COMPUTER AIDS TO CHEMISTRY

IMPORTANT NEW & FORTHCOMING BOOKS IN CHEMICAL SCIENCE

COMPUTATIONAL METHODS FOR CHEMISTS
A. F. CARLEY and P. H. MORGAN, University College, Cardiff
REDUCTIONS IN ORGANIC CHEMISTRY
M. HUDLICKY, Virginia Polytechnic Institute and State University
BIOHALOGENATION: Principles, Basic Roles and Applications
S. L. NEIDLEMAN and J. GEIGERT, Cetus Corporation, Emeryville, California
ORGANOSILICON AND BIOORGANOSILICON CHEMISTRY: Structure, Bonding, Reactivity and Synthetic Application
Editor. H. SAKURAI, Tohoku University, Sendai, Japan
COMPUTER AIDS TO CHEMISTRY
Editors: G. VERNIN and M. CHANON, Université de Droit d'Economie et des Sciences, Marseilles, France
HANDBOOK OF AQUEOUS ELECTROLYTE SOLUTIONS
A. HORVATH, ICI, Mond Division, Runcorn, Cheshire
BIOINORGANIC CHEMISTRY
R. W. HAY, University of Stirling
ORGANIC PHOTOCHEMISTRY
J. D. COYLE, Open University
PHYSICAL PHOTOCHEMISTRY
R. DEVONSHIRE, University of Sheffield
INORGANIC PHOTOCHEMISTRY
A. HARRIMAN, The Royal Institution
PHOTOCHEMICAL AND PHOTOELECTRICAL CONVERSION OF SOLAR ENERGY
A. MILLS, University College, Swansea and J. DARWENT, University of London
MATHEMATICAL AND COMPUTATIONAL CONCEPTS IN CHEMISTRY
N. TRINAJSTIC, Rugjer Boskovic Institute, Zagreb
CHEMICAL MONITORING OF OCCUPATIONAL TOXICITY
V. FOA, *et al.*, University of Milan
HANDBOOK OF LABORATORY WASTE DISPOSAL
M. J. PITT, Loughborough University of Technology and E. PITT, University of Aston in Birmingham
COMMUNICATION, STORAGE AND RETRIEVAL OF CHEMICAL INFORMATION
JANET E. ASH, *et al.*
CELLULOSE CHEMISTRY AND ITS APPLICATIONS
T. P. NEVELL, UMIST and S. HAIG ZERONION, University of California

COMPUTER AIDS TO CHEMISTRY

Editors:

G. VERNIN, M.Sc., Ph.D.
Director of Research, National Centre of Scientific Research
Marseilles, France

and

M. CHANON
Professor of Chemistry, Université de Droit et des Sciences
Marseilles, France

ELLIS HORWOOD LIMITED
Publishers · Chichester

Halsted Press: a division of
JOHN WILEY & SONS
New York · Chichester · Brisbane · Toronto

First published in 1986 by
ELLIS HORWOOD LIMITED
Market Cross House, Cooper Street, Chichester, West Sussex, PO19 1EB,
England

*The publisher's colophon is reproduced from James Gillison's drawing of the ancient Market
Cross, Chichester.*

Distributors:

Australia and New Zealand:
Jacaranda-Wiley Ltd., Jacaranda Press,
JOHN WILEY & SONS INC.
GPO Box 859, Brisbane, Queensland 4001, Australia

Canada:
JOHN WILEY & SONS CANADA LIMITED
22 Worcester Road, Rexdale, Ontario, Canada

Europe and Africa:
JOHN WILEY & SONS LIMITED
Baffins Lane, Chichester, West Sussex, England

North and South America and the rest of the world:
Halsted Press: a division of
JOHN WILEY & SONS
605 Third Avenue, New York, NY 10158, USA

5 42.8
C 7 3
1 4 2 / 6 5
Sept. 1987

© 1986 G. Vernin and M. Chanon/Ellis Horwood Limited

British Library Cataloguing in Publication Data
Computer aids to chemistry. —
(Ellis Horwood series in chemical science)
1. Chemistry — Data processing
I. Vernin, G.   II. Chanon, M.
542'.8   QD39.3.E46

Library of Congress Card No. 86–10452

ISBN 0–85312–774–3 (Ellis Horwood Limited)
ISBN 0–470–20338–2 (Halsted Press)

Phototypeset in Times by Ellis Horwood Limited
Printed in Great Britain by The Camelot Press, Southampton

Table of Contents

5

Foreword

The development over the past twenty years of big computers has accelerated calculations in many fields—quantum chemistry, molecular mechanics, computer-assisted chemical synthesis are only some typical examples. A new surge in the contribution made by computers to progress in chemistry is the result of more recent commercialization of the high-performance personal computers (PCs). Thanks to its relatively low price and easy availability, such equipment has become commonplace in most chemistry laboratories throughout the world. The purpose of this book is to show chemists, with the aid of judiciously selected examples, how such recently developed and easily available techniques can be of invaluable assistance.

The eight chapters deal with such varied topics as organic synthesis, chemical education, chemical kinetics, analysis of chemical data, crystallography, structural analysis through data banks, and chemical information. Although far from exhaustive, this list is quite representative of the types of aid that microcomputers can offer to chemists. The presentation of the mathematical basis of these computational methods is reasonably condensed, allowing the average chemist to understand their development and applications. The contributors are experimental chemists applying computational methods to their own fields of research or teaching which is at least one reason why this text is so naturally assimilable.

This excellent book fills a gap in chemical education and will without doubt be of valuable help to a large population of chemists in research, teaching and industry. I am quite confident it will do just this.

Marseille. France
May 1986

Jacques V. Metzger

Introduction

Chemistry is an evergreen subject. This is so because chemistry has roots in conceptual fields which range from quantum mechanics to the molecular organization of life, and because chemistry deals with such a wide variety of problems in diverse fields, such as analytical, synthetic, physical, and industrial, living as well as inanimate matter, solid state as well as solution and gas phase systems, and from the simplest molecules to the most complex structures. Consequently, advances in nearly any area of scientific endeavor has an effect on chemistry, and vice versa.

One of the more recent revolutions to affect the field of chemistry is the advancement of spectroscopic techniques such that they have become ubiquitous in the everyday life of the chemist. Although far from over, this spectroscopic revolution is already undergoing its own revolution as a consequence of a new advance affecting the lives of all chemists. The new advance is, of course, the effect of the computer sciences on chemistry, made possible by the commercial availability of computers with ever-increasing capabilities and steadily-decreasing prices. Thirty years ago, computers were already known and used, but perhaps 90% or more of the chemists then working functioned quite well without them. However, this is no longer the case; within the next twenty years, quite the opposite will no doubt hold true.

The goal of this book is to show why the swing to computer aids to chemistry is occurring as well as examples of how. Since we accept the multifaceted nature of the field of chemistry, we have selected some representative, yet quite different, facets which exemplify the changes computers have made in the activity of chemists.

Of the eight chapters, three (Chapters IV, V, and VI) treat typical quantitative data. They have been written so as to be understandable to any chemist willing to understand how a computer makes such fundamental tools as X-ray Crystallography, Kinetics, and Data Analysis/Experimental Design easier to use. Chemists often complain of being inundated by information from all directions and thus being unable to keep abreast of current topics either in general or specific areas of chemistry. Three chapters, one devoted to information access (Chapter VIII) and two treating the specific areas of spectrometry databases (Chapter VII) and synthetic organic chemistry (Chapter I) bring good news in these areas. The computer can ease the burden of keeping abreast of the wave of chemical information not only by efficient information management, but also in the 'pretreatment' of information for a given chemical speciality.

Finally, each of us is familiar with our own process of learning chemistry, both as a student and throughout our careers. Two chapters describe how computers are affecting this aspect of our lives as chemists. Chapter II summarizes some of the exciting uses of computers as complements to the teaching and learning of chemistry. Chapter III deals more specifically with computer graphics and its effects far beyond activities concerned only with teaching. Chemistry has often been the battlefield and proving ground for mathematical models of all sorts. It was often fashionable to denigrate the chemist's intuition (which is somewhat impressionistic) in favor of the precision of mathematicians. The cross-fertilization of these approaches was strongly inhibited, since even though the chemists had an idea of the phenomena they had extracted from their laboratory experiments, they often felt unable to understand the mathematics involved in the elaboration of various models. With modern computer graphics this situation changes, since every mathematical model can be transformed into something tangible and observable. Thus, the problem of communication is solved, and the chemist can say to the mathematician, 'Although I do not fully understand the mathematics involved, your model does (or does not) fit my knowledge and understanding of my experiments, in such and such respects'.

Early in his history (actually in *pre*-history), Man realized that tools, adapted to specific tasks, extended the usefulness and efficiency of his own hands and fingers. Now, at the close of the twentieth century and on the verge of the twenty-first, it becomes apparent that a tool is now being introduced to extend the usefulness and efficiency of Man's own *brain*. This is a fantastic advance, and chemists should not attempt to avoid it. However, a good tool in awkward and inexperienced hands does a poor job. Thus, although each chapter attempts to show what computers have brought to a subject, we also try to illustrate in which ways the human brain must direct and decide, drawing comparisons between the abilities of man and the computer as often as possible.

This book is illustrative rather than exhaustive, as the extent of the topic precludes even an attempt at exhaustivity. One still may ask, 'Why no chapters dealing with quantum chemistry, infrared and n.m.r. databases, drug design, instrumentation interfacing, robotics, etc.?' Our only answer is the constraint of size of this book, and our own prejudices. We have chosen what we hope to be representative and useful applications of computer aids to chemistry.

ACKNOWLEDGEMENTS

Firstly, we wish to express our thanks to Professor J. Metzger for his valuable encouragement and for his dynamic interactions in his laboratory.

We wish to thank also all the contributors and B. W. Zoellner, now of Northern Arizona University, who translated some chapters.

Our thanks are also due to Mrs G. Gebbenini who typed Chapter I and Mrs G. Vernin for her valuable assistance in the artwork and in some stages of production of this book.

Finally, thanks are due to the University of Aix-Marseilles for financial support, and to our publisher Ellis Horwood, for his encouragement and assistance.

July 1985   G. Vernin and M. Chanon

CHAPTER I

Computer-Aided Organic Synthesis (CAOS)

René Barone and Michel Chanon

Faculté des Sciences de Saint Jerôme LCIM, Marseille, France

I.1 INTRODUCTION

In 1967, Corey and Wipke [1] began the great adventure of teaching organic chemistry to a computer, with the expectation that in teaching this rather bright and untiring student, fresh blood could be infused into the old and respectable field of synthetic chemistry. Now, almost twenty years later, more and more groups are entering the land opened by these pioneers. Thus, it is interesting to consider how very differently these two explorers have managed their time since then. Corey is sometimes said to have diminished his efforts in this direction (which is inexact, see refs 14 and 49(b)) and, rather, has concentrated on discovering, in each successive year, another new series of reactions and strategies [2]. The average chemist, reluctant to allow the computer into his laboratory, would paraphrase this great master by saying, 'I don't need computers to find new synthetic pathways'. Further, the average chemist can support his argument by stressing the number of publications originating from Corey's group, or, even more strongly, by citing the number of outstanding young synthetic chemists who received their training with Corey and did not enter the field of CAOS. On the other hand, W. T. Wipke has followed a totally different track: he has invested more than seventy-five man-years in the original SECS program. About twenty industrial firms have found the SECS program useful enough to buy it and five academic laboratories currently use it. W. T. Wipke is now perfecting a completely new approach whose aim is to overcome some of the problems that will be specifically dealt with later in this text. So, these two pioneers in the field each

19

give different messages to the newcomer. One says, 'Synthetic chemists can do very well without computer aid', while the other answers 'Yes, synthetic chemistry is a formidable challenge for artificial intelligence, but even if we don't solve everything we can still build useful new tools for the chemist'. In the discussion which follows we hope to show how the same field allows such opposite judgements. Before doing so, we shall briefly introduce the leading research groups active in the field. Keeping in mind our double-entry approach (computer scientist and synthetic chemist) we will first attempt to show how the computer can be made to communicate with the chemist, then, with the communication issue settled, we will explore what can be taught to the computer, starting from the easiest things and going on to more and more difficult matters.

We shall attempt to delineate the main directions of action involved when a computer-aided synthesis program is constructed. It is indeed important in any endeavor to recognize the dimension and the basis axis of the representation space [3] of a given activity. The second step then being the evaluation of which axes man clearly leads and which ones are better tackled *via* the computer (Section 5.4). So, rather than being the captive of a monodimensional approach ('synthesis is better dealt with by man *versus* synthesis is better dealt with by computers') we attempt to identify the directions in which the computer indisputably leads the match and those in which man is still far ahead. These latter give us, by the same token, the challenges that the 'artificial intelligence workers' must face for the future and a strategy of development *whose goal is using the 'synergy of the chemist–computer couple'*.

Computer aids to synthesis have been reviewed by various authors. In 1967, Corey [5] furthered Woodward's [6, 7] approach of systematic and in depth reflection concerning the question: '*How does a chemist choose a pathway for the synthesis of a large organic molecule, given either the great diversity of organic structures and reactions, or, in contrast, the critical importance of each step to ultimate success?*'. The first report concerning the application of computers to this question did not appear until 1969, when the same author summarized the work performed by his group in writing the OCSS program. Some of the difficulties mentioned in that article still lack solutions [8]. The analytical approach [31] has, however, stimulated work far beyond the CAOS field, [32, 273–278]. Although these overviews are by far the most cited in the field, some other early contributions deserve mention: Vleduts [9(a)] and Sarett [9(b)] suggested earlier that computers could be applied to organic synthesis. In 1969, J. E. Dubois, in a very systematic and thorough way examined the synthetic paths available to obtain increasingly sterically hindered ketones [10]. After these forerunners, the fascination exerted by the field manifested itself in a number of reviews in different languages, among which we have selected six for their readability and/or completeness. The best known is by Bersohn [11]; an ACS symposium on the subject gives a detailed account of the state of the art attained by the leading research groups in 1976 [12]; a recent and very enjoyable review appeared in a somewhat exotic journal and as such could be easily overlooked if no stress were placed on it [13]. A concise but up-to-date and precise description of the field may also be found in the May 1983 issue [14, 15] of *Chemical and Engineering News*. Gund reviewed computer-aided synthesis in a medical journal [153]. In addition to these basic contributions, many other reviews bear testimony to the world-wide attraction that the topic exerts on a growing number of synthetic chemists: Japan [16], Yugoslavia [17], GDR and FRG [18, 56], Finland [19], USA [20, 153, 158], Sweden [21], UK [22], Belgium [23], Czechoslovakia [24], Canada [25], France [26], Netherlands [27, 28], Italy [29], and China [30].

In this chapter we will use some of the abbreviations proposed by workers in the field; the meaning of each is given in Section I.6 at the end of the chapter (p. 87).

I.2   RESEARCH GROUPS INVOLVED IN COMPUTER-AIDED SYNTHESIS

We could not find a better title for this section but we must warn the reader that this title is somewhat misleading because we introduce here only the programs identified in publications. It should be clear, however, that an essential part of the work in CAOS is done by experienced chemists continuously using and improving a given program without publishing much about it (see, however, SECS at Merck in refs. 155–156). A well-built algorithm is something interesting but what *brings it to life and keeps it alive is continuous use by researchers actively involved in laboratory work* [130]. Many industrial firms, while using the leading programs, continuously point out basic flaws in treatments and challenge algorithm writers to correct them. *This positive interaction is one of the strongest driving forces for the continuous improvement of existing programs.*

One may classify the programs whose goal is computer-aided synthesis according to a wide variety of criteria. If one thinks in terms of the first year of publication the ancestor then becomes OCSS (Corey-Wipke) [1]. This ranking, illustrated in Table 1, gives an approximate idea of the experience gained by a group in the field, but this must obviously be adjusted by an estimation of the average size of the group involved. We estimated this in man-years of work and, with a few exceptions, the leaders in the field agreed to give a *rough* evaluation of their program in these terms (Table 2). We insist on the adjective 'rough' because these figures may vary widely depending upon the quality of researchers involved, the fact that one may or may not include the contributions made by program users, the quality of the scientific environment, etc. Even though approximate, we give these numbers to show that, after fifteen years, computer-aided synthesis is still a very young science. If one compares the sum of contributions ($\simeq 150$ man-years) whose goal is to forge a new tool to make the synthesis of any target easier, to the efforts invested in attaining the famed target vitamin-B12 (about 210 man-years) [4] one realizes that there *is still a tremendous lack of time invested in the computer approach*.

Table 1 History

Number	Authors	Prog. Name	1st Pub.	Ref.
1	COREY-WIPKE	OCSS	1969	1
2	COREY	LHASA	1971	5, 31–49
3	UGI-GASTEIGER	(a)	1971	18c, 50–82
4	HENDRICKSON	(a)	1971	83–95
5	BERSOHN	—	1971	97–111
6	WEISE	AHMOS	1973	112–122
7	GELERNTER	SYNCHEM	1973	123–130
8	BARONE-CHANON	SOS	1973	131–143, 210b
9	POWERS	DINASYN	1973	188–189
10	BROWNSCOMBE	EXTRUS	1973	209
11	BROWNSCOMBE	HEXARR	1973	209
12	WIPKE	SECS	1974	144–160
13	UGI-GASTEIGER	CICLOPS	1974	57 (see N°3)
14	BENEDEK	SIMUL	1974	161–166
15	DUBOIS	SYNOPSYS (b)	1975	167–181
16	WHITLOCK	—	1976	185–187, 281
17	DONOVA	HEDOS	1976	17b
18	POWERS	REACT	1977	190
19	GOVIND	REPAS	1977	190 (p. 81 ref. 4), 191
20	DJERASSI	REACT	1977	12 (p. 188), 204–205
21	PENSAK	LHASA (Dupont)	1977	12 p. 1
22	KAUFMANN	PASCOP	1978	195–202
23	MOREAU	MASSO	1978	203
24	GELERNTER	SYNCHEM 2	1978	127–130
25	UGI-GASTEIGER	EROS	1978	18c, 65 (see N°3)
26	YONEDA	GRACE	1978	206–208
27	GASTEIGER	PSYCHE	1979	72
28	WEISE	GSS (c)	1979	117
29	BARONE-CHANON	SAS	1979	137, 141
30	STOLOW-JONCAS	LHASA (Educ)	1980	210a
31	AGNIHOTRI	CHIRP	1980	211–212
32	JORGENSEN	CAMEO	1980	213–220
33	GUND	SECS (Merck)	1980	155–156
34	HENDRICKSON	SYNGEN	1981	93 (see N°4)
35	HIPPE	SCANSYNTH	1981	221–224
36	HIPPE	SCANPHARM		245
37	HIPPE	SCANMAT		245
38	ZEFIROV	FLAMINGOES	1981	225–232
39	GHOSE	(d)	1981	96
40	SCHUBERT	ASSOR	1981	74
41	ZIN	—	1982	233a
42	ERDOS	ASR	1983	233b
43	KAUFMANN	PSYCHO	1984	200b
44	SEIDEL	—	1984	235a
45	HARA	PFP	1984	234
46	WIPKE	SST	1984	159
47	CENSE	MICROSYNTHESE	1985	235b
48	BARONE-CHANON	TAMREAC	1985	138b

(a) for basic ideas. (b) Ref. 171 is chosen to class Dubois' group, but in 1969 (ref. 10) a description of reaction was published which was totally different from the IGLOO model which is used in the DARC/SYNOPSYS system (Ref. 176). (c) The original name of GSS was SYNAB/SYNPLAN. (d) Only an algorithm is described according Hendrickson's system.

Another consequence of this 'man-year' evaluation is to aid in the recognition of where the 'hard-core' of the field is. Even if, for sake of completeness, we gathered about 200 references to cover Table 1, 80% of the information would be concentrated in the 22 references that we have tentatively identified as 1, 5, 11, 12, 13, 14, 38, 49(b), 69, 93, 108, 119, 130, 142, 155, 159, 171, 174, 202, 205, 219, 222. These references must be read and reread. However, this does not mean that the 20% of the remaining information is without interest: this occurs simply because computer-aided synthesis has not yet solved many problems [155] and innovative approaches may originate anywhere.

A convenient pictorial representation has been recently given in *Chemical and Engineering News* [15, 26(a), 26(b)] and Fig. 1 is our adaptation, in a more complete form.

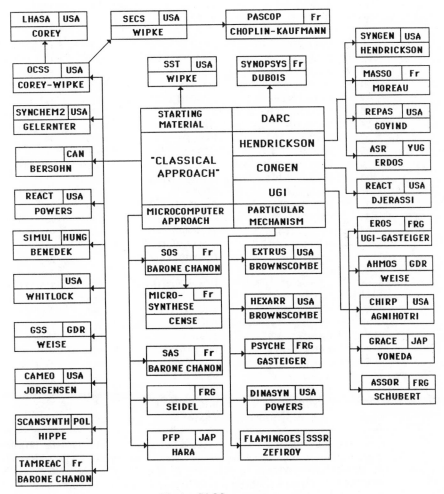

Fig. 1   CAOS programs.

Table 2a   Generalities

Name	Size	Computer	Language	Field	Ref.
LHASA	30 000 lines	VAX	F	A	45
SYNCHEM 2	2, 4 Mb (a)	IBM, VAX vms	P	A, B	127–130
BERSOHN		IBM 370/165	A	A	99, 105
AHMOS	150 Kb	IBM 360	P	C	245
		EC 1040, 1055			
SOS	30 Kb	IBM 1130	F	A, B, D	132, 134
		Apple II, IBM-PC	B		142, 143a
DINASYN				E	188, 189
EXTRUS	1 170 lines	BURROUGHS 550	F	F	209
HEXARR	3 000 lines	idem	F	G	209
SECS	130 Kb	PDP 10, PDP 20			
	(32 bit	UNIVAC 1108	F	A	146, 152
	words)	IBM 370			
		FUJITSU			
EROS	300 Kb	IBM 360–370	P	H	65
		AMDAHL 470 V6			
		Telefunken TR 440			
SIMUL	—	—	—	A	163, 166
WHITLOCK	—	—	L	I	185
WHITLOCK	—	IBM 7094	F	J	186
REPAS	—	—	—	A	190, 191
REACT	2 500 lines	IBM 360–370	F	K	190
(Powers)	of code				
REACT	—	—	—	—	12 p. 188,
(Djerassi)					205
PASCOP	400 Kb	UNIVAC 1110	F + A	A, L	196
			+ IGL		
MASSO	50 Kb	IBM, PDP 11	F	A	203
		VAX 11			
GRACE	—	HITAC 8800	—	M	208
PSYCHE	—	IBM 360/91	P	N	72
		AMDAHL 470 V6			
GSS	300 Kb	IBM 360–370	F + A	H	117, 245
		EC 1040, 1055			
SAS	10 Kb	IBM 1130	F	A	137
		Apple II	B		
CHIRP	—	—	—	K	211
CAMEO	45 000 lines	HARRIS, VAX	F	A	213, 215, 245
SYNGEN	—	DEC 11/23	F + A	A, O	245
FLAMINGOES	1 000	BESM-6	F	P	245
	operators	ISKRA 226	B	P	
SCANSYNTH	(b)	RIAD 32	F	A	245
SCANPHARM	(c)	ODRA 1305	F	A	245
SCANMAT	(d)	MERA 400	F	A	245
		SM-3, SM-4,			
		ODRA 1305			
ASSOR	600 Kb	CDC, Cyber 175	P + F	A	74, 245
		IBM 370/145			
ASR		ES 1022, 1040	P	A	233b
PSYCHO	400 Kb	UNIVAC	F	A	200b
SYNOPSYS	—	—	—	A	176
SEIDEL	—	—	—	—	235a
SST	26 000 lines	DEC 20	F	A	159
MICROSYNTHESE	40 Kb	APPLE II	A	A	235b
TAMREAC	40 K	PDP 11/44	F	Q	138b

SIZE:
(a) including dynamically allocated work storage region, (b) disk 32 Mb, CPU 256 Kb, max overlay: 32 K, number of overlay: 9, (c) disk 8 Mb, CPU 40 Kb, max overlay: 40 K, number of overlay: 22, (d) disk 2, 45 Mb, CPU: 32 Kb, max overlay: 20 K, number of overlay: 21

LANGUAGE:
F = FORTRAN, P = PL/1, L = LISP, A = Assembly, B = BASIC

FIELD:
A = Organic Chemistry, B = metabiotic pathways, C = Polar organic reactions, D = heterocyclic chemistry, E = DNA synthesis, F = unimolecular pericyclic reactions, G = general cyclic transition state, H = all chemistry, I = linear organic molecules, J = functional switching problem, K = industrial chemistry, L = phosphorus chemistry, M = heterogeneous catalysis of gas phase reactions, N = pericyclic reactions, O = heterocycles with N only, P = polycyclic systems, Q = transition metal reactivity.

A more quantitative classification would attempt to represent each program in a three-dimensional achievement space. One can select as 'basis axes': strategies, exhaustiveness and organization of reaction data-bank and, in-depth treatment of transforms. An N-dimensional representation [3], including such parameters as flexibility toward utilization and modification by chemist users, efficiency measured as the number of good synthetic paths with respect to what is found in the literature, etc., would perhaps be better. At this point, however, it is difficult to evaluate the foregoing criteria for all the programs gathered in Table 1 because of lack of precise information. In addition, the representation axes selected would not be completely independent. For example, if a program is very highly comprehensive in terms of proposed transformations (including even innovative reactions) [69, 93] then it usually proposes thousands of precursors and would snow the chemist under with propositions *unless it is coupled with very efficient pruning* of the synthetic tree (Section 4.2). REPAS, based on Hendrickson's half-reaction approach, faced with the synthesis of acrylic acid without further specification would generate more than 10 000 reaction pathways [190] Tables 2a and 2b gather some of the characteristics displayed by the published programs.

I.3   BASIC OPERATIONS IN COMPUTER-AIDED SYNTHESIS

3.1   Connect chemist's and computer's capabilities

3.1.1   *Representation of molecules*

The preliminary step in any use of a computer in organic chemistry is to transform the chemist's two-dimensional visual description of chemical substances into representation systems more amenable to algorithmic computer processing. Many different solutions were offered for this problem even before the beginning of projects aimed specifically at computer-assisted synthesis. A good account of the overall field may be found in references 146 and 237–244. The choice of a particular representation scheme depends upon the functions to be performed, the available hardware and software and the

Table 2b Generalities

Name	a	b	c	d	e	Keywords
SYNCHEM2	20	1	1	2	B	Knowledge-based domain-specific expert system-solving problem, heuristic search, knowledge-based management
AHMOS	8	0	1	3	B	First direct simulation program
SOS	8	7	42	2	C	Microcomputer, heterocyclic chemistry, mechanisms
SECS	75	20	5	5	B	1st program to handle chemistry, 3D modules, aromatic chem., pattern transform for heterocyclic chem. and to achieve industrial continued use. Initiated in 1967 with the 1st synthesis program OCSS on a PDP-1 with 24 K 18-bit words. A one person project until latter Howe and Cramer joined WTW. That program predicted a synthesis of Patchouli alcohol that was later accomplished.
PASCOP	42	1	2	3	B	Empirical, interactive, graphical, retrosynthetic, strategic.
MASSO	1	1	4	0	C	Mechanisms, half reactions, rearrangements, creativity, new reactions
GSS	10		1	3	B	The first universal synthesis simulation program by means of synthon substitution
SAS	$\frac{1}{6}$	0	1	0	C	Analytical synthesis, disconnection approach, skeletal dissection
CAMEO	45	2	3	7	C	Mechanistic reasoning, forward synthesis, algorithms for organic reactivity
SYNGEN	7	0	1	2	C	Skeletal dissection, constructions reactions, digital reaction description, canonical numbering, logical driven program
FLAMINGOES	—	—	2	3	C	Non empirical approach, cyclic bond transformations, combinatorial algorithm, back-track procedures, mathematical statement of the main algorithms
SCANSYNTH	10	1	1	4	B	Own original metalanguage for description of chemical structures and reaction extremely powerful (low memory reservation) backward and forward
SCANMAT	3	0	4	4	B	Powerful method of canonicalization, originally developed methods of identification of topological fragments, feasibility of simulation of any chemical reaction, extended set of strategies to control the simulation process or the prediction process
SCANPHARM	3	1	1	4	B	Powerful method of canonicalization, originally developed methods of identification of topological fragment, feasibility of simulation of any chemical reaction, extended set of strategies to control the simulation process or the prediction process, inclusion of data base of industrial technologies, enables both backward and forward development of syntheses and gives more realistic results
ASSOR	6	1	1	5	B	Mathematical model, deductive approach, elementary reactions, reaction simulation, synthesis planning
PSYCHO	5	0	0	6	B	Empirical, synthetic, interactive, ergonomical, graphical, evolutive.
SECS (Merck)	3	1	0	0	B	
MICRO-SYNTHESE	5	2	10	3	C	Microcomputer, expert system, friendly, cheap, computer-assisted instruction.
TAMREAC	3	0	1	2	C	Mechanisms, organotransition metal reactivity, homogeneous catalysis, forward synthesis.

a = Number of man-years. b = Number of industrial firms using the program. C = Number of academic laboratories using the program. d = Size of the group presently working on the project or its ameliorations. e = Origin of the searchers involved in the project: C = mainly chemists. B = chemists + computer scientists.

desired balance between manual and machine processes. Here we will study only the representations used effectively for the objective of 'computer assistance to synthetic chemistry'. Several algorithms performing the automatic interconversion of a given structure into different representations have been published (see ref. 246).

The modes of representation are usually classified into two main groups: those that are extensions of classical nomenclature (usually linear) [242, 243] and those that are essentially graph theoretical exercises [247]. These two broad classes allow hundreds of variations [248]. To clarify some of these variations we will describe how the same molecule is described in the SYNCHEM, LAHSA, DARC and EROS programs.

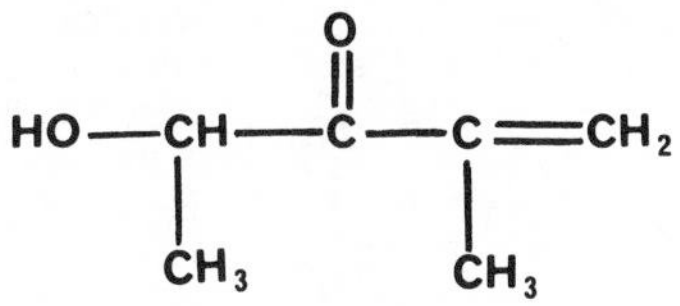

$$HO-CH-\overset{\overset{O}{\parallel}}{C}-\underset{\underset{CH_3}{|}}{C}=CH_2$$
$$\underset{CH_3}{|}$$

Scheme 1

In SYNCHEM (Wisswesser), the molecular structure is represented in two different forms [124]. Aldrich Chemical Company provides a list of available compounds (11 000) in Wiswesser Line Notation [242] and SYNCHEM compares this list against the availability of the precursors which it proposes. On the other hand, for manipulating molecular structures, SYNCHEM uses a connection–matrix representation (TSD) adapted to make any transformation into the WLN system easier. WLN is a linear string consisting of a set of symbols which represent a complete topological description of a compound. It has symbols which represent atoms or groups of atoms, a syntax to describe interconnections, and rules (about 300) for ordering the symbols to provide a unique and unambiguous representation of the topology of a molecule. For the foregoing molecule, the Wisswesser Line Notation is: QY1&VY1&U1. The OH substituent is cited first and represented Q; Y, which stands either for

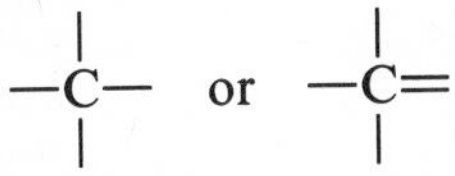

$$-\overset{|}{\underset{|}{C}}-\quad \text{or} \quad -\overset{|}{\underset{|}{C}}=$$

follows and 1 represents the methyl group. The & indicates that the first fragment has been described and that we are going to move on to the next fragment, made up of a CO (V) another C= (Y) and a CH_3 (1). After the second & of separation, we finish with the description of the double bond (U) and CH_2 (1). No information is provided about the stereochemical relationship linking these substituents. WLN allows a very fast retrieval of a given compound in a large data-bank, but is not very well adapted to computer-aided synthesis. SYNCHEM 2 itself has elaborated its linear representation into a new system: SLING. WLN's importance comes from the large amount of interesting information stored in this form by important firms (ISI, Aldrich, Searle, Dow Chemical, Hoffmann-La Roche, etc.).

Connectivity matrices or connectivity tables are the primary form of representation used in computer-aided synthesis. The most used derive from Morgan's type [250], improved by Wipke [145, 147] and Moreau [251] to include stereochemical information. In a progressive approach we will first write a connection table without any priority considerations to illustrate the very simple principle of such a description. Then we will see what the introduction of priorities brings to the numbering and the importance of the canonicalization operation.

In the connection–matrix representation, our ketone is described both by an atom table and by a bond table.

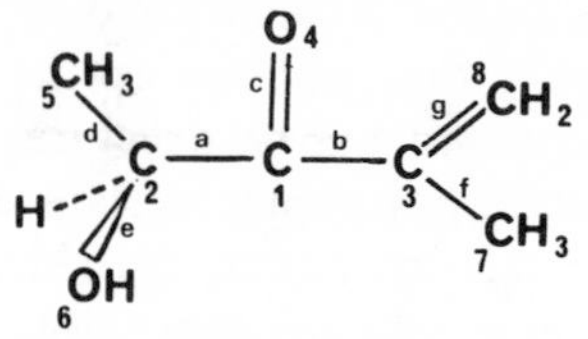

Scheme 2

Atom table

Atom No.	Type(a)	Stereo(b)	NBS(c)	NATCH(d)	ATBD(e)	BD(f)
1	1	0	4	3	134	abc
2	1	1	3	3	165	aed
3	1	0	4	3	178	bfg
4	4	0	2	1	1	c
5	1	0	1	1	2	d
6	4	0	1	1	2	e
7	1	0	1	1	3	f
8	1	0	2	1	3	g

(a) C = 1, O = 4; (b) stereocenter = 1 no stereocenter = 0; (c) number of valence used; (d) number of bonds other than to H; (c) and (d) differ when there are double or triple bonds; (e) number of atoms connected to the atom considered; (f) bonds involved.

Bond table

Bond input number	B type(a)	B stereo	AT. 1	AT. 2
a	1	0	1	2
b	1	0	1	3
c	2	0	1	4
d	1	0	2	5
e	1	1	2	6
f	1	0	3	7
g	1	0	3	8

(a) 1 = single bond, 2 = double bond, 3 = triple; (b) stereo 0 = none, 1 = At. 2 up with respect to 1; 6 = At. 2 down with respect to atom 1; 4 = At. 2 either up or down with respect to atom 1.

For the representation of stereochemistry [145, 147], the list of attachments (ATBD 165) is ordered so that viewing down the bond from the first attachment to the central atom, the other attachments are arranged in a clockwise manner. Implicit hydrogens are permuted to the end to prevent 'holes' in the list.

The connectivity table, so written, is unambiguous because it applies to only one chemical substance (assuming at this level that enantiomers have been taken care of [145–147]). It is, however, not unique because alternative numberings of the connection table would result in different representations for the same chemical substance (*n*! equivalent matrices all interconvertible by row/column interchange). To pass from this non-unique representation to a unique one is called '*canonicalization*'. The practical consequence of 'canonicalization' is that whoever in the world applies one given set of canonicalization rules to a given structural formula, only one representation is obtained. Bersohn has clearly explained [104] all the advantages associated with such an operation: fast identification of the canonicalized structure or part of it with any other structure or substructure stored in the computer, fast identification of equivalent atoms in a given structure, a basis for stereochemical description [99], etc. Although there are advantages, this is not a 'must' for a computer-aided synthesis program so long as no comparison of proposed precursors with commercially available compounds is involved: molecular representations in LHASA were not canonicalized in the earlier versions [38] (in LHASA 11, however, they are canonicalized [49(b)].

There is an amazing variety of canonicalization rules: these rules, indeed, come to a hierarchical (set of priorities) treatment of structural features. Different groups have proposed different hierarchies (structural priorities). The most used is probably the one originally proposed by Morgan [250] as an extension of Gluck's treatment [253] and improved to include all the aspects of stereochemistry [145, 147]. This hierarchy appears here as a process of successive partial orderings. The main ordering factor is the consideration of 'connectivity values' for each atom of the structure. This is reminiscent of the familiar chemist's vision of ramification expressed with the terms of quaternary, tertiary, secondary, and primary carbons: here quaternary carbons are ranked at the top of the hierarchy. The elaboration, expressed by a set of rules, is necessary for those cases where the competition for higher ranking involves atoms with less clear-cut differences (e.g., two quaternary carbons). These rules are:

All non-hydrogen atoms in the molecule are first numbered from 1 to *n*.

(1)   One atom is selected and assigned the number 1.
(2)   The atoms connected to atom 1 are numbered 2, 3, etc...
(3)   The unnumbered atoms connected to atom 2 are now numbered.
(4)   This procedure is followed until all atoms have been numbered.

These rules reduce the number of ways in which a compound may be described. The algorithm uses properties of the compound to further reduce the number of descriptions to be considered. These are the connectivity of each atom (that is, the number of non-hydrogen atoms attached to it), the atomic symbol, and the bond code. An iterative (or step-by-step) procedure is used to enhance the connectivity values as follows:

(1) Each atom in the structure is given a connectivity value.
(2) The number of different connectivity values, k, in the compound, is determined.
(3) A new connectivity value is calculated for each atom by adding together the connectivities of all atoms directly attached to it.
(4) A new value of k is calculated from these values and compared with the previous values.
(5) If the new value is greater, the new connectivity values are assigned and step (3) is repeated.
(6) If the new value of k is less than or equal to the old one, the connectivities determined at the end of the last cycle are retained and the process is ended.

This process is illustrated for our ketone (Scheme 3).

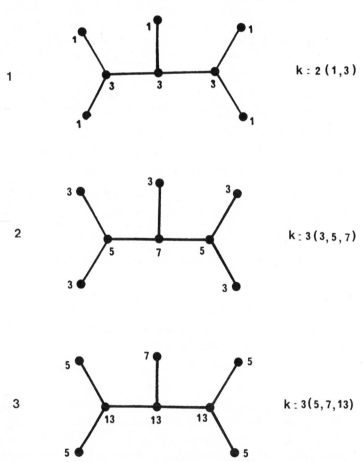

Scheme 3

In stage (3) the value of k has not increased; the connectivity values of stage 2 are therefore used in the determination of the actual numbering. The ranking of the carbonyl carbon atom at the top is not obvious because in the zeroth order approach three atoms are equivalent. The higher ranking of C=O originates from the fact that it is connected to one quaternary carbon and one tertiary, whereas its competitors (other atoms with $k = 3$) are connected respectively to a quaternary C, a primary C, a primary-like O atoms, and a quaternary C, a secondary-like C, and a primary C.

Even when this zeroth diagram has been obtained, ambiguities appear for atoms to be ranked 2 and 3, respectively. So, to decide between the two rankings we have to write both and proceed with the numbering of the other atoms.

Scheme 4

In this ranking we see that C has priority over O (alphabetical order in Morgan) and also that single bond C—C has priority over C=C. If we now write the linear strings associated with A and B:

(A)
$$\begin{array}{cccccccc} 1 & 2 & 3 & 4 & 5 & 6 & 7 & 8 \\ C & C & C & O & C & O & C & C \end{array}$$

(B)
$$\begin{array}{cccccccc} 1 & 2 & 3 & 4 & 5 & 6 & 7 & 8 \\ C & C & C & O & C & C & C & O \end{array}$$

and assign 0 to C (priority in alphabetical hierarchy) and 1 to O (second ranked in this hierarchy) we find:

(A) 0 0 0 1 0 1 0 0   (B) 0 0 0 1 0 0 0 1

The smallest number is retained (i.e., B), so the Morgan canonical numbering is:

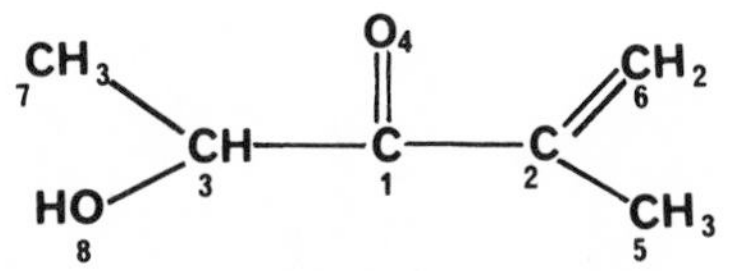

Scheme 5

For some structures, Morgan's hierarchy encounters problems with oscillatory behavior in the automatic numbering; Moreau recently proposed an improvement which suppresses these problems [251]. One may propose other priorities to get another 'view' of the molecule; this could be of use for some structures. After all, in set theory some objects demand several irreducible representations whereas others do not [252]. The possibility of using several different canonical numberings within the same program is, however, too time consuming: Bersohn [104] reported that for some problems, the program spends as much as 40% of its time canonicalizing the molecular structure representations. Even if this performance is improved [106] it still consumes much computer time.

The comparison of the Wisswesser and Morgan systems shows immediately that the former is much to be favored in terms of storage requirements and retrieval speed in a large population of structures. Some remedies have been offered to this inconvenience. Compact connection tables, such as those devised by Gluck [253] lead to a major saving in memory. More recently, Fujiwara and collaborators [254] proposed a hierarchical view of chemical structures in terms of blocks (ring assemblies) and atoms. By introducing a block, a chemical structure is described in an intermediate form of tree structure called *a block cutpoint tree* (BCT). A 'block dictionary' completes the system. This method is used for quick substructure searches.

Dubois's group has developed over twenty years a consistent system of structural representations [169]. The DARC code resembles a connection table in that it expresses or implies the nature of each atom and bond. The canonicalization step is entangled with the writing of the connection table because, from the beginning an implicit hierarchy is introduced by choosing a focus FO (e.g. C=O in ketones, but which may be a ring or any other structural feature) which then organizes the whole description of the environment. The rest of the environment is structured concentrically around the focus. This organization starts with a strictly ordered description of the first set of atoms directly connected to the focus [167–168]. Our ketone would be centered on the focus C=O, but then, to select which atom is A_1 and which is A_2 one would again have to go further in the graph to rank them according to DARC hierarchical rules [168]. The number of bonds starting from A (two links in each case) does not allow one to rank the A's, neither does the kind of bond linking A to the focus, nor their nature (both C). However, when we reach atoms of B rank, one is at once top ranked because it is double bonded to A. This priority ensures that A_1 is associated as shown in Scheme 6.

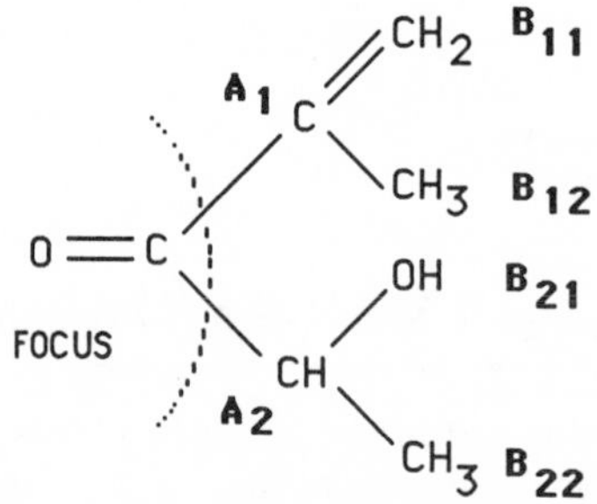

Scheme 6

The numbering B_{21} and B_{22} follows the rule that O (heavier) has priority over C. A simplified graph results: DEL

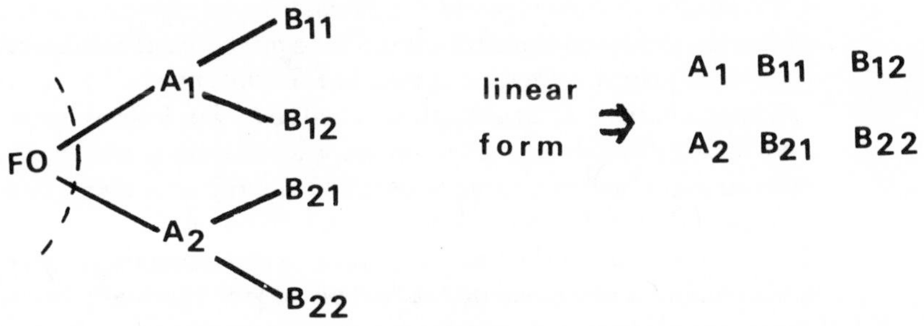

FO — A$_1$ — B$_{11}$, B$_{12}$; FO — A$_2$ — B$_{21}$, B$_{22}$

linear form ⇒

A$_1$	B$_{11}$	B$_{12}$
A$_2$	B$_{21}$	B$_{22}$

Scheme 7

Scheme 7, built from unambiguous rules of priority [168], corresponds to a complete canonicalization of an implicit connectivity table. Therefore, an unambiguous linear representation may be extracted for a condensed representation. It first describes the focus (FO):

$$(1100/2:11/8:1)$$

where 1100 describes the DEX of FO according to rules described later,

2:11 double bond which ends on C (11)

8:1 the atom numbered 1 in the focus is an oxygen (atomic number = 8).

Then the DEL (Descriptor of Limited Environment), which describes only two ranks of atoms concentric with the focus, is given:

$$(2220/2:11/8:21)$$
DEX DLI DNA

DEX (Descriptor of Existence) summarizes the graph under its linear form telling which positions of the graph are occupied (1) by an atom other than H, and which are empty (0):

A$_1$	B$_{11}$	B$_{12}$	B$_{13}$	1	1	1	0	
A$_2$	B$_{21}$	B$_{22}$	B$_{23}$	1	1	1	0	
				2	2	2	0	DEX

DLI (Descriptor of 'Liaison', French for bond), where 2 stands for double bond and 11 defines the node of the graph where this double bond ends. DNA (Descriptor of Nature of Atoms) tells us that the position 21 of the structural graph is occupied by an oxygen (atomic number 8). The general description of all the descriptors is open and precise rules have been written to describe stereochemistry, conformations or any other structural information thought relevant to the problem [171].

One may wonder what causes the amazing efficiency of DARC system in terms of structure searches [181]. It may be viewed as a set of implicit connection tables *putting stress at will on a given structural feature (a view of the molecule from various angles)*. The inconvenience of having several connection tables is largely outweighed by two facts. One is that the topological nature of the representation is used, at best, to get a very compact representation of connection tables. The second is that the system accepts a caricaturization of the compound: one may focus on one part of the structure (as large as one wants) and just neglect the rest if it is considered as not being significant for the searched objective. This is typically what is done in *pattern recognition methods: the selection of invariant features among sets of objects to be recognized* [255]. The great versatility introduced follows from the fact that polyhierarchical approaches are allowed (i.e., one may choose the focus at will). This is an important point for the future molecular descriptions: *hierarchies are important in obtaining consistent descriptions and allowing memory saving representations. For some properties,* however *(particularly biological properties and chemical syntheses), rigidity of hierarchy could be moderated by the creation of several different and consistent hierarchies.*

Hendrickson's description of molecules further illustrates this point [83]. It results from an in-depth study of the chemical changes involved in synthetic pathways in aliphatic and alicyclic systems. Here the focus is on the carbon skeleton of the compound and the functionalities (nature and position) that it bears. The simple notation adopted for representing molecular structure consists of a list of f numbers, which are arbitrary labels given to the various carbon atoms:

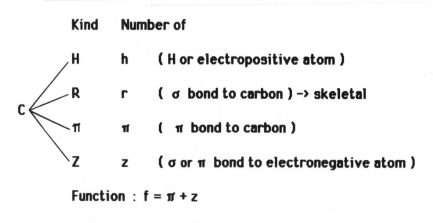

Kind	Number of	
H	h	(H or electropositive atom)
R	r	(σ bond to carbon) -> skeletal
π	π	(π bond to carbon)
Z	z	(σ or π bond to electronegative atom)

Function : $f = \pi + z$

Oxidation state : $x = z - h$

Functional Oxidation State : $x' = 2f - \pi$

Scheme 8

The f numbers of C forming the skeleton of our ketone are:

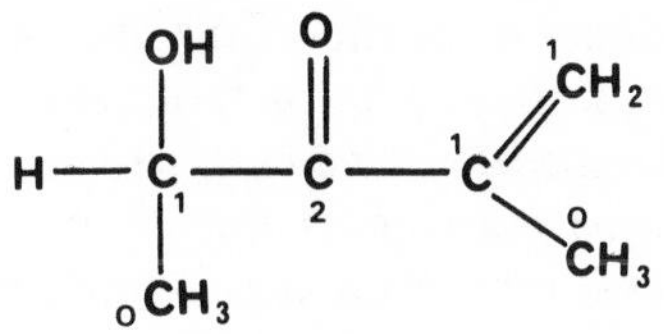

Scheme 9

This skeleton may therefore be represented:

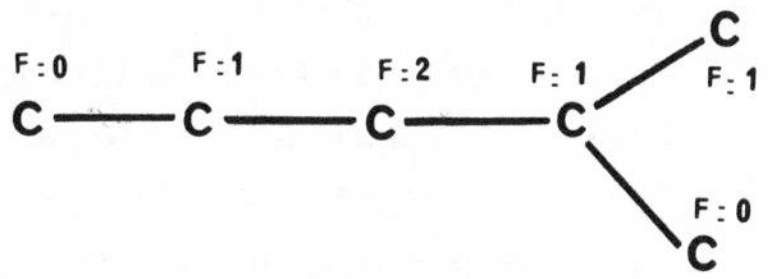

Scheme 10

Hendrickson has then settled the canonicalization of these skeletons in a very logical way [94, 95] based *on the maximization of the linear binary number corresponding to the adjacency matrix associated with the molecular skeleton*. The adjacency matrix is an $n \times n$ matrix of the n skeletal atoms in which the elements are 1 or 0, respectively, for atoms bonded or non-bonded. For our ketone, two possible numberings out of the 6! possible ones are:

	1 2 3 4 5 6 T			1 2 3 4 5 6
1	/ 1 1 1 0 0 3		1	/ 1 1 1 0 0
2	1 / 0 0 1 0 2		2	1 / 0 0 0 0
3	1 0 / 0 0 0 1		3	1 0 / 0 0 0
4	1 0 0 / 0 0 1		4	1 0 0 / 1 0
5	0 1 0 0 / 1 2		5	0 0 0 1 / 1
6	0 0 0 0 1 / 1		6	0 0 0 0 1 /

Scheme 11

In terms of linear string representations, one simply takes the succession of rows above the diagonal. Therefore (A) and (B) are respectively described:

(A)   1 1 1 0 0/0 0 1 0/0 0 0/0 0/1   A > B

(B)   1 1 1 0 0/0 0 0 0/0 0 0/0 1/1

        ↑

We do not need to read the complete binary strings to chose between (A) and (B) because it is clear that the maximum value of the linear string is determined by the position indicated by the arrow. The method proposed by the authors mainly rests upon selecting the starting atom (no. 1) as the one of greatest connectivity or valence. Once the starting atom has been selected one simply seeks the maximum for each successive matrix row, assigning numbers to the skeleton successively as each row is maximized. In this canonicalization, functionalities appear as second order factors which determine the ranking of atoms only when the C-connectivity considerations have been insufficient for the determination. In the conclusion of his most recent report on the subject, [95] Hendrickson notes that for acyclic graphs, the full connectivity is uniquely represented by the T-list (row sums from the maximal adjacency matrix) requiring only $2n$ bits in length (n number of C). Complemented by the R-list [95] for polycyclic graphs, it becomes competitive with the WLN notation for rapid numerical search techniques.

Still other orders of priority are adopted in the EROS description of molecules [53, 57, 61]. For this system several canonicalization methods [53, 57, 61] have been proposed; only Gasteiger's will be given here.

The first step is to define equivalence classes using the criteria of NOON (Number of Outermost Occupied Neighbors) to quantitatively describe a molecule as a town beginning at the center (top ranked) and moving further and further out until the suburbs are reached. For our ketone it gives:

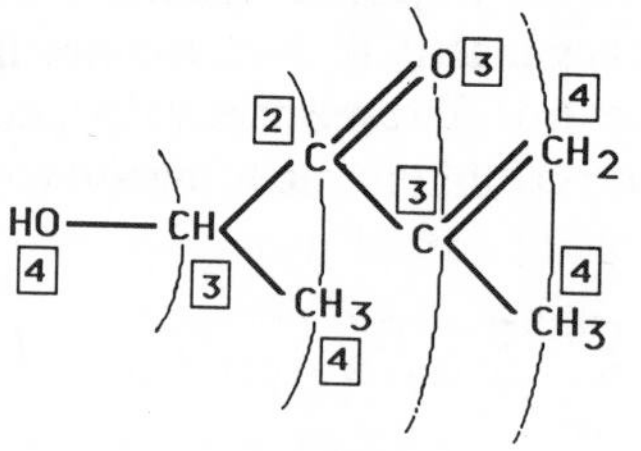

Scheme 12

For every atom one counts concentrically the number of neighbor spheres necessary to accommodate all atoms of the molecule. We have represented this operation for the atom O of OH (we neglect H) by dashed lines which show that the NOON of O is 4. The NOON of other atoms are determined in the same way and indicated on Scheme 12. The center of the molecule is the atom with the minimum NOON: here the C in C=O. Atoms which have the same NOON are equivalence classes and these classes are ordered according to increasing NOON. We have here three equivalence classes (NOON = 2, 3 and 4) and the next step is to rank the atom within each equivalence class. A precise set of priority rules allows this operation. The first is that the atom with the higher atomic number gets priority: therefore in the equivalence class of NOON 3 the first ranked atom will be O. To distinguish between the two carbons left in this equivalence class one uses the rule: the atom which has an α-atom with higher atomic number than the other gets priority. Therefore C bearing OH gets the overall ranking 3 and the other one 4. One then applies the rule of atomic number to find the lowest number in the equivalence class of NOON 4: this indicates O for the number 5. Number 6 is given by the rule 'the atom which has more bonds to the α-sphere gets priority'. And to distinguish between the two CH_3 groups left one uses the rule 'the atom which lies closer to an atom already numbered gets priority'. The overall numbering is therefore:

	1	2	3	4	5	6	7	8	9	10
1	0	2	1	1	0	0	0	0	0	0
2		4	0	0	0	0	0	0	0	0
3			0	0	1	0	1	0	0	0
4				0	0	2	0	1	0	0
5					4	0	0	0	1	0
6						0	0	0	0	1
7							0	0	0	0
8								0	0	0
9									0	0
10										0

Scheme 13

For the sake of clarity we have numbered only the first three ranked H atoms (9, 10, 11). The connectivity matrix of bonds may now be written under a canonical form. In Ugi's approach it displays two specific features. First, the diagonal entries are numerically equivalent to the number of free valence electrons belonging to the atoms defined [63]. This diagonal divides the matrix into an upper part (bond orders) and a lower part. The lower part (and this is the second feature) can represent any property (bond length, bond energy etc.), that exists between the pairs of atoms defining the considered box. Such a matrix is called a BE matrix (bond electron matrix).

The Cahn–Ingold–Prelog [256] treatment of stereochemistry allows us to expect that computer adapted descriptions of stereochemistry should also rely heavily on the kind of hierarchy adopted for ranking the substituents. Lack of space precludes treatment of this point here. Fortunately, these descriptions have been clearly discussed and the interested reader may consult the references directed to this specific point for programs SECS [145, 147], EROS [52, 54, 59, 71], SYNCHEM [126], BERSOHN [99, 101, 109], SYNOPSYS [170, 177], LHASA [40], and PASCOP [196(b), 198(b)]. By this overall sampling of representations (see also Table 3), we hope to have shown that beyond the variety, there is an intrinsic unity provided by connection tables. Furthermore it is now clear that several condensed forms of these tables may be used as unique and unambiguous linear representations within a given set of rules. Other hierarchies or graph numberings [184] may be proposed in the future but there appears to be little room left for fundamental changes.

One important technical point has still to be solved for good communication between the chemist and the computer at the level of molecular description. It concerns the *clean graphical representation of a molecular*

Table 3 Description of molecules

Name	Code	Stereo	Ref.
LHASA	1	Y	35, 38, 40
BERSOHN	1	Y	99, 102, 109
SYNCHEM 2	2	Y	126, 128, 130, 245
AHMOS	3	N	245
SOS	1	N	132, 134
DYNASYN	—	N	188, 189
EXTRUS	4	N	209
HEXARR	4	N	209
SECS	1	Y	145, 146, 147, 152
EROS	5	Y	65, 69
SIMUL	3	N	163, 166
WITHLOCK	6	N	185, 186
REPAS	7	N	190, 191
REACT (Powers)	3	N	190
REACT (Djerassi)	3	N	12 p. 188, 204, 205
PASCOP	1	Y	196
MASSO	3	N	203
GRACE	5	N	208
PSYCHE	8	Y	72
GSS	3	N	117, 245
SAS	3	N	137
CHIRP	5	N	211
CAMEO	1	Y	213, 245
SYNGEN	7	N	93-95
FLAMINGOES	3	N	230, 245
SCANSYNTH	9	N	222
SCANMAT	3	N	245
SCANPHARM	3	N	245
ASSOR	1	Y	74, 245
ASR	7	N	233b
PSYCHO	1	Y	200b
SYNOPSYS	10	Y	176
SEIDEL	—	—	235a
SST	1	Y	159
MICROSYNTHESE	1	Y	235b
TAMREAC	1	N	138b

Code used: 1 = atom, bond tables + binary sets of important structural features. 2 = incidence matrix + canonical linear string (SLING). 3 = connectivity table. 4 = list of the 6 atoms of the studied reaction. 5 = B.E. Matrix. 6 = set of significant substructures. 7 = HENDRICKSON formalism. 8 = atom and bond lists + n, π electrons. 9 = CONOL II. 10 = DARC.

structure from its connection tables. The problem here is to avoid overlapping atoms and bonds and to adopt similar graphical representations for identical or similar structures. For this question, the reader should consult the recent article by Shelley [258]. Even this technical point is still progressing fairly rapidly; recently, LHASA's group has implemented a new graphical interface to reduce even further the communication barrier between the chemist and the computer [14, 49b]. From the beginning, this point was addressed with care in LHASA as testified to by the esthetic quality of drawings in the synthesis of patchouli alcohol [1].

We have devoted some time to the representation of molecules because this first step in man/machine communication is of prime importance for computer-aided synthesis. Presently, the chemists using the programs need no longer worry about the technical aspects: with most programs *they only need to draw their target very simply on a tablet and the software takes care of all the matricial representations*. They should realize, however, that these representations *are not neutral*: they must be manifold because they will serve contradictory aims such as in-depth structural descriptions (pruning based on stereoselectivities) but also simplified or compact description to allow fast comparison with large populations of memory stored structures.

3.1.2 Reaction representations and storage

By the close of the preceding section the first stage of the communication between the chemist and the computer is completed. The chemist can tell the computer which structure (called the *target*) for which he wishes some assistance in synthesis. We must therefore move to the study of the various techniques and philosophies of reaction descriptions.

The most classical approach mimics the organic chemist's way of thinking. This method usually recognizes characteristic structural features in the target which automatically give the chemist a lead to a given reaction or set of reactions leading to the construction of a part of the target. Within this 'close to the chemist' approach, two subclasses may be distinguished: (a) the 'overall transformation' one and (b) the mechanistic one. At a higher level of abstraction we will then describe Hendrikson's and Ugi's reaction approaches.

3.1.2.1 The 'classical' approaches of transformations

The adjective 'classical' may hint at a rather non-innovative approach to the problem. We will precisely show that, at this point, just the reverse is true, using the LHASA example. This apparent paradox is easily understandable: *if the field of synthetic organic chemistry has accumulated innovations for more than eighty years, the biggest pool to tap for adding innovation to the programs is probably still the study of what chemists have done during all this time*. We will describe in the simplest possible case how the computer is taught to perceive a significant structural feature. For more complex situations we will return to the chemist's language in describing 'what' can be taught to the computer without describing 'how'.

The preparation of an amine may be written:

$$C - X + NH \longrightarrow C - N + HX$$

This equation describes a synthetic pathway but we are interested in the retrosynthetic approach:

$$C - N \xrightarrow{\text{transform}} C - X + NH$$

Scheme 14

Therefore, for a computer, the recognition of a possible synthetic pathway and its communication to the synthetic chemist involves three successive steps:

(1) Perception in the target of structural features significant for a synthetic approach. This very important step tries to mimic as much as possible the chemist's view of a compound.
(2) When a structural block suggesting a possible synthetic route is recognized the computer must move one step further ahead trying to evaluate whether the structural environment is favorable or unfavorable for this suggestion (this point will be elaborated upon later), and
(3) In the target, the computer replaces the recognized structural fragment by its precursor (i.e., replaces C–N by C–X + NH). Although we will not treat this technical concept in detail it should be clear that in some complex structures this step (display of the precursor) is far from obvious. Another problem may arise because constitutional symmetry present in target or precursors may lead to redundancies in the tree of synthetic pathways [64].

To partly illustrate how LHASA [34, 36] mimics the operation of recognizing important synthetic features in a given structure, we must return to the atom–bond connection table that we introduced in the section dealing with Morgan's method of canonicalization (although canonicalization is *not* compulsory for substructural perception):

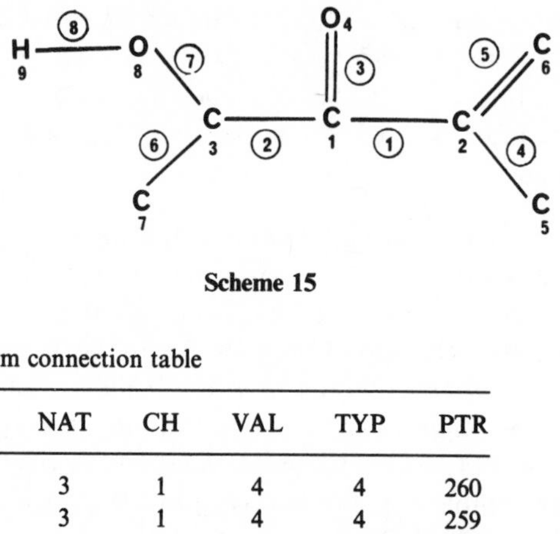

Scheme 15

Atom connection table

At	NAT	CH	VAL	TYP	PTR
1	3	1	4	4	260
2	3	1	4	4	259
3	3	1	4	4	257
4	1	1	2	2	255
5	1	1	4	4	253
6	1	1	4	4	251
7	1	1	4	4	249
8	2	1	2	2	247
9	1	1	1	1	245

where At = atom number; NAT = number of attachments; CH = charge (1 = neutral); VAL = valence; TYP = atom type 4 = carbon, 2 = oxygen, 1 = hydrogen; BD = bond number; BO = bond order; PTR = pointer field.

The foregoing array is self-explanatory except for what is called PTR. This PTR number connects the atom table to the bond table. Indeed, in reading 260 at the end of the row for atom 1 we can follow the process in going to the bond table and searching for the entry 260. This leads us to atom 2 and thus we see that the bond connecting atom 1 to atom 2 is numbered ① and its bond order (BO) is 1. To continue the affiliation, PTR tells us that atom 1 is also connected to atom 3, which is indicated by the value 258 that we found both in PTR of bond ① (last column) and first column of bond ②. The last column of this bond ② tells us that we continue the numbering by travelling to atom 4 through atom 1; this atom 4 is oxygen, which is the last atom connected to carbon number 1. The column PTR therefore indicates ∅. The whole bond table is built with this principle of direct affiliation with the atom table and internal affiliation in the numbering of bonds as shown below.

	AT	BD	BO	PTR
245	8	⑧	1	∅
246	9	⑧	1	∅
247	3	⑦	1	246
248	8	⑦	1	∅
249	3	⑥	1	∅
250	7	⑥	1	248
251	2	⑤	2	∅
252	6	⑤	2	∅
253	2	④	1	∅
254	5	④	1	252
255	1	③	2	∅
256	4	③	2	∅
257	1	②	1	250
258	3	②	1	256
259	1	①	1	254
260	2	①	1	258

Because bonds join two atoms they are stored in duplicate in adjacent locations in the table. This organization of data allows the mechanism of structural feature perception to be founded on the notion of a *set*. An atom is either a carbon (1) or is not (0), a bond is either a double bond (1) or is not (0). Therefore any Boolean condition related to the structure of the compound represents this truth or falsehood of information *in a two-word data structure called a set*. The following sets illustrate the perception of heteroatoms O, of a carbon bearing electron withdrawing atoms, of the stereocenter and of doubly bonded atoms at the level of a linear representation of atoms array [38].

OXYGEN	0	0	0	1	0	0	0	1	0
EWA	1	0	1	0	0	0	0	0	0
STRCNT	0	0	1	0	0	0	0	0	0
DBONDAT	1	1	0	1	0	1	0	0	0

The bond table may be used to represent the perception of carbon heteroatom bonds, axial substituents on a cyclohexane ring, bonds in a ring, etc. Set notation permits all operations of Boolean algebra to be employed (and, or, exclusive or, and not). For example, if the program wants to identify a C=O in the foregoing structure it will make an 'and' between the sets DBONDAT and OXYGEN (because we know that only C and O are present in the target).

OXYGEN   0 0 0 1 0 0 0 1 0

DBONDAT 1 1 0 1 0 1 0 0 0

Only atom 4 fits both sets so the C=O bond is identified. Set oriented data are associated with a great gain in computational efficiency because they allow the extremely valuable property of parallel processing. The LHASA program contains more than 200 structural sets, such as:

HETBND	Heterobonds	ATOMI	Atoms with single bonds
ARMATM	Aromatic atoms	STWDGB	Stereo wedged bonds
STRBDS	Strategic bonds	STDOTB	Stereo dotted bond
RNFUSB	Fusion bonds	WTHBND	Bond with withdrawing character
RNBDGA	Bridgehead atoms	etc.	

For any structural data whose size is variable and determinable only during the run time (for example, the order in which the atoms and bonds are placed around a ring in polycyclic structures) a more elaborate perceptual algorithm called a *list structure* has been used. Lists are collections of information doublets, one part being the datum and the other being the address of the next related datum. This approach has been particularly useful in the perception of functional groups.

Although we have used the simplest problem of perception to show 'how it works', the reader should realize that the algorithms which permit the perception of some structural features are reminiscent of Chinese puzzles. The problem of perceiving all the rings present in a structure is such a puzzle for some natural products. In 1974, three different algorithms had already been implemented and used in LHASA to perceive such rings (ref. 34, p. 56). The nature of the challenge is illustrated by the fact that several of the groups working in the field of computer-aided synthesis have at least one publication dealing with this problem. LHASA [35, 37, 41, 45] EROS [67], DARC [173], Hendrickson [95], AHMOS [120], CAMEO [214], Bersohn [98, 100], and SECS [149, 193]. From a chemical point of view, two important consequences result from the current thinking about these structural situations: one is the distinction between *real rings* and *pseudo-rings*, the other is the notion of *strategic bonds*. The chemist recognizes at once the difference between a real and a pseudo-ring as illustrated by the following examples [1], where the circled numbers indicate real rings.

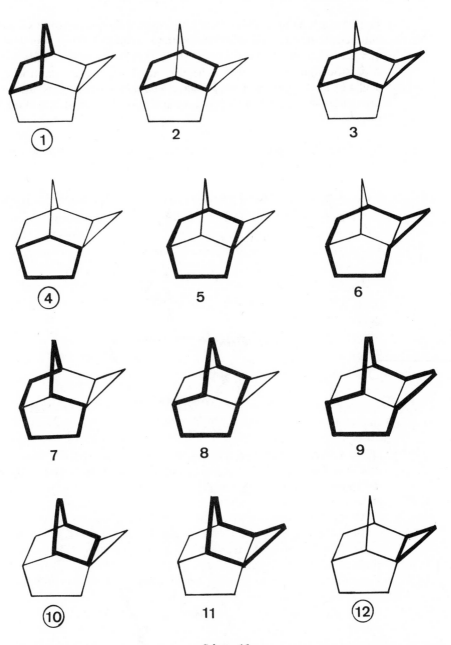

Scheme 16

For each bond in a molecule, the smallest ring containing that bond is called a real ring. Pseudo-rings are the pairwise envelopes of real rings with the restriction that the size of the envelope be seven or less [45]. The second important consequence is recognizing and teaching the computer to perceive strategic bonds in polycyclic systems [41]. A strategic bond may be recognized in that the transform operating on such a bond generates especially simplified precursors. 'Simplified' requires that the precursor has a minimum number of side chains, a minimum number of chiral centers, and a minimum number of bridged rings in the total number of rings. This is usually understood intuitively by the chemist but the computer has to be taught analytically to do so. Corey established six simple rules allowing the perception of such strategic bonds (there may be several in a given target) [41].

(1) A strategic bond is usually part of a 4, 5, 6 or 7-membered ring.
(2) It is directly connected to another ring, provided that this ring is not a 3-membered ring
(3) It must be part of a real ring having the maximum number of bridgehead atoms, such as is shown:

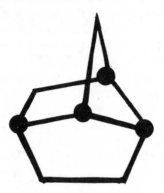

Scheme 17

(4) To avoid precursors having rings of a size greater than 7, one discards as strategic bonds those at a junction of rings whose overall envelope is a larger-than-7-membered ring:

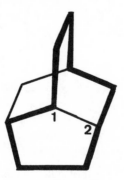

Scheme 18

1–2 is not a strategic bond since a transform operating on it would generate the 9-membered ring drawn in bold.
(5) Bonds belonging to an aromatic ring are not strategic

(6)   Bonds belonging to a ring which possesses a chiral center may or may not be strategic. They may be if the chiral atom is one of the ends of the bond, they may not be for all other situations.

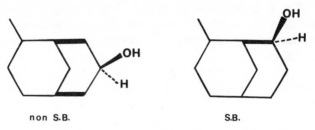

Scheme 19

These rules were given for C–C bonds only. For C-heteroatom bonds only rules 4, 5, 6 and part of rule 2 still hold. We defer the discussion of the utility and limitations of these strategic bonds to the section dealing with strategy; at this point, however, we assume that the computer is now able to perceive them in any target. At this stage the computer is also able to perceive those bonds which are not members of rings but are directly attached to rings: they are gathered in a set called appendages. Jorgensen later introduced the distinction and recognition of 'ring' and 'branch' appendages [42]. *A ring appendage* is a group of atoms linked to a ring in such a way that the link is not itself part of another ring; the non-ring end of the link may be either a carbon or a heteroatom. If it is a heteroatom it must have at least two carbons attached. *A branch appendage* must include at least three carbon atoms and originate on *a non-ring atom* that has a total of three or more attachments other than H (a double bond counts as two attachments). These distinctions are important in stereochemical strategies. LHASA also perceives aromatic rings according to the $4n + 2$ rule.

If one wishes the computer to generate reasonably sophisticated sequences in natural product synthesis, stereochemical strategies and evaluations must play a major role in the analysis. The first step in this direction is for the program to have access to the representation of the stereorelationships within the target molecule. Both LHASA [34, 49(b)] and SECS [146] are able to perceive:

(1)   *cis/trans* relationships between pairs of ring substituent atoms.
(2)   *cis/trans* relationships between pairs of double bond substituent atoms.
(3)   Double bonds with E or Z character.
(4)   Axial and equatorial substituent atoms [47] on rigid ring systems. This includes the elimination of possible ambiguity when an atom may be axial on one ring and equatorial on another (by reference to a *specific ring*)
   — *cis* and *trans* ring fusions
   — R and S configurations of all stereocenters.

The constraint placed on the chemist is to consider each asymmetric center *individually* during input, ensuring that each center is properly defined by the attached indicator bond. Attempted three-dimensional representation of a bond as dotted or wedge-shaped causes an entry to be made at the appropriate position in the internal bond table.

The recognition of functional groups involves atom-by-atom matching between some part of a current target structure and a table listing the structural requirements for functional groups [35, 45, 49(b)]. More than 60 functional groups were perceived by the first versions of LHASA; this program has recently been modified in collaboration with Pr. A. P. Johnson to allow it to recognize an unlimited number of different functional group types (see note 8 in ref. 49(c)). An important aspect of such a detailed perception in LHASA [49(b)] and SECS [152] is that when a reaction in a given part of the target is being proposed, information is provided to decide *whether this reaction will affect another part of the target*. Furthermore, LHASA was recently increased in efficiency by introducing a set of programs designed to assist the chemist in the selection of protecting groups. PRO-TECT is a program which accesses a data base of reactivities of 228 protective groups versus 108 prototype reaction conditions [49(c)]. In these projects, *a transform* is viewed as a description of what a particular chemical transformation is and the factors which affect it when the transformation occurs or does not occur. Specific languages, ALCHEM in SECS [152], and CHEMTRAN in LHASA [49(c)], have even been created to describe precisely every transform. In LHASA 11, the treatment of transforms covered by the program (1100 reactions) makes of each of them a mini-review oriented towards organic synthesis [49(b)]. For every listed transform, the following information is given [156]: name, reference, substructure (which describes the necessary structural environment for the reaction considered), priority (which indicates the plausibility of the transform), character (which describes the kind of structural modification made by the transform), conditions, scope and limitations. Since the transform library is not part of the SECS program, new transforms are easily added. A precursor to this in-depth approach of a given reaction was proposed by Dubois for the special case of ketones [10, 171] and shows the amount of work needed to exhaustively describe a reaction.

One important feature of Bersohn's program is that it 'compares' the target with potential starting materials. Improvements in this direction have recently been made by writing an algorithm for finding the *intersection of molecular structures*. The method initially constructs a list of those atoms in one molecule which have a matching atom in other molecules and whose neighbors match the corresponding neighbors in the other molecules, and so on [111].

The perception of all relevant structural features in a target opens the door to retrosynthetic analyses based on overall transformations as well as to mechanistically oriented approaches. The SOS program has shown that mechanistic approaches to a target can lead to innovative solutions [133–136]. LHASA 11 transforms usually include information on the mechanism of reaction to make the evaluation of structural and medium effects on it easier [49(b)].

3.1.2.2 Theoretical approaches to reactions   For his sequence of constructions with no refunctionalizations [93], Hendrickson has defined a reaction
by the *net structural change* from substrate to products as measured by
changes in σ and f at a given carbon, as defined in Section 3.1.1. This results
from an overview of all possible reactions that a carbon center may undergo.
The three variables taken are the number, σ, of C–C bonds linked to a
carbon, the number, h, of hydrogen atoms linked to the same carbon and the
value, $x = f - h$, of its oxidation state. Therefore a diagram may be drawn
for all possible reactions [83, 90]: Scheme 20 quantifies the net structural
change [83] occurring in any transformation:

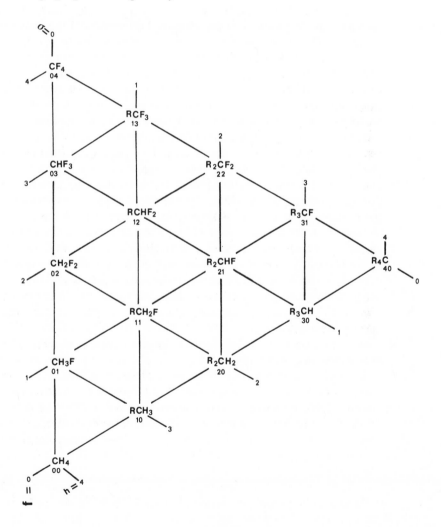

Scheme 20

As a consequence, one notes that a compound of the kind R_2CHF may be
prepared by at least six different methods (through $RCHF_2$, R_2CF_2, R_3CF,

R_3CH, R_2CH_2, or RCH_2F), and actually by at least seven because of the possibility of a substitution reaction in which none of the three basic values (σ, h, x) changes. One sees also that CF_4 or R_4C may be obtained by only two routes. Out of these reactions one may extract only those corresponding to construction reactions forming C–C bonds. The interesting point of this diagram is that it covers, in a consistent way, the one-step reactions leading to a substrate without omitting any (Camp and Power, adding the site characteristics for heteroatoms, obtain a pyramidal form of the diagram and thus 132 reactions in place of the 70 provided by the triangle [190]). It also obviously states which node by node pathways are available to transform one kind of carbon (say HCF_3) to another one (say R_2CHF), and which appears to be the simplest route. Hendrickson further elaborated this diagram [91], but this basic form is sufficient to understand his reaction treatments. Here, in place of having to perceive a specific structural feature present in the target or one of the precursors, the computer has only to recognize *a generalized structural situation*. This structural situation, by its succession of $f\pi$ and f values associated with each carbon of the skeleton, suggests a large number of possible synthetic routes. At this point, there is apparently no need for a set of actual reactions and the computer may be more creative (perhaps too creative) than the classical programs relying only upon existing reactions. One may remark that with a large research group it would be perfectly possible to systematize the already known reactions according to this overview of chemical transformations, the reward of this huge task would be the realization of a more realistic bank of reactions for the Hendrickson program but, in addition, to identifying the holes to be filled in synthetic chemistry [87]. Hendrickson himself [85], by a critical examination of the Diels–Alder principle of reaction identified such gaps for a reaction which had been studied for more than fifty-five years. T. Brownscombe [209] has computer generated all possible unimolecular pericyclic reactions (see also ref. 72). *Such thorough examinations of important reactions should be developed more in the future.*

Ugi and his group have developed a mathematical approach to reactions apparently quite far separated from the chemist's usual representation of transformations. The system is based on the proposition that all chemical reactions correspond to interconversions of isomeric ensembles of molecules (IEM) within a family of isomeric ensembles of molecules (FIEM) [51, 52, 55, 58, 61, 65]. One may illustrate this proposition for the reaction:

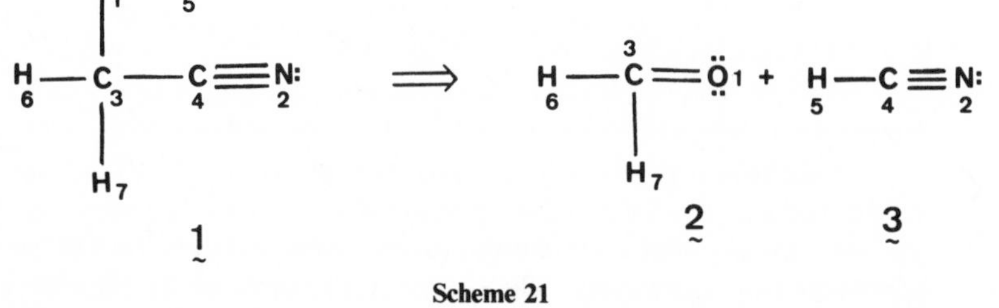

Scheme 21

The BE matrix of **1** is:

	1	2	3	4	5	6	7
1	4	0	1	0	1	0	0
2		2	0	3	0	0	0
3			0	1	0	1	1
4				0	0	0	0
5					0	0	0
6						0	0
7							0

One may determine the numbering in reagents **2** and **3** as if we had kept track (a kind of labelling) of each atom during the reaction. This results in a BE matrix for the reagents:

	1	2	3	4	5	6	7
1	4	0	2	0	0	0	0
2		2	0	3	0	0	0
3			0	0	0	1	1
4				0	1	0	0
5					0	0	0
6						0	0
7							0

Comparing this matrix with the one first written, one sees that in order to obtain the second matrix, one has only to add the following matrix to the first matrix:

	1	2	3	4	5	6	7
1	0	0	+1	0	−1	0	0
2		0	0	0	0	0	0
3			0	−1	0	0	0
4				0	+1	0	0
5					0	0	0
6						0	0
7							0

This matrix is called the R-matrix (reaction matrix) [53]. Returning to the Hendrickson triangle, one would expect that for the formation of this target, seen as an RCH_2F type, there would be at least seven main routes to reach it. Some of them would form **1** together with another compound (H_2O, NaCl, CO_2-etc.) so that the BE matrix containing the target and a standard set of by-products would be submitted to the action of R-matrices to locate all the isomeric ensembles of molecules (IEM). Without further information one could guess at this point that so many R matrices will arise to cover all the known transformations that the main problem with EROS would be a combinatorial explosion. An outstanding series of studies by Bart and

Garagnani has shown that this is not the case. Analyzing a set consisting of about *1900 C–C bond formation reactions*, they showed that only *thirty general R-matrices* are sufficient to describe them all. Furthermore, over 92 %: of all organic reactions are covered by ten general reaction matrices [260]. This pattern of hierarchization and consistent organization of the ensemble of synthetic reactions was confirmed on studying organic name reactions [261] and rearrangements [262]. Combined with an efficient method of canonical numbering [79] of reaction matrices, these results hint at a bright future for the DU model in reaction information handling systems. *This kind of hyperbolic distribution seems to be quite a general phenomenon.* We observed it for reaction schemes in the preparations of heterocycles [131], and Lynch [257] did also for functional group interconversions.

One positive point about the R-matrix is that it allows synthetic (forward) as well as retrosynthetic representations. IGOR exhaustively generated the irreducible matrices under chemical boundary conditions and found 100 irreducible R-matrices which correspond to distinct reaction principles [77(b)]. Since the known reactions of organic chemistry belong to twelve distinct reaction principles, it should follow that there are 90 virgin reaction principles. *This announcement constitutes a formidable challenge for synthetic chemists even if many of the 88 reaction principles are chemically unfeasible* [76]. Another advantage of R-matrices is that they give hints to the mechanism involved because a complementary look at the diagonal and non-diagonal terms of the R-matrix provides a view of the electron displacements involved in the reaction [61, 77(a)]. Similar to this 'electron pushing' approach, but closer to the chemist's representation, is the description of reactions as adopted in MASSO [203]. This approach is a combination of Hendrickson's half-reaction definitions [87] and of an electron pushing reaction view. Any construction reaction consists of two partial synthons, each of three or fewer linear carbon atoms functionalized as shown:

$$\overset{\gamma\ \ \beta\ \ \alpha}{C-C-C} + \overset{\alpha\ \ \beta\ \ \gamma}{C-C-C} \longrightarrow \overset{\gamma\ \ \beta\ \ \alpha\ \ \alpha\ \ \beta\ \ \gamma}{C-C-C=C-C-C}$$

Scheme 22

Each partial synthon may be considered separately and independently and the construction at each may be called a half-reaction, as illustrated below:

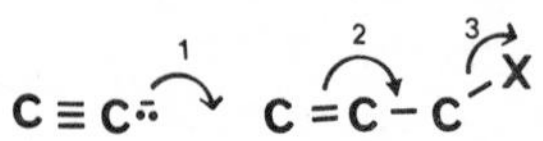

$$C\equiv C\overset{1}{\cdot\cdot} \quad C=\overset{2}{C}-C\overset{3}{\cdot}X$$

Scheme 23

First half-reaction: formation of C–C bond at arrow 1. Second half-reaction: migration of the double bond (arrow 2) and loss of X (arrow 3). The sum of the two half-reactions makes an overall reaction. The movement of electrons, indicative of a mechanistic scheme is *not* compulsory. Moreau [203], introduced it, but at the beginning, Hendrickson stressed that the half-reaction representation does not imply any mechanistic considerations. Since the reaction is created from two half-reactions, the parts of the structures being modified during the reaction are called half-spans. One may choose these to be as large as one wishes but for most practical purposes, Hendrickson limited their size to three atoms (numbered α, β, γ).

Thus any half-reaction on one synthon may be coupled to a whole family of partner half-reactions on the other synthon to produce a bond. Moreau [203] has applied this principle in the writing of MASSO and has extended it to carbon heteroatom links, generating a series of about 90 half-reactions. Some are negative (when one notes that the α-carbon in the synthon displays a decrease in its electronic population ($\Delta x_{C_\alpha} = +1$) and some are positive (for the opposite situation ($\Delta x_{C_\alpha} = -1$)).

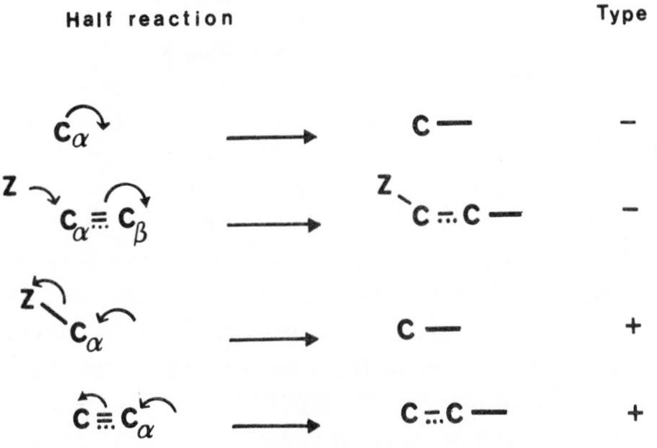

Half reaction   **Type**

Scheme 24

The formation of a bond results from the combination of two half-reactions of opposite type. Therefore, to suggest methods for forming a bond between C_1 and C_8 in the following target:

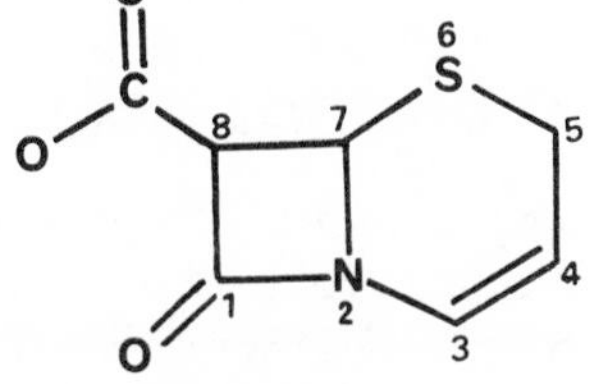

Scheme 25

The program will automatically find the structural fragments centered on C_8 and search in its files for all possible half-reactions compatible with the functionality of these fragments. It will do the same with the C_1 end of the bond and then combine complementary half-reactions to propose schemes, such as:

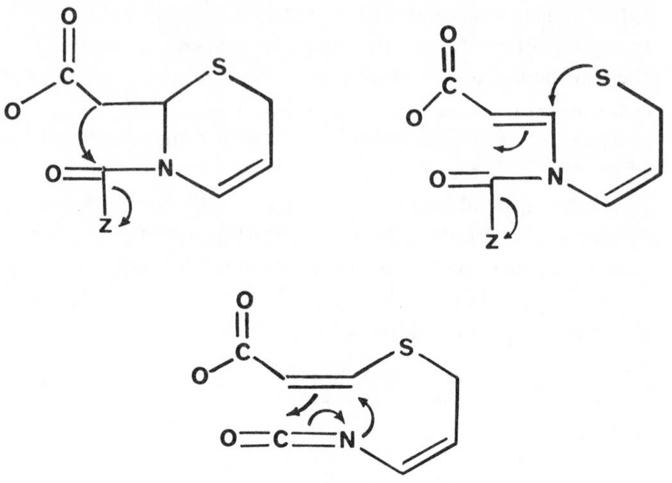

Scheme 26

Some proposed schemes are unsound, but the propositions may be highly innovative since *one of the benefits of the half-reaction approach is to overgeneralize all the reaction schemes already known in synthetic chemistry, plus the creation of some not yet explored.* The attractiveness of Hendrickson's model was also investigated by Ghose to propose an algorithm for computer assisted design [96]. In this work, problems arising when a half-span bears several functionalities are thoroughly discussed, as well as strategies taking into account the lack of selectivity expected from the combination of certain half-reaction couples.

Although lack of space precludes our developing all the non-empirical approaches to synthetic reactions, we cannot close this section before calling the attention of the reader to the recent highly flexible and coherent representation of synthetic reactions proposed by Dubois and co-workers [174]. Table 4 tentatively summarizes the various methods adopted for describing reactions in CAOS.

3.2 Retrosynthetic and synthetic approaches

In all the foregoing examples the program was designed to answer the question 'What synthetic routes do you propose for obtaining this target molecule?'. To answer this question, the program would analyze, in a retrosynthetic way, the possible first-rank precursors, then investigate each of the interesting precursors as a target and so on. *In the process of emulating the methods of a good synthetic chemist, however, it became clear that an efficient plan has to perform both synthetically and retrosynthetically.* [49(b)] Indeed, it would sometimes be important to be able to indicate what kinds of secondary products may be formed. Therefore the computer must be taught to answer

Table 4 Description of reactions

Name	Code	Number	Biblio	Ref.
LHASA	T-F-1	1100	—	36, 45, 49b
BERSOHN	T-F	1000	—	103
SYNCHEM2	T-F	1000 (a)	f	124–130
		200 (b)		245
AHMOS	M-2	8	—	115, 245
SOS	T-M	20 (c)	f	133, 134
		17 (b)		143b
DYNASYN	3	3	—	188, 189
EXTRUS	4	1 (c)	—	209
HEXARR	4	1 (c)	—	209
SECS	T-F-5	3000	f	146, 152, 245
EROS	U	3 (d)	—	18c, 65
SIMUL	T-F	—	—	163, 166
WHITLOCK	T-F-6	—	—	185
WHITLOCK	T-F	—	—	186
REPAS	H	—	—	190
REACT (Powers)	T-F-7	300	—	190
REACT (Djerassi)	T-F	—	—	12, 204, 205
PASCOP	T-F-5, 8	750	a, b, c, f	196–199, 245
MASSO	H	90	g	203, 245
GRACE	H	—	—	206, 208
PSYCHE	4	1 (c)	—	72
GSS	T-F-9	8000	e	117, 245
SAS	10	0	g	137
CHIRP	U	9	—	211, 212
CAMEO	T-M	(e)	a, b, c, d, f	213–220, 245
SYNGEN	H	32 (f)	g	93, 245
FLAMINGOES	11	—	—	225–232, 245
SCANSYNTH	T-F	468	b	221, 245
SCANMAT	12	40	g	245
SCANPHARM	12	40 (g)	g	245
		200 (h)	g	245
ASSOR	U	—	g	18c, 74, 245
ASR	H	—	—	233b
PSYCHO	T-F	20 (i)	a, b, c, f	200b
SYNOPSYS	13	—	—	171, 175
SEIDEL	—	—	—	235a
SST	no reaction	0	g	159
MICROSYNTHESE	T-F	130	—	235b
TAMREAC	T-M, F	64	f	138b

code used: T = Trilogy: research of substructure + tests + transformations to build the precursors.   F = description of precise substructures, functions, etc..   M = mechanism.
H = Hendrickson formalism. U = Ugi formalism. 1 = CHMTRN language. 2 = Elementary donor-acceptor reactions. 3 = 3 classes of reactions (protection of nucleotides, single strand condensation, duplex joining). 4 = pericyclic reactions. 5 = ALCHEM language. 6 = linear compounds only. 7 = reactions are subroutines. 8 = CLASS language. 9 = synthon transformations automatically extracted from reaction equation. 10 = skeletal dissection. 11 = cyclic rearrangement. 12 = reaction generators. 13 = DARC/IGLOO system.
Number: a = organic chemistry. b = metabiotic pathways. c = mechanisms. d = R-categories. e = infiny due to mechanistic approach. f = reaction generators. g = generators. h = real industry. i = program is still under development.
biblio: a = current contents. b = current reactions. c = chem. abstr. d = derwent. e = SPRESI doc. system. f = others. g = not necessary.

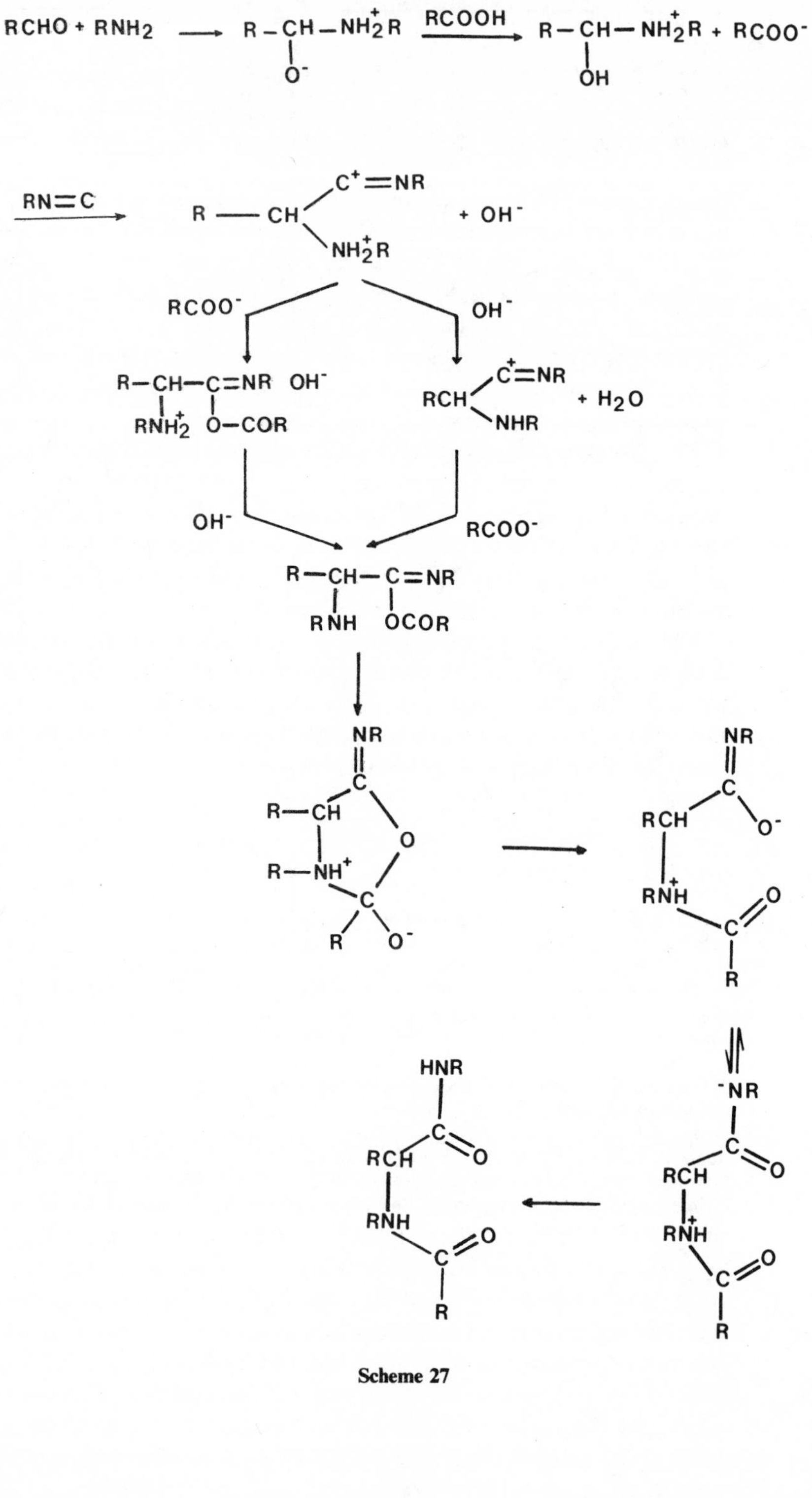

$$RCHO + RNH_2 \longrightarrow R-CH-NH_2^+R \xrightarrow{RCOOH} R-CH-NH_2^+R + RCOO^-$$

Scheme 27

the question 'What product should we expect if we mix the set of reagents A + B + C?

Yoneda in Japan addressed this question very early [206] in the course of computer aided synthesis, but the paper appeared in a Japanese language review which dealt primarily with heterogeneous catalysis, thus it went unnoticed for a long time. GRACE [208] is able to generate elementary reaction networks for simple reactions including free radicals, ions, and even active sites in heterogeneous catalysis. Similar to EROS, but independent of it, GRACE uses square symmetric matrices to represent reactants and products; a linear input, however, is possible [207]. Submitting the reaction of ethylene with H_2 in the presence of a heterogeneous catalyst to GRACE provided two reaction schemes: one corresponding to the associative adsorption mechanism for hydrogenation, the other corresponding to the dissociative adsorption mechanism for isotope exchange.

The German Democratic Republic project AHMOS/GSS was probably the second to address this question [117]. The AHMOS synthetic program is capable of predicting the products of simple reactions. To do so, it employs the nucleophilic–electrophilic properties of the reagents [113]. For example, given the starting reagents RCHO, RNH_2, RN=C, and RCOOH it would propose the scheme displayed in scheme 27.

Although the mechanistic scheme is not totally sound (for example, OH⁻ is viewed as a leaving group, and the treatment of acid catalysis is not totally integrated; compare with ref. 263), the final product obtained out of a mixture of these four reagents is correctly predicted. Working in a forward direction the program was able to propose a new synthesis of the following target:

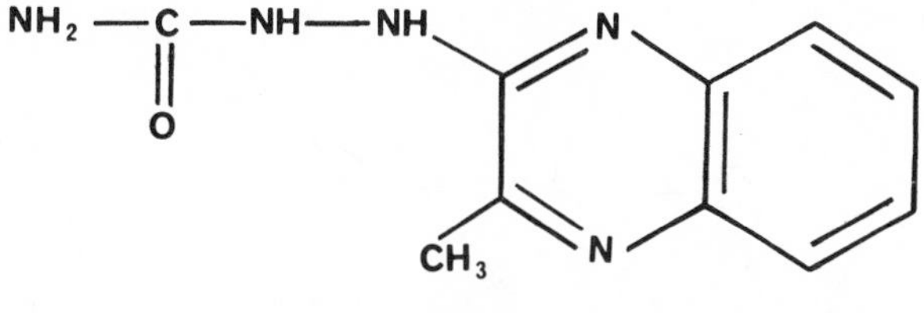

Scheme 28

This synthesis was experimentally checked (55% yield starting from *ortho-phenylenediamine*) [121].

The SOS forward approach was initiated in two distinct areas of chemistry because of external queries. The first 'mechanistic treatment of catalysis' (TAMREAC) was the result of Marseille's chemistry department challenging us to initiate research at the frontier between organic and inorganic chemistry. The second was a problem that G. Vernin, analyzing very complex mixtures of heterocycles involved in flavor [268] chemistry, submitted to us. For the first project, we taught the computer the basic reactions involved in the mechanisms of transition metal induced catalysis [264–267]. This accomplished, we then asked the computer 'what succession of steps could arise when one mixes ethylene and a transition metal complex?' The computer proposed a great number of possibilities, among which one was recognized by M. L. H. Green but not yet considered in the literature, although theoretically sound [138(a)]. The critical intermediate:

Scheme 29

was proposed because it provided an explanation of the experimental result that no trimers of ethylene are found in the reaction products [138(a)]. Independently and at about the same time, Schrock made a similar proposal [182]. This project is being developed in the direction of more efficient pruning of proposed steps with regards to the currently emerging understanding of organometallic reactivity [138(b)]. Prior to this work, Schleyer and Gund [148] had used the capabilities of SECS to generate structures according to chemist's specifications to study the possible mechanisms of rearrangement of pentacyclotetradecanes to diamantane:

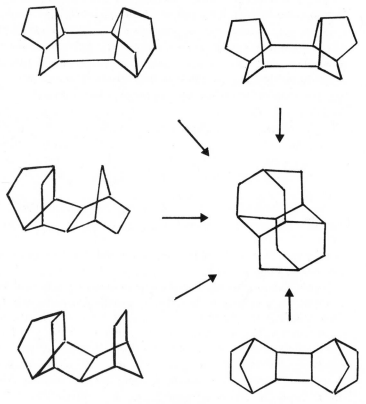

Scheme 30

Graphs of isomeric hydrocarbons were generated from a chosen precursor by 1,2-alkyl shifts adopting certain assumptions to simplify the number of possible intermediates ($\gg 40\,000$). As SECS contains a module which calculates rough strain energies, the strain energy of each intermediate was calculated, and low energy pathways were proposed for these rearrangements [148].

The second forward project corresponds *to the construction of a complementary tool for the analysis of highly complex mixtures*. Such complex mixtures are formed in cooking and are responsible for the delicacy and the flavor of a meal. The chemistry of cooking involves the reactions between sugars, amino acids and their degradation products. Several model reactions have been proposed to rationalize this chemistry [268]. One of them begins with the mixture:

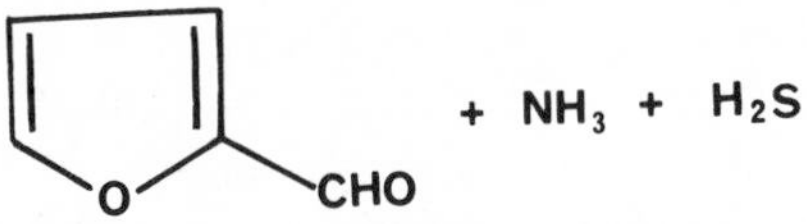

$$+ \; NH_3 \; + \; H_2S$$

Scheme 31

Teaching the computer all the important forward reactions of these reagents and between them and their first rank products leads to the generation of a highly complex mixture [139, 140]. The comparison of these structures with experimentally identified compounds in actual flavor mixtures shows that most of the identified structures were, indeed, predicted by the computer. The computer predicted many more products than were originally identified, among which G. Vernin has been able to experimentally discover some not yet identified. Of course, to be fair, part of the overproduction of structures also originates in a too loose treatment of reactivity rules associated with the considered substrates. This example *shows that computer aided synthesis has interest far beyond the realm of simply synthetic chemistry*.

Previous to this study, Djerassi's group had reported work which further supports the preceding proposition. The REACT program, an extension of CONGEN, is designed to carry out representations of chemical reactions carried out in the synthetic direction [204, 205]. A given compound is subjected to the repetitive application of a set of reactions with the double purpose of simulating the biosynthesis of natural products and of suggesting solutions to a structure elucidation problem. For example [205], the C_8 compound:

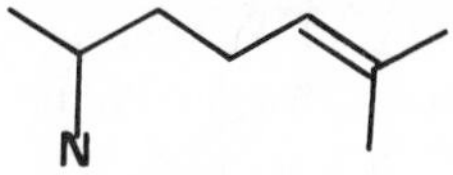

Scheme 32

was subjected three times to the sequence of reactions displayed in the following tree (each node in the tree constitutes a new product).

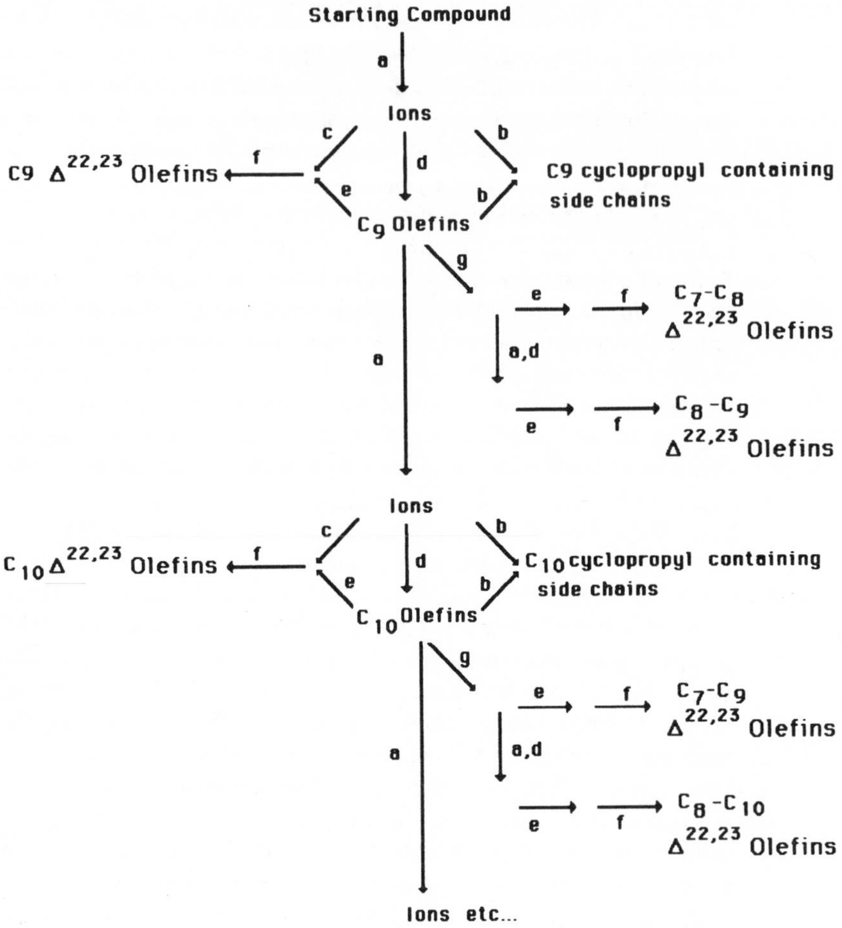
Starting Compound
a
Ions
c
b
d
C9 Δ22,23 Olefins
f
e
b
C9 cyclopropyl containing side chains
C9 Olefins
g
e
f
C7-C8 Δ22,23 Olefins
a
a,d
e
f
C8-C9 Δ22,23 Olefins
Ions
c
b
d
C10 Δ22,23 Olefins
f
e
b
C10 cyclopropyl containing side chains
C10 Olefins
g
e
f
C7-C9 Δ22,23 Olefins
a
a,d
e
f
C8-C10 Δ22,23 Olefins
Ions etc...

Scheme 33

In the tree, the reactions are: (a) C-methylation of a double bond, (b) rearrangement to cyclopropyl systems, (c) quenching to form saturated side chains, (d) proton elimination and formation of a double bond, (e) reduction to form saturated side chains, (f) oxidation, (g) degradation to shorter side chains *via* loss of allylic methyl groups. Combining the set of side chains so generated with the seven most common steroidal skeletons, these authors generated 1 778 possible sterols. Separating the structures by molecular weight reduces the number of candidate structures to be considered in a given problem. Thus, in a GC/MS experiment, the maximum number of structures for a sterol of molecular weight 292 is 264. This number drops further if any other spectroscopic data are available (see Chapter 7). REACT is even more impressive in *the modelling of labelling experiments with the goal of the elucidation of biosynthetic pathways.* Being able to monitor the transformation of the precursors to the products, the program follows the isotopic labels throughout a reaction sequence, and allows, therefore, a systematic investigation of all of the possible aspects of a proposed experiment so as to avoid any ambiguous experiment. Doing so, it constitutes *a powerful new tool for efficiently choosing the labelled precursor which will provide the maximum information*. Istin and Grognet recently used SOS to perform a comparable study on the metabolites to be expected in the biological degradation of sulfiride, an organophosphorous pesticide [143(b)].

DINASYN [188] was devised for optimizing the *synthesis of bihelical deoxyribonucleic acid*. Only three different classes of reactions are used to transform the four mononucleotides (adenine, cytosine, guanine, thymine) into a DNA strand; these are protection of *mononucleotides*, single strand condensation, and duplex joining. Using an evaluation of the time required for reaction, separation, and analysis, Powers calculated the time required for a different total synthesis of the alanine gene actually performed in twenty man-years by Khorana's group. Optimization of the synthesis path of DINASYN predicted *that only 50% of this time would have be required if the optimized synthesis had been followed*.

CHIRP [211] uses a limited number of R-matrices (hydrogenation, cracking, oxygenation, alkylation, hydration, halogenation, dehydrogenation, dehydration, and deshydrohalogenation) [212] to find alternate reaction paths, products and by-products *for a process of industrial interest*. The system has the capability to carry out the reaction to the desired number of stages. The first stage is the application of the R-matrices to the starting reactant set. In the next stage each of the generated products undergoes the same reaction that the parent reactant set underwent and so on.

CAMEO [219] mimics the mechanistic reasoning of organic chemists and predicts the products of organic reactions, given starting materials and conditions. It is able to deal with (1) base catalyzed and nucleophilic chemistry [213] (2) electrophilic chemistry in which the key reactive species are carbonium ions (3) ylide chemistry and the organometallic chemistry of lithium, magnesium and lithium cuprates [215], (4) thermal pericyclic processes, cycloadditions, electrocyclic reactions and sigmatropic rearrangements [218, 220(a)], and (5) reactions of unsaturated electrophiles including nucleophilic aromatic substitution [220(b)]. During the perception phase, it recognizes rings, functional groups (over 100), aromaticity, and stereochemistry. It also recognizes the potential nucleophilic and electrophilic sites. The perception is even further refined by estimating the pK_a values for sites with acidic hydrogens, and checking for unstable functional groups, tautomers and overly strained rings [214]. One important choice behind CAMEO results from the belief that hundreds of reactions simply involve different combinations of the same few fundamental mechanistic steps. *So, rather than creating tables for innumerable specific reactions it seems preferable to thoroughly understand the behavior of the fundamental steps and the competition between them.*

Faced with the problem of possible products when 2-bromocyclohexanone (1) is treated with sodium methoxide in methanol, CAMEO first perceives five potential nucleophilic sites [219]:

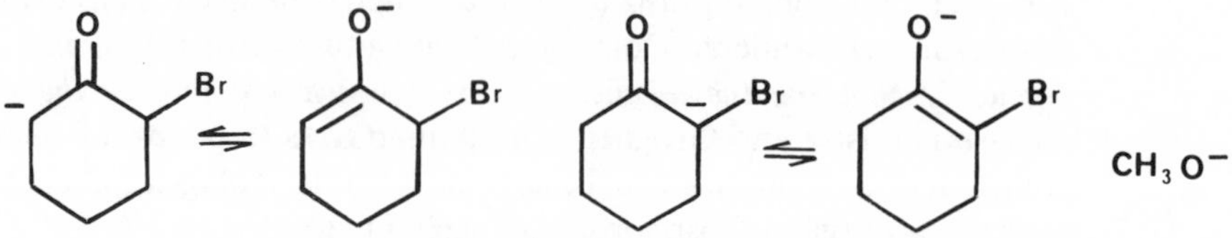

Scheme 34

It then recognizes the electrophilic centers in the starting material (C=O and C—Br). The next step combines electrophiles and nucleophiles along S_N2, Ad_N, E_2 and ElcB pathways. Both intramolecular and intermolecular situations are dealt with. The products obtained are either saved or eliminated by the chemist and then themselves submitted to further elementary steps. The synthetic tree below is constructed after four steps.

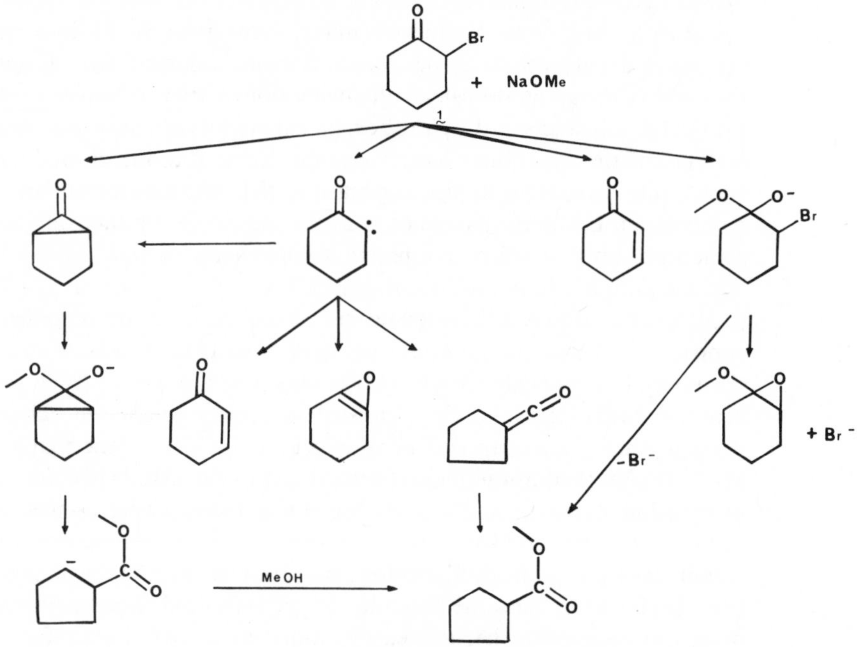

Scheme 35

The Favorskii rearrangement product is correctly predicted, as is the formation of several by-products [219]. The computer recognizes three possible mechanisms as explanations of the Favorskii product. CAMEO has also been adapted to deal with organosilicon chemistry [217]. One distinctive feature of this treatment is that it includes quantitative data which allows consideration of factors such as base strength, steric hindrance, and the pK_a rule for the evaluation of S_N2 feasibility with respect to the relative basicities of attacking and leaving groups. One important parameter has, until now, not been given much consideration: medium effects [269] and their consequences on selectivity. In the mechanistic model handling thermal pericyclic reactions, the quantitative analysis of reactivity may be elaborated one step further [218, 220]. To determine the regiochemistry and the likelihood of such reactions for specific cases, a frontier molecular orbital (FMO) approach was used. After estimation of the FMO energies using a specifically designed algorithm which takes into account substituent effects, the ends of the π systems with the largest MO coefficients in the FMO's are determined. The controlling HOMO–LUMO pairs and the likelihood of the cycloaddition are then estimated from the HOMO–LUMO energy gaps, while the regioselectivity is estimated by matching up the ends with the largest coefficients in the HOMO–LUMO pair. *Endo* stereoselectivity is also treated along these lines. In addition, the enthalpy change for every reaction is computed in this model, which is the most sophisticated in CAMEO. This, of course, is not meant to say that this program will predict with complete confidence what is going to occur in the actual experiment, because as chemists know quite well, accurate quantitative theoretical treatments of reactivity are still to be realized. However, the chemist will have at hand what can be expected from an up-to-date and logical analysis. In the same spirit, Jorgensen's group has experimented with electrophilic aromatic substitution [216] and, more generally, electrophilic chemistry (generation of carbonium ions and their subsequent rearrangement, elimination, addition to multiple bonds and quenching by nucleophiles) [218]. The several sets of modules constituting CAMEO is an outstanding example of the way in which computer-aided forward synthesis *stimulates the search for fundamental principles governing organic reactivity* [270].

We have treated here only the programs specifically devoted to the forward approach in synthesis. It should be mentioned that the best retrosynthetic programs, SYNCHEM 2, [127] LHASA [14], SECS, EROS [61], and SCANSYNTH [222], all have forward and retro capabilities, although the latter are more developed. For example, SYNOPSIS includes a module: MECOPSYS dealing with the forward approach. An important improvement being implemented in LHASA is the bidirectional approach to sequence generation, in which paths are grown concurrently backward from a target and forward from a starting material [14, 49(b)]. Kauffman's group has developed a forward module (PSYCHO) in the framework of his PASCOP project [200(b)].

I.4 ELABORATED OPERATIONS IN COMPUTER-AIDED SYNTHESIS

4.1 Strategy and tactics

At the conclusion of the foregoing section the computer is technically fit to receive and clearly redraw any compound, perceive structural features, and propose many reaction paths from a data file of reactions. We have seen that, whatever the technical solution employed to determine these points, a great amount of work has already been invested. *Nevertheless, at this stage, the overall set of available algorithms would compare rather poorly with any average chemist.* The main weakness may be summarized simply as: '*erudition without judgement*'. As a result, when a program is asked a question, it may provide an *astronomical number of answers* (this point has been well illustrated in refs 93, 105, 110). What must still be taught is 'strategy and tactics'. Actually, strategy and tactics were already somewhat present in the foregoing sections because the mode of molecular representation, the depth and nature of structural feature perception, and the choice of exhaustive versus hierarchical files of reaction data all reveal implicit strategic choices by the group involved in the project. *Strategic choices constitute the heart and soul of the problem of computer-aided synthesis and the direction in which the greatest amount of progress is still needed.* Let us first define the difference between strategy and tactics. *Strategy* is meant to represent the method of thinking, which among an infinitely ramified tree of decisions, allows one to select only those branches leading *to success* (*in this context, 'least expensive' synthesis of the target*). We will return to this definition later to classify the strategies described in the literature. This classification will strongly rely on a graphical representation which can be compared with the actual action of pruning a tree while gardening.

The aim of *tactics* is an overall increase in yield within a given strategy by an optimum choice of successive protections and deprotections, functional group additions (FGA) and interchanges, (FGI) selectivity, etc. This final point clearly shows that there will be a strong overlap between the field of strategy and that of tactics. For example, an attractive overall strategy may very well be a failure because some practical details of tactics were not thought of (for an example, see ref. 126). On the other hand, an apparently dull overall strategy may reveal itself to be more successful than a seemingly brighter strategy simply because the tactics were perfectly mastered in the former. *This point must be stressed*: the classical definition of strategy (Karl von Clausewitz, Basil H. Liddell Hart) is: 'The art of using military forces to reach results decided within a given politic'. The hierarchy (1) politic, (2) strategy, and (3) tactics transpires from this definition. *In terms of chemistry there are cases when this hierarchy is turned upside down.* For example, a chemist involved in increasing the optical yield of a given step (*tactics*) as part of an overall synthetic scheme (*strategy*), whose goal is the production of a molecule which was selected as a target because it had some economic interest (*politic*), may discover such an interesting method for increasing optical yield that politics cause him to initiate a project where the synthesis of enantiomers of several targets is systematically studied.

4.2 Strategy and pruning

Before giving some examples of the ways strategy has been tackled by different groups some preliminary comments are needed. They may be made more lively by presenting them as common-sense applications of quotations extracted from an ancient book, *The Art of War* by Sun Tzu, written about 500 years BC. 'The elements of military art are: first, measurement of space; second, estimation of quantities; third, calculations; fourth, comparisons;

and fifth, chances of victory. The measurement of space are derived from the ground...' It appears that the first element has been underestimated in computer-aided synthesis. In several reviews synoptically describing the characteristics of the various programs, most of the programs are described *as covering the whole of organic chemistry*. This has even been specifically stated 'it is felt that a global system for designing syntheses should cover the entire spectrum of organic compounds and should be applicable to the diverse variety of synthetic problems, from laboratory scale synthesis to industrial processes' [18(c)]. We believe rather, that the first step in any strategy is 'appreciation of the ground', *namely, recognition of the type of target* that we wish to treat. One cannot adopt the same strategy for an industrial target made of 12 atoms [190] and that for vitamin B12 [4], for an aromatic [84] or for a heteroaromatic compound [18(d), 141] or a prostaglandin derivative [2], for a phosphorus-centered compound [200] and a carbon only skeleton, or for linear polypeptides [50] and polyenes [46]. This proposition is supported by our own experience in heterocyclic chemistry, but also by the fact that Kaufmann's group had to invest much work to adapt one of the best 'general' programs of synthesis (SECS) to deal efficiently with phosphorus-centered organic compounds. *There is a an actual need for general rules of pattern recognition related to the overall 'synthetic shape' of molecules.* One could imagine that the classical division of chemistry [272] into hydrocarbons, halo compounds, oxygen compounds, nitrogen compounds, carboxylic acids, phosphorus compounds, sulfur compounds, selenium compounds, silicon compounds, boron compounds, organometallic compounds, heterocyclic compounds, and biological compounds [272] is sufficient for this purpose.

A simple example can show that even if the foregoing classification helps, it is insufficient, since the two following structures:

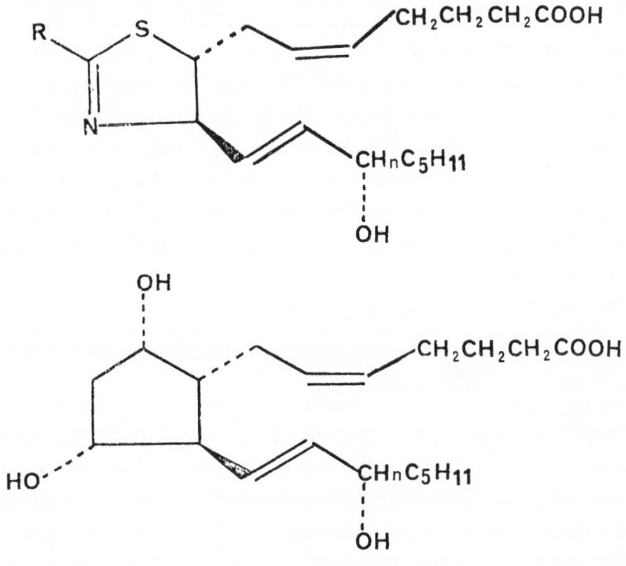

Scheme 36

would respectively be classified as heterocyclic (Vol. 4) and as hydroxy-carboxylic acids (Vol. 2) in *Comprehensive Organic Chemistry* [272]. Both, however, involve one common structural feature: appendages for which a specific strategy has to be devised [42]. Therefore what we really need is a *classification of the type of synthetic problems which have been solved or which constitute present challenges*. One could identify the former by using the Science Citation Index as a first approximation, the latter are more difficult to localize. Then, a focal- or multifocal-centered DARC-like approach [181] could possibly allow one to place the emphasis on the difficult parts of the target, thus allowing an early recognition of the need for a mono-hierarchy or poly-hierarchy of structural features. This would allow one to find a *balanced solution to the dilemma of 'a unique, huge program for all the possible targets or a multitude of specific little programs adapted to every classical class of compounds'*. One practical consequence of having this pattern recognition step programmed would be that, for a given target, only the optimal part of the program would have to be called upon. This is an important consideration in terms of microcomputer applications to organic synthesis [142] but also for the question of time management on large computers. In Sun Tzu, one also reads [271]: 'One attack may lack ingenuity, but it must be delivered with supernatural speed.' We take this sentence to defend the point that the 'Eureka Syndrome' may, in some cases, be quite positive and not at all naive as it is sometimes characterized [152]. Why do some good chemists, looking at a target, feel it quite straightforward to synthetize? They probably recognize in it a sequence of successful steps that they have repeatedly read about in the literature (see functional group oriented strategies in ref. 49(b)). This looks like a biased approach but it could be taught to the computer at the price of studying large populations of reactions to extract the most successful ones. We performed such a study in heterocyclic chemistry [131] and realized that more than 90% of heterocyclizations may be accounted for in terms of four or five basic schemes (see also ref. 257). Our point is simply 'before diving into the loops of refined strategies or exotic reactions, try a search to see if the target is not one of those that can be rapidly treated within the Eureka Syndrome approach'. Although these kind of targets are viewed as 'unglamorous ones' they could well represent more than 70% of the overall manpower devoted to synthetic work. Powers [190] has shown that for some of these targets there are still enough complications of other sorts to overcome before reaching the industrial stage without searching for further burdens.

We may now turn to the main strategies which have been implemented in different programs. *To do so, we will rely heavily upon the graphical representation of a synthetic tree since the definition we gave earlier for the word strategy tells us that the problem is to select a few from among the multitude of branches in a tree*. This 'pruning' may be based on positive or on negative considerations. By *positive* we mean that criteria are elaborated to select some branches out of the many, while by *negative* we mean criteria elaborated to screen out certain branches. The following tree [11] illustrates the different kinds of strategies which have been implemented:

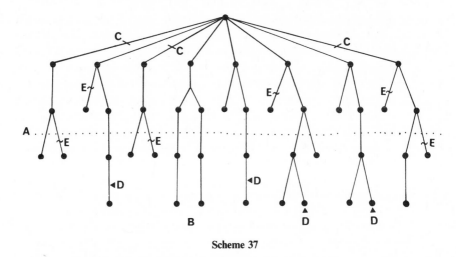

Scheme 37

Type A: Search for only those branches having less than *n* nodes between starting material and the target.

Type B: Search for branches having a given shape.

Type C: Early pruning.

Type D: Pruning of all the branches which are not in a given direction.

Type E: Diffuse pruning: wherever in the tree, when a given characteristic (usually structural) is met.

The various kinds of strategies will be better served either by a *depth-first* or a *breadth-first* tree search [92]. The first corresponds to a walk from the target to the end of one branch, whereas the second would first explore all the first rank precursors of the target then, all of the second rank, and so on. Another consideration of crucial importance is the difference between *interactive and non-interactive programs*. In *interactive programs* the pruning results from both pre-codified rules taught to the computer and from the chemist's knowledge and decisions taken at the very moment when the program is working. To make the intervention of the chemist easier, PASCOP even includes a language of communication (STRATOS) [197, 198] whose syntax rules are simple enough to be learned rapidly by any chemist. In the *non-interactive mode*, the strategy results from completely precodified rules and the chemist cannot intervene during the operation of the program. Both modes are available on PASCOP [199], LHASA [14], and SECS [245].

We shall now give one example of each of classes A–E pruning, but we do not wish to develop this part in great detail, since it has been excellently treated in Bersohn's and the LHASA group reviews [11, 14, 49(b)]. Furthermore, several books specifically deal with the art of synthesis viewed as an extension of Corey's original [5] retrosynthetic (also called disconnection) approach [18(d), 273–279].

Type A pruning has been specifically used in non-interactive programs. The three most sophisticated programs of this type are those of Gelernter, Ugi and Bersohn; we wrote a non-interactive version of SOS for heterocyclic compounds [135] but have given up this kind of computer-aided synthesis because we had difficulties in managing the quantity of precursors generated. For example, Bersohn's program, implemented on an IBM 370-165 and working on a steroid target, would generate 200 000 intermediates in 15 minutes [105]. The kind of constraint which has been first adopted in the programs of Gelernter and Bersohn is to *retain only those proposals in which there is a limited number of steps* between the target and readily available compounds. In SYNCHEM 2 the list of available compounds is 5 000 entries selected from the catalogs of several chemical supply companies [130]. In a less systematized form, pruning of type A is obviously present in all interactive synthesis programs. In such programs the chemist user may intervene to discard a precursor displayed on the screen on the basis of his own chemical judgement.

Type B pruning follows from the general remark that a convergent synthesis is preferable to a series of consecutive steps [280]. The following two shapes of tree branches made up of the same number of steps (average of 90% yield for each step) lead to a different maximum total yield.

$$
\begin{array}{l}
A \rightarrow B \rightarrow C \\
\hspace{2.5em}\searrow\hspace{-0.5em}\nearrow G \rightarrow \text{ TARGET } 65\% \\
D \rightarrow E \rightarrow F
\end{array}
$$

$$A \rightarrow B \rightarrow C \rightarrow D \rightarrow E \rightarrow F \rightarrow G \rightarrow \text{ TARGET } 48\%$$

Scheme 38

Hendrickson has further elaborated these notions by *quantifying the various possible shapes* of synthetic trees [89, 93].

Early pruning (type C) follows from the fact that if one can cut a branch at an early stage of the search (near the target), much computer time will be saved. This was realized from the beginning by Corey [5, 8] and is behind the notion of strategic bonds [41]. In terms of fighting the target, this is simply an expression of Sun Tzu's statement: 'always select and advance to the spot where the resistance is the weakest. Avoid or by-pass a strong defense and assault a weak spot'. If the statement is easy, the actual recognition of strategic bonds or sets of bonds in a structure is not necessarily so [222]. Several *heuristic* rules have been written [41]. The adjective heuristic is important, since it means that these rules simplify problem solving but *cannot guarantee success*. These rules which constitute the body of what Corey calls *topological strategies* [49(b)] are:

Recognize strategic bonds (Section 3.1.2.1) in the target as early as possible in the retrosynthetic analysis. This entrains positive pruning because the branches of the tree which include those transforms creating these bonds will be favored with respect to the others. It also helps diffuse pruning (type E) because each precursor is in due time considered as a target. It must be made clear however, that recognizing strategic bonds or sets of bonds is equivalent to ranking structural features, and we have already stated that there *cannot be an unique hierarchy covering the 6 000 000 already-known structures*. Bersohn's automatic program [105] classifies reactions into eight priority classes, as follows, (1) reactions that introduce functionality, (2) reactions that build a molecular skeleton, e.g., carbon–carbon bond forming reactions and heterocyclic ring-forming reactions, (3) isomerizations including epimerizations, (4) reactions that protect functional groups, (5) reactions that remove the protection from functional groups, (6) reactions that alter functionality, (7) reactions that remove functionality, and (8) fragmentation reactions such as ozonolysis. In the retro-search these reactions are used in the order given. Bersohn, being conscious of the drawbacks of such a unique ranking, tempers it to avoid negatively-oriented pruning which may eliminate branches of the tree of synthetic value. For this, the program has a section whose task is to recognize familiar situations and call into use reactions of low priority which are necessary. Within priority class 2, those reactions which simultaneously form several bonds (carbon-type synthesis of three-membered rings, Diels–Alder, etc.) are placed first. Early pruning is so efficient that Hendrickson has devised a special representation of the target, placing the stress on the carbon skeleton [85]. This caricature, which necessarily implies a *hierarchization of structural features*, as it coalesces trivial structural distinctions so as to manipulate fewer items, favors both early pruning and diffuse pruning [88]. It really must do so, since the main problem of the so-called mathematical models of synthesis (see Section 3.1.2.2) is the mastering of the combinatorial explosion. These models not only propose known reactions for every transform but also new ones not yet explored (see ref. 77(b)). Our group introduced the idea of skeleton construction in an early article where it was shown that allowing the computer to search for all combinations of two or three simultaneous bond constructions would propose innovative ideas of synthetic strategies [137]. Negative early pruning is highly efficient because it kills so many possibilities in one shot. The price, of course, *is high risk* because, as noticed by Bersohn, a route may appear as mediocre near the target and become simply superb when one reaches steps far from the target. Furthermore, heuristic rules (like those of grammar) suffer from exceptions [281]. Many proofs can be given for this; we give only two of them. The first one is linked to the fact that aromatic rings are often considered as a whole building block in synthesis; nevertheless elegant strategies by Vollhardt [282] and Subba Rao [283] prove that this needs not necessarily be so. The second one involves the rapid synthesis of longifolene which has served as a challenging test case [284a] for synthesis methodology and planning for a long time. Oppolzer reported an original synthesis of this target in which the key step involves the formation of two C–C bonds, one of which (C(8)–C(8a)) would not be classified as strategic using Corey's criteria (Section 3.1.2.1) [284b]:

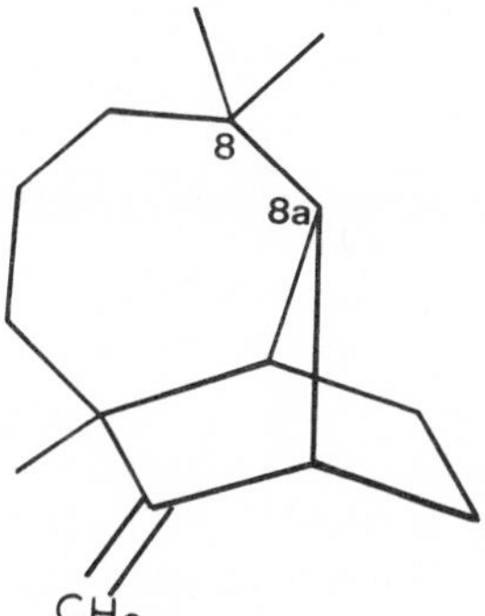

Scheme 39

This synthesis clearly involves a non-strategic bond formation in its key step and, as such, would have been overlooked by LHASA. The fact is that negative pruning may be looked upon either as prudent advice based on past experience (in terms of Sun Tzu's 'attack cities only when there is no alternative') or as a challenge. For example, Deslonchamps [285], in his illuminating article on synthetic strategies, directly challenges the proposition of trying to avoid the synthesis of non–real rings (see Section 3.1.2.1). He proposes that the medium-sized rings (see ref. 286) could be made much more available in the future and therefore become interesting precursors. No such drawback is associated with positively oriented pruning. A striking example of such pruning is the search for *structural features which suggest transforms with one pot formation of several bonds.* Such examples of *holosynthons, holoretrons* [49(b)], *and holotransforms* are provided by the Diels–Alder and related reactions (hetero DA, quinone DA, etc.) and other internal cycloadditions including dipolar cycloadditions [287–289]. The prefix holo (from the Greek *holos*: whole) draws the attention to the fact that for such transformations the chemist has to look at the target as a whole in contrast to the majority of synthetic reactions in which only one bond is formed at a time. Corey has shown that quite elegant solutions to synthetic problems can result if one can successfully apply two or three powerful simplifying transforms within a strategy [8, 40, 49(b)]. The most spectacular holotransform that we are aware of is the transformation in which the open-chain polyolefin, squalene, undergoes (enzyme-catalyzed) polycyclization to produce the tetracyclic substance, lanosterol [290 (a), (b), (c)].

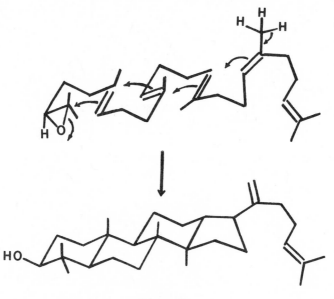

Scheme 40

Four bonds are formed in such a reaction, and what is amazing in the biosynthetic pathway is that seven asymmetric centers are formed with such stereoselectivity that only one of the 128 different stereochemical isomers is formed. Johnson was able to support a process involving concerted formation of two rings [290(d)] in related examples associated with non-enzymic olefinic polycyclizations. Such a characteristic, when present, is the trump card in the game of obtaining high stereoselectivity. Polymerizations are trivial examples of holotransforms which display stereoselectivity only with appropriate catalysts (Ziegler–Natta [290(e)]). We are presently gathering other examples of holotransforms; following a discussion of these kinds of transforms, Marko (Louvain-la-Neuve) brought some other typical examples to our attention:

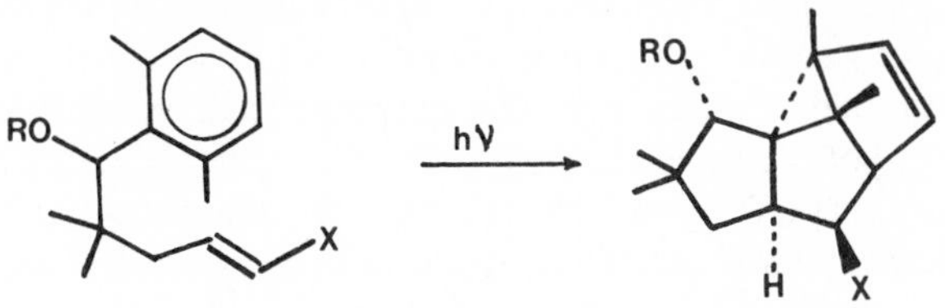

(290f)

3 bonds and 3 rings in one pot

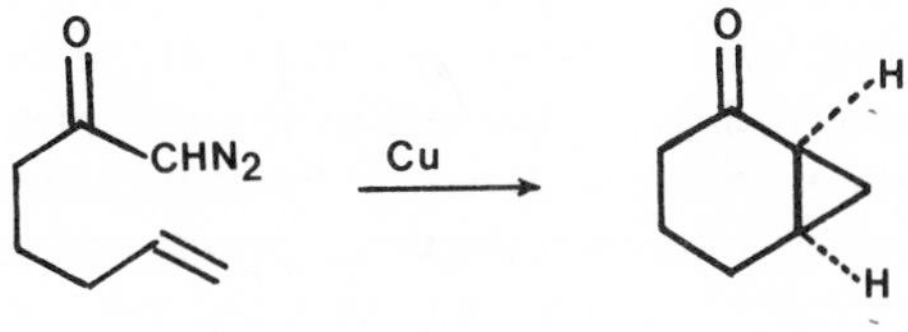

(290g)

2 bonds and 1 ring in one pot

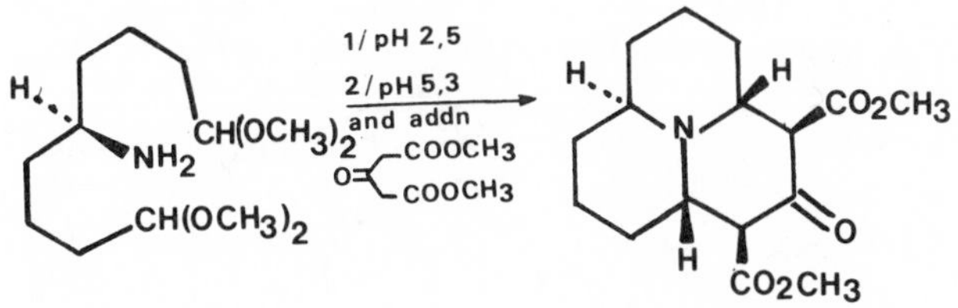

(290h)

4 bonds and 3 rings in one pot

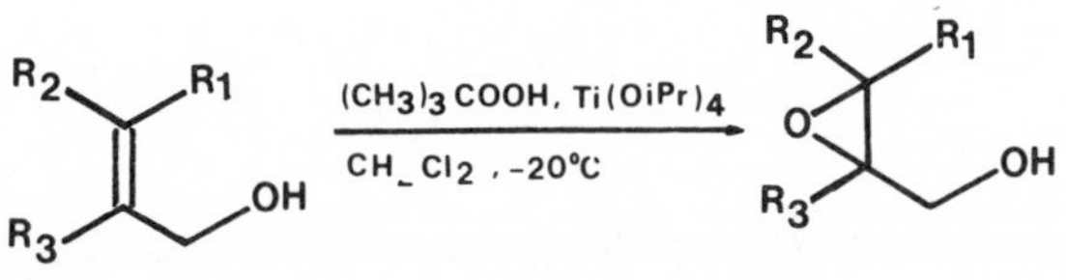

(290i)

2 bonds and 1 ring in one pot

Scheme 41

This point leads us to *Type D pruning*, which covers both transform based strategies (T-goal in ref. 49(b)) and structure goal strategies (S-goal (49(b), 152)), Recognizing the importance of such kinds of transforms, Corey's group has written algorithms specifically designed to teach the computer to look ahead in the search up to 25 steps from the target (14, 40, 49(b)). When a transform such as the Diels–Alder [40], double bond formation [46], halolactonization [49(a)], or Robinson annelation [48] is perceived through a depth-first search, then the branch containing it is given high priority. One may even adapt the involved precursor in such a way that it undergoes this kind of reaction better [49(b)]. This positive kind of pruning is equivalent to a directional pruning, which has also been advocated by Sridharan propos- ing 'a search in a planning space taking leaps along the synthesis sequence' [125]. Another way of directing the pruning is to favor any branch reaching commercially available compounds or any pool of selected structures (aro- matic, chiral, etc. [22(d)]). SYNCHEM 2 first introduced this approach and, here, the untiring computer is able to compare every possible precursor with a list of 5 000 commercially available items, doing a task which would make months of boring work for a group of chemists. Borsohn [105], remarked on the dramatic pruning efficiency which could be associated with such a strategy. Hippe [222] has improved it by considering not only 'identity with a set of reactants stored in the computer' but also 'similarity'. Another even more elaborate algorithm has recently been proposed by Wipke [159]. This algorithm does more than only favor branches of the tree which attain preferred precursors. It even creates new ones, allowing the junction of the target to precursors with higher numbers of atoms (degradative synthesis) but which are readily available. This SST approach is meant *to simulate the fact that the organic chemists sometimes make an 'intuitive leap' to a specific starting material from a target without consideration of the reactions needed for interconversions*. This intuitive leap has been programmed for the follow- ing four situations: (1) target = starting material (identical match the same as in SYNCHEM 2, (2) target > starting material (superstructure match), (3) target < starting material (substructure match) and, (4) none of these (simi- larity match). The last case was elegantly dealt with by an automatic caricature of the target called *abstraction* as illustrated below:

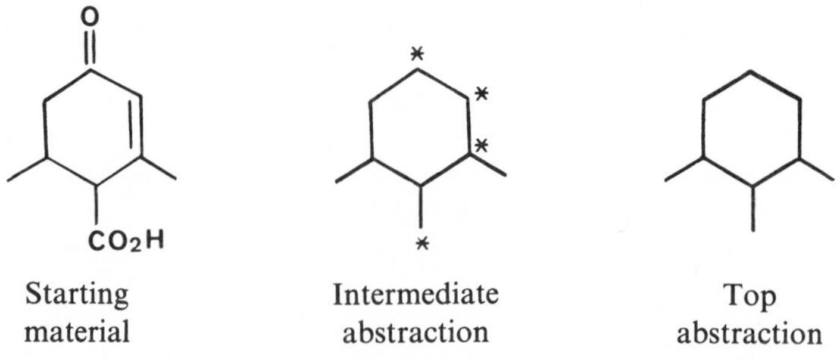

| Starting | Intermediate | Top |
| material | abstraction | abstraction |

Scheme 42

This concept of abstraction (introduced in a slightly different way by Hendrickson [93]) is applied both to the target and to the starting material library. It has three positive results: (1) it saves much time for the search for matches; the original set of 11 000 starting materials drops to 1 436 top level abstraction structures, (2) it represents a 60-fold reduction in the search space and, (3) it connects the target to starting materials which could have been missed by the classical retrosynthetic approach. It has been shown to allow for the discovery of over 90 % of a published collection of starting materials. As such it constitutes the major breakthrough in the last ten years in the field and suggests that very efficient programs for substructure searches, such as DARC, should play an important role in the future of computer-aided synthesis [181].

The last type of pruning has probably been the most developed. Being based on rules of evaluation, it assigns figures to the different nodes of the tree. This is a particular case of the Problem Solving Graph (PSG) developed by Gelernter (1962), which associates numerical weights to each node in the PSG (here the synthetic tree) [291]. Incidentally, in terms of games, this clearly puts the game of computer-aided synthesis much closer to chess than to backgammon. This is relevant information when one recalls that in July 1979, the program BKG 9.8 (written by H. Berliner) defeated the world champion player Luigi Villa 7 to 1 in a match whose prize was $5 000 [292]. Up to that point, the chess programs had not been able to defeat the great masters. These rules may be reactivity or theoretically based [123]. The term reactivity covers both yes–no situations (reactive versus non-reactive, illegal structure pruning [152]) and those which express less well defined differences (regio, stereoselectivities [47, 49(b), 152]) and their consequences in terms of yield [97] and tactics [152]. *Any chemist, having tried to quantitatively study these notions, knows that there still remains many open problems in this area.* He therefore expects limited success even with the best of these situations. Keeping this reservation in mind, the teaching of structural scope in ketone preparation [175], Diels–Alder yields [38], stereoselectivity of nucleophilic addition to ketones [144, 150], electronic substituent effects [129, 152], and steric effects [129, 150], orientation rules in aromatic substitutions [129], endo or exothermicity of reactions [66, 76], relative reactivity towards reducing or oxidizing agents, comparative stability of carbonium ions [129], leaving group tendencies [129], and conformational preferences [47, 49(b)] have been taught to the computer. Therefore, these features and their consequences in terms of yields are taken into account by several programs which give an evaluation function of every node proposed in synthetic schemes. The theoretical approach of diffuse pruning (involved also in early pruning) rests on the notion of *the space of states* applied to a synthetic tree. Any set of objects may be represented in an N-dimensional space [123]. One may therefore imagine an N-dimensional space which describes at best the target and all the possible precursors (space of states). If the representation is fair, one should be able to 'see' in an N-dimensional space which objects (precursors) are closest to the target. Ugi's matrix representation of FIEM is particularly suited for such an approach.

A $n \times n$ BE-matrix can also be represented as a b vector with n^2 components:

$$b = (b_{11}, \ldots, b_{1n}; b_{21}, \ldots, b_{n1}, \ldots, b_{nm})$$

where the entries b_{ij} are the cartesian coordinates of a point P(B) in the space of states. The same is true for an E-matrix representing the chemical system after reaction, and a point P(E) may therefore be defined in the n^2-dimensional matrix space. One may show that the R-matrix corresponds to a vector R from P(B) to P(E). Ugi and Dugundji [53] call the distance between P(B) and P(E) *the chemical distance* between B and E (in chemical terms between reagents (educts) and products in the n^2-dimensional space). Since this distance is measured by R (reaction matrix), one should be able to calculate all the chemical distances corresponding to any reagent–product couple, and therefore any step of the synthetic tree will have an associated chemical distance. The sequential sum is characteristic of the total synthesis. The best synthetic routes may correspond to sequences with *minimum chemical distances (PMCD)* [69]. Although the idea is attractive, computing time on a target such as a steroid would make it prohibitively expensive at this time. A possible method of managing the search could be to simplify the space of states in a way comparable with that introduced by Hendrickson [93]. This author [92] himself proposed the use of 'the distance function' to calculate the minimum number of steps needed to convert a given substrate to a given product. This 'distance' is a function of the net change in the functionality of each atom from substrate to product. The distance function is given by:

$$N = 1/2 \sum_i (|\Delta h_i| + |\Delta z_i|)$$

Scheme 43

where N is the minimum number of steps and Δh_i and Δz_i refer, respectively, to the overall change in the number of hydrogens and heteroatoms on each carbon i. LHASA recently implemented this concept in its possibilities [49(b)].

All the types of pruning that we have seen are complementary and may be used simultaneously or consecutively. When more and more examples have been treated, different general shapes of synthetic trees should evolve and enrich the general view of strategies. For this reason the PASCOP ability to perform storage, retrieval, and updating [201] of synthetic trees should be extended to other programs.

The foregoing section illustrates how the strategic problems oblige computer chemists to tread lightly between the Scylla of infinitely branched trees and the Charybdis of overlooking too many relevant synthetic routes [192]. The field has evolved greatly since the first days when the synthetic trees were explored mainly retrosynthetically. An overall picture of the next decade interactive program emerges: *it will work forward as well as backward, will include an extensive basis of starting materials, and will save much computer time by a sound hierarchization (classification and caricature) of reactions, mechanisms and targets. The most chemical part of it (reactivity, selectivity) will probably have to rely very much on the chemist's knowledge as assisted by a well-constructed database* [49(b), 176].

4.3   Tactics

Strategy has now provided the chemist with four or five attractive routes to his target. He now has to critically examine every step very carefully to try to foresee where difficulties are going to arise: which kind of reaction would be the best suited for building a given structural feature (e.g., olefins [46]), when should the functionalization steps be introduced, what kind of media should be avoided in each step, are there any structural features forbidding a step [47, 171], what kind of by-products could be expected for every step, which functions should be protected [44, 49(c), 186, 259] and when should they be unprotected, what sequences of functional group interchanges (SEQFGI) should be realized [43, 186], etc.? This enumeration only shows that an attractive strategy has many possibilities for failure and explain why so much time has been devoted to these matters in LHASA and SECS. For these systems the thoroughness of transforms [49(b), 152, 156] *makes our division into strategy and tactics obsolete*: transforms play a definite role in diffuse pruning *and* in tactics. SYNCHEM 2 has even devised a special system of interactive access and control modules for building and managing the knowledge base. The Knowledge Interchange System (KIS) [128] makes it easier for the user to revise and elaborate the information associated with the defective reaction scheme to reflect what has been learned as soon as the lesson of the failure is understood. This illustrates an important new trend in computer-aided synthesis: *after a painstaking time when the communication between man and machine had to be taken into effect, and strategies to master the combinatorial explosion have been mastered, the programs have now reached a stage where they learn from all the past successful syntheses and the ones currently being performed.* Gelernter's group describes the benefits of such a knowledge-base enhancement via a training sequence in an article which raises much hope for the future of the field [130]. Hundreds of structures have now been examined and the ones displayed in Fig. 2 are only a small sample of what has been done by the various programs. The case of phosphacarnegin [200] deserves special mention because here the target has been examined theoretically and experimentally [202] by the same group. This double situation of designer and user allowed the PASCOP team to deduce directly which were the most important improvements to be made within their program. (Of course, the group where such a positive situation is most often encountered is Corey's [39, 49(b)].)

Table 5 lists keywords describing the evaluation and strategies associated with the main CAOS programs.

I.5   EVOLUTION OF COMPUTER-AIDED
ORGANIC SYNTHESIS

5.1   Microcomputers and super computers

When computer-aided synthesis began, a single possibility was offered: computers. Now both supercomputers and microcomputers are available. Bersohn has discussed what supercomputers could mean to computer-aided

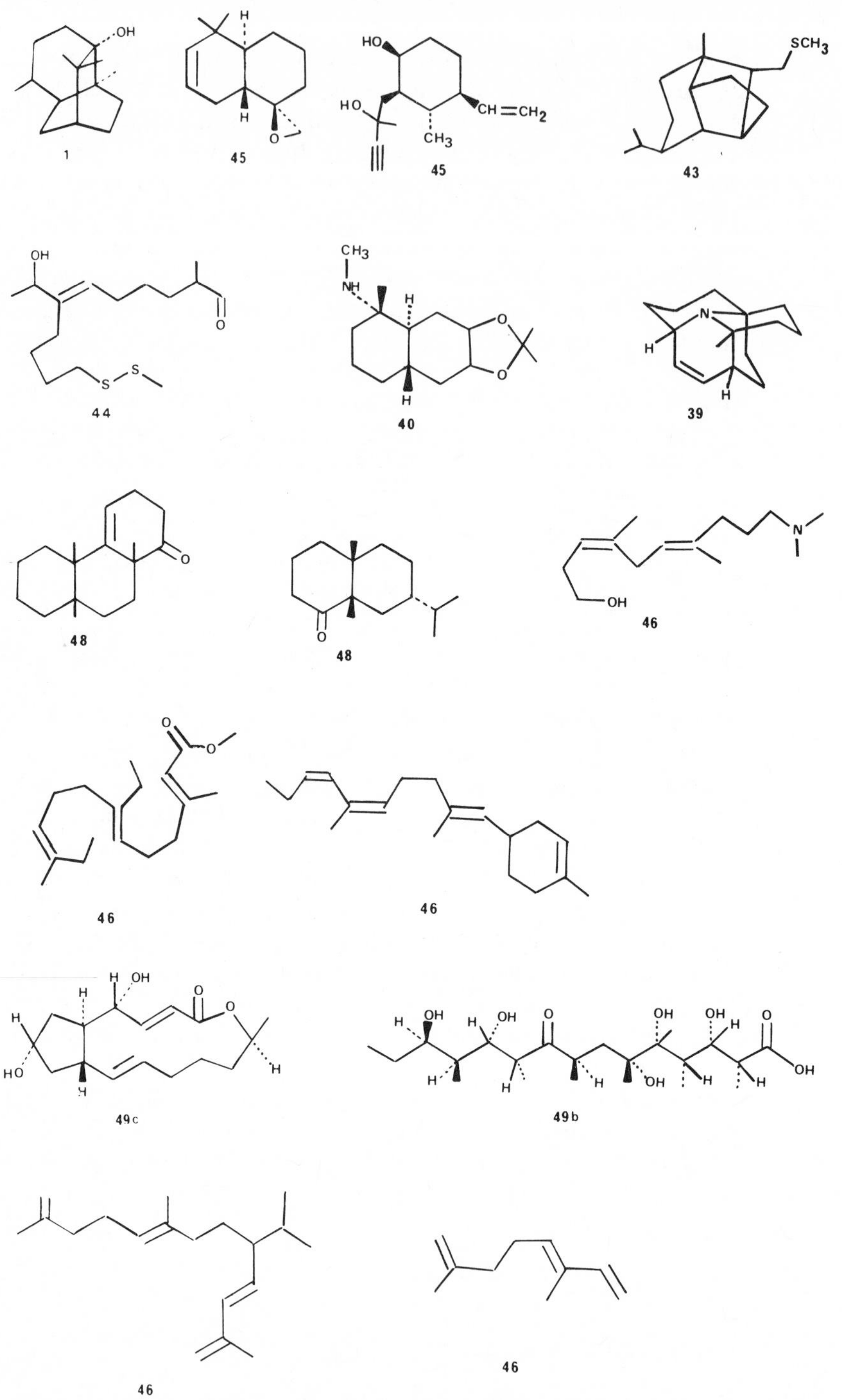

Fig. 2　Representative structures studied by CAOS. (the numbers stand for the literature reference).

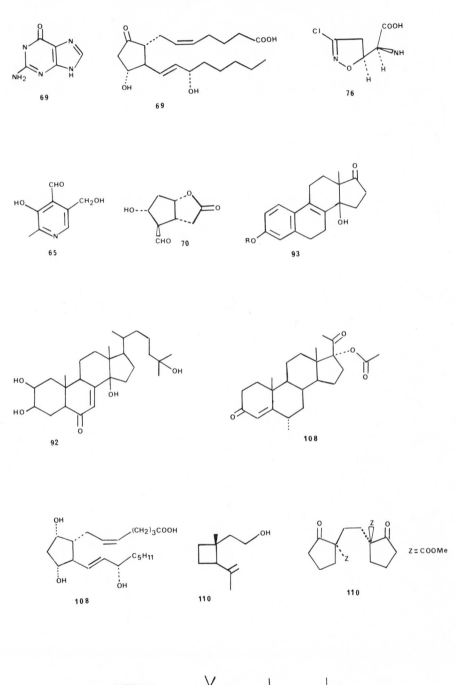

Fig. 2   Representative structures studied by CAOS. (*continued*)

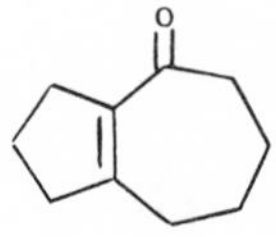

126

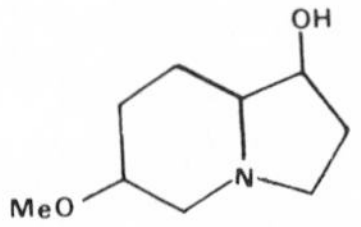

126

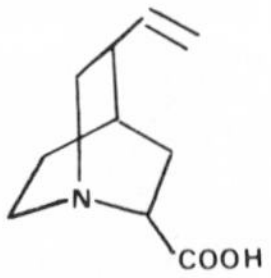

126

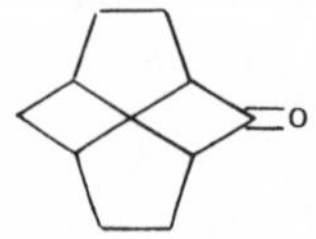

126, 127

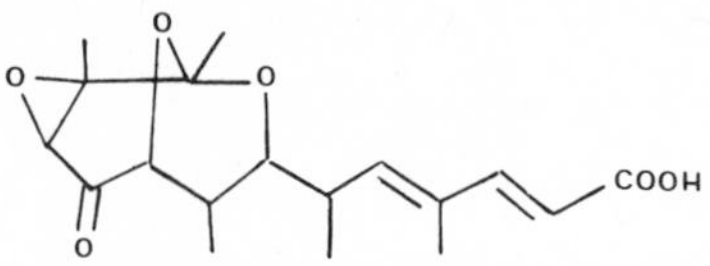

126

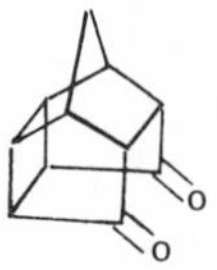

130

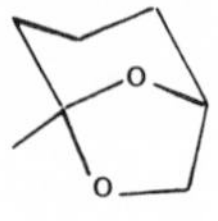

130

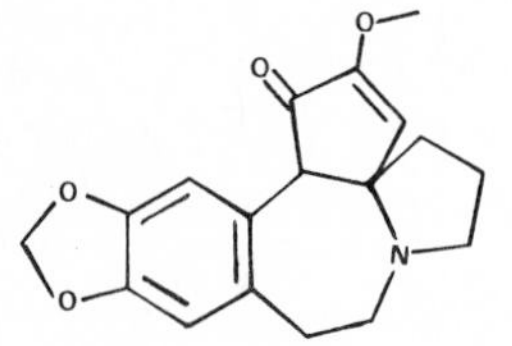

128

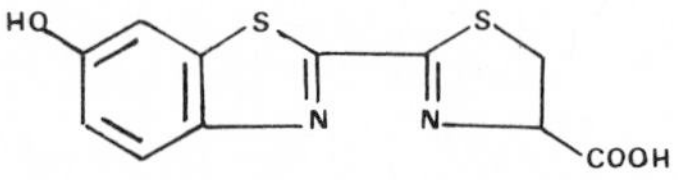

132

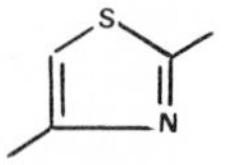

133, 135

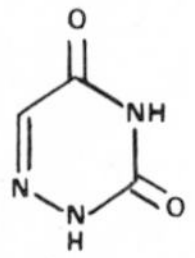

134

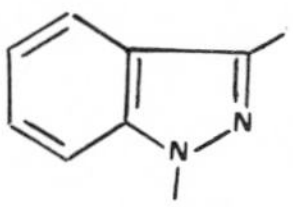

136

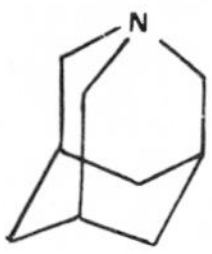

137

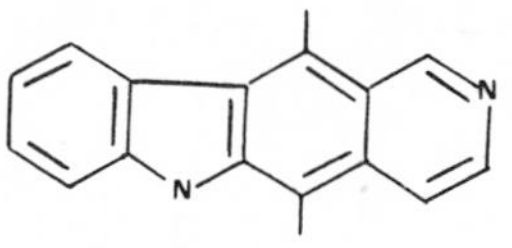

141

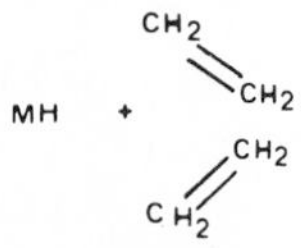

138 a

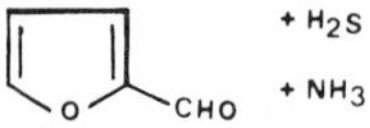

140

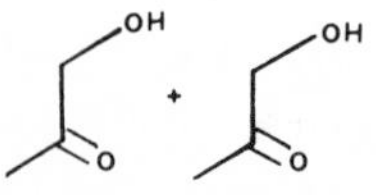

139

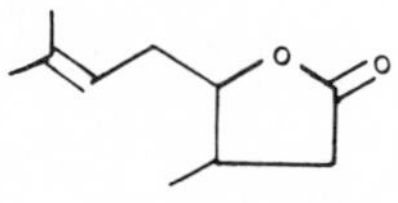

143 a

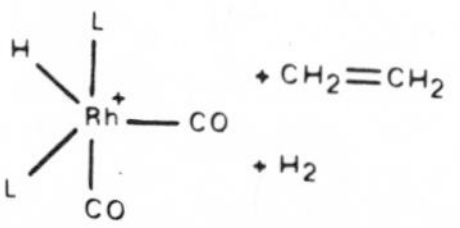

138 b

Alanine tRNA

189

Fig. 2 Representative structures studied by CAOS. (*continued*)

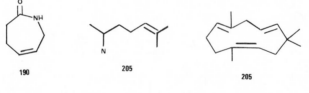

Fig. 2   Representative structures studied by CAOS. (*continued*)

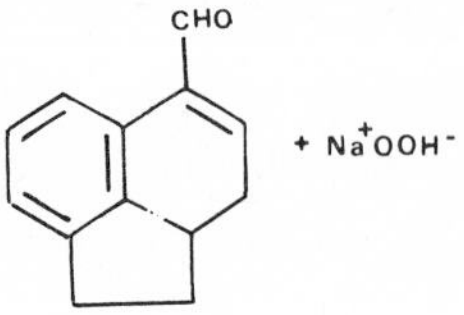

213

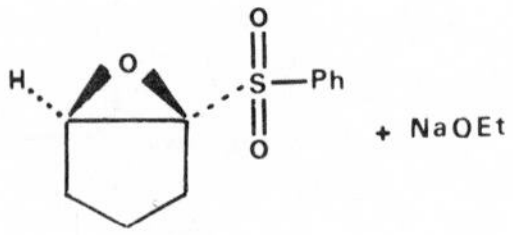

213

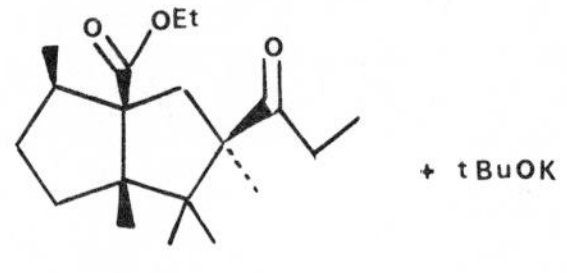

213

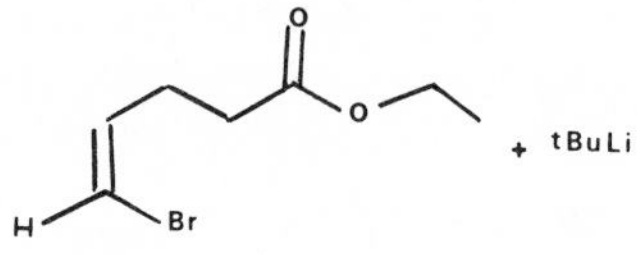

215

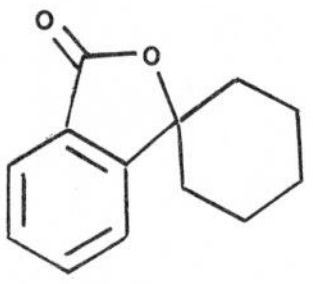

215

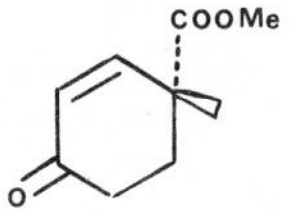

217

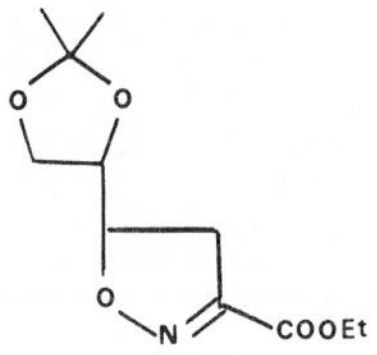

218

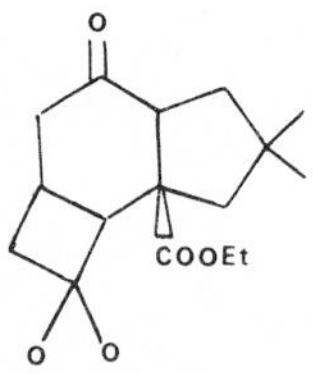

218

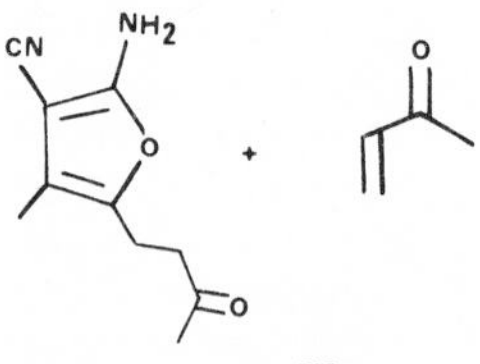

218

Fig. 2 Representative structures studied by CAOS.

Table 5 Evaluation and strategies

Name	Evaluation and Strategies	Type	Ref.
LHASA	Functional-group, transform-based, topological, stereochemical strategies, strategic bond disconnections for polycycles, appendage reconnections and disconnections, important ring forming reactions.	I-R	40, 41, 42, 43 45, 46, 48, 49b
BERSOHN	Yields, Definition of available compounds, classification of reactions, limits of the number of facilitatives reactions, number of steps.	A-R	105, 108, 110
SYNCHEM 2	Pretransform and post-transform tests (electronic, steric, atomic, structural features)	A, I-R	128, 129, 130
AHMOS	Heuristic scales of nucleophilicity, etc.	A, I, -S	113–115
SOS	Tests atoms, bonds (BASIC-like language: IF-THEN-ELSE)	I-R, S	13, 132, 134
DINASYN	Reaction, separation and analysis times are evaluated by means of mathematical models.	S	188, 189
EXTRUS	Thermodynamic evaluation, limitation concerning	I-S	209
HEXARR	Atom nature, range of exothermicity.		
SECS	Evaluation: complete, including 3-D and electronic. Strategies: user directed, starting material symmetry, simplicity.	A, I-R	152
SECS (Merck)	Atom or bond inclusion, exclusion, aromatic chemistry permitted, excluded, FGI permitted, excluded, user designed strategies	I-R	245
EROS	Search of breakable bonds, energy of reactions from thermochemical data.	A-R, S	18c, 61, 65, 66, 69, 76
WHITLOCK	To make carbon-carbon bond. Eliminates marginally functional groups. Emphasizes the chemical difference in non identical primary functional group. Eliminates quaternary centers, if any. Makes maximum use of existing functionality. Exploits the symmetry elements of the molecule. Tries to form a bond closest to the center of the molecule.	A-R	185
REACT (Powers)	Evaluation: Feasibility by transform constraint, heuristic scoring at transform level, stoichiometry with raw material, costs side reactions. Search Strategies: Depth-first guided by transform scoring function, breadth-first, user driven by interaction through a teletypewriter.	I-R	190
REACT (Djerassi)	Reaction site constraints refering to features of the molecule which affect the reaction.	I-S	12 p. 188, 204, 205
PASCOP	Chemical and topological evaluation, interactive strategy by directions in specialized language, automatic strategy	I-R	196–199
MASSO	No evaluation.	I-R	203
GRACE	Tests on atoms, bonds, reactives sites, reaction complexity	A-S	208
GSS	Evaluation: probability measures derived from structural environment descriptors. Strategies: Simple general strategies like maximal structural reduction/enlargement, similarity with specific structures, application to special types of transformations.	A, I-R, S	245

(continued)

Table 5 *(continued)*

Name	Evaluation and Strategies	Type	Ref.
SAS	No evaluation.	A-R	137, 141
CHIRP	Thermodynamical evaluation. Eliminates unstable compounds on chemical valence criteria. Prevents formation of redundant product sets. Eliminates reactions which involve replacement of an atom by an equivalent atom set.	S	211
CAMEO	Specific conditions for each mechanism, HOMO LUMO energies, Frontier Molecular Orbital energies, ΔpKa rules, reactivity of aromatic rings, carbonium ion stabilities, base strength, steric hindrance, unstable functional groups, ylides, silanes reactivity...	I-S	213–220
SYNGEN	Simplification by dissecting skeleton is best convergent modes from real starting material skeletons. For each bondset generate functionality digitally to all possible routes from real starting materials allowing only sequences of constructions. Finds all shortest convergent syntheses.	A, I-R	93
FLAMINGOES	Evaluation: valencies, reaction generated must belong to the set of prescribed types of cyclic bond transform Strategies: choice of possible reaction centers, criteria for rejection of uninteresting results.	A, I-R, S	245
SCANSYNTH	Strategic bonds, self-adapting function for automated modification of the generated structure	A, I-R, S	221
SCANMAT	Strategies to control the simulation process (canonicalization of molecules, chemical, heuristic, tautomeric and thermodynamic controls).	A, I-R, S	245
SCANPHARM	Idem.	A-R, S	245
ASSOR		A, I-R, S	245
PSYCHO	Topological and Chemical Evaluation.	I-S	245
MICRO-SYNTHESE	Tests on the target with a BASIC-like language.	I-R	235b
SST	Search type, user abstraction, aromatic construction, aromatic degradation, required atoms, evaluation cut-off, chirality, price, atom count, functional evaluation of proposed starting materials.	I-R	159
TAMREAC	Tests on atoms, bonds, ligands, metal.	I-S	138b

Type: A = Automatic. I = Interactive. R = Retrosynthetic. S = Synthetic.

synthesis [110] and SIMULA shows the important benefit (up to 24-fold speed enhancement) to be expected from *parallel processing* [160]. At the opposite size extreme, we proposed that microcomputers could have a role to play in CAOS [142, 143]. A referee for this article [142] was strongly against the idea. The combinatorial explosion and the quantity of information needed in major programs supported that opinion. Nevertheless, *one may profitably compare the structure of computer-aided synthesis to theoretical chemistry* and even perhaps to some spectroscopic methods. At the beginning of quantum mechanics, computers were unavailable but the basic equations connecting Hamiltonians and wave functions were already understood. It was also clear that if the Hamiltonian 'describing' H_2 was easy to write, the calculation of wave functions was something quite different. Some hoped that progress in computer science would unravel these practical difficulties. Some played another card and began to develop approximations starting from HMO, through semi-empirical to more or less elaborate *ab initio* methods. It is interesting to remark that, in this adventure, the researcher who won the highest awards was not the one using the greatest amount of computer time … On the other hand, if it is recognized that for some problems HMO provides a good idea of what is going on, for others problems one must use *ab initio* methods. The reader will then understand when we suggest that SOS as adapted to microcomputers [142] compares well with HMO and EROS with *ab initio* methods. One has to remember that LHASA (in between HMO and *ab initio*) works on a VAX–11/750 computer whose cost is about \$95 000, whereas SOS works on an Apple II which costs about *90 times less*. In the world of semi-empirical methods, Dewar recently wrote that MNDO is, with some restrictions, comparable with *ab initio* methods that need at least *1 000 times more computing time* [183].

Furthermore it is very probable that by the end of 1995 *microcomputers will have increased their computational and retrieval abilities by at least one order of magnitude without a corresponding increase in price*. The structure of information related to chemical structures and synthons [122], or to synthetic reactions used shows that usually there is a small sized hard-core of frequently used items surrounded by successive layers of less and less common and current items. If one considers also that clever simplifications such as those proposed by Hendrickson [93] and Wipke [159] could considerably cut the amounts of computer time needed, then perhaps the idea of microcomputers as aids in synthetic chemistry for some classes of targets is not ridiculous. In terms of comparison with spectroscopy, we would advocate that a good chemistry department needs both a high resolution multinuclear NMR but, in addition, smaller, less expensive ones.

5.2   Entrance of the idea in chemical laboratories

The growth of computer-aided synthesis should also depend critically upon the way the idea enters chemical laboratories. If we pretend to build tools for chemists that chemists do not use (for whatever the reason) the conclusion will be obvious. To catalyze this entrance in laboratories one may: try to teach synthetic chemistry using more computers [142, 210], teach it in a way reminiscent of what is taught to the computer [18(d), 274], teach the computer programs most of the synthetic successes of the last thirty years [127], use microcomputers *so that decentralization occurs*, and design the use of programs in such a way that users may easily participate in the improvement of them as tools (by addition of new reactions, addition of new starting materials, simple implementation of personal strategies, etc.)

A deceptively prospective issue which is sometimes raised is 'could computer programs compete with the great masters in synthesis?'. We call it 'deceptively prospective' for several reasons. *The first one* is that to speak of competition one should have a fair measure of what is meant by the winner. Suppose that in ten years one proposes five different targets to the 'wisdom' of a well taught program and to three great chemists (assuming that they accept this kind of competition). To know who provided the best answers one would have to conduct at least four or five years of experiments to determine who actually proposed the best route. *The second reason* is that computer-aided synthesis will probably still keep its best efficiency in the interactive mode [156] so the competition, if any, would rather be between a chemist computer couple versus a great master without a computer. The *third reason* and probably the most relevant from a practical point of view is that more than 80% of the actual synthetic problems *do not* demand the intervention of a great master to be efficiently solved. *Fourth*, the improvement of SECS by experienced chemists at Merck [156] shows that rather than seeing these programs as potential competitors, it seems far more positive *to exploit the synergy computer-experienced chemist to forge a tool able to increase the overall productivity of the research group.*

5.3 How to properly 'feed' a CAOS program

There are still many poorly solved problems in computer-aided organic synthesis [155, 156]. These emphasize the route in which the most progress should occur. The first problem has to do with the question 'How is one to keep a program not only alive but in good health?'. It seems that all the situations between starving and over-fattening 'diets' exist in the field. Not much about an healthily balanced 'diet' is available. Over-fattening is fairly easy when one considers that about 25 000 reactions are available [110] and that about 200 'new reactions' each month can be extracted from the current literature [110]. The reader can easily imagine what kind of H-bomb combinatorial explosion one would have if no preorganization of the data were performed. *Any organization of this information supposes a consistent system for classifying synthetic reactions.* What one could imagine would be a kind of 'periodic table' of reactions allowing one to recognize whether a paper describing a 'new reaction' is only adding one more example to an already occupied position or if it really fills an empty position. For this reason systematic approaches [77(a), 293-300] aiming at classifying synthetic transformations and mechanisms are of utmost importance [68]. Indeed, only if a good organizing principle is provided, will firms accept the investment of the huge amounts of money needed to feed the system with what is already known and then keep it alive monthly. It is noticeable that even with 500 transforms, SECS at Merck is seen as needing many more to produce a reasonably thorough analysis of a structure [156]. The two best known classifying systems of synthetic reactions are probably Ugi's and Hendrickson's. One of their really positive aspects is that they not only permit the classification of all the known reactions but also all those to be discovered in the future. On the other hand, Bersohn's group recently described a system whose goal was to collect relevant information on synthetic reactions from the English-language journals on tape and translate the information into a standardized form [107]. Presently the oldest data bank of reactions, SPRESI contains about 15 000 reactions [117, 119]. With such a large data set one problem is to automatically discern synthon substructures from non-stoichiometric appendants. Weise has proposed an algorithm to solve this problem [119]. Several reaction data banks are already available. In the framework of DARC-SYNOPSYS (DARC-RMS), KETO REACT [176] recognizes the difficulty in an exhaustive compilation of reactions of general interest and therefore concentrates on the limited area of aliphatic and acyclic ketones. This documentary fund contains 136 references covering 2 788 specific reactions classified according to a consistent model. The controlled passage from specific reaction data banks to generic reaction data banks is presently under study [178–181]. CDRS, produced by Derwent, covers vol. 1–30 of 'Synthetic Methods of Organic Chemistry', the *Journal of Synthetic Methods* from 1975, and is completed by Patents and coverage of various journals [301]. As such it contains 55 000 reactions and this number increases by about 3 000 every year. ISI has created Current Chemical Reactions (CCR) [302a] to meet the organic chemist's need for information about new synthetic methods and modifications of known reactions or syntheses. Approximately 3 000 papers selected from 100 primary chemistry journals are summarized per year. CCR is cumulated annually and has been available since 1979. Verlagsgesellschaft publishes the weekly *Chemischer Information* service which gathers all the new information on organic, organometallic and inorganic reactions. The emphasis is on chemical transformations but theoretical aspects are also reported when thought relevant to the theme of reactivity. Molecular Design Ltd installed REACS at Eastman Kodak's synthetic division before marketing the system industry-wide. In this system [302(b)], chemical reactions and associated data are catalogued and retrieved through graphical displays. In 1982, it already provided access to more than 20 000 structures and 15 000 reactions. On this data bank, a search can be made for reactions with a specific reactant or product, catalyst or solvent, range of associated conditions, a yield or other defined characteristics in a specific range. The system is highly flexible; for example, one may search for all reactions of a reactant with an ether linkage where that linkage does not react. REACSS also has the ability to recognize reacting centers in a reaction. Johnson's group, at the University of Leeds, recently developed ORAC, a program concerned with the indexing and retrieval of information concerning reactions [22(d)]. It allows the chemist to quickly retrieve the best literature precedents for a desired chemical transformation. The ORAC data base is being created by an international consortium of synthetic chemists who are entering information obtained both from the current literature and also from personal card indexes which have been built up over many years.

The pharmaceutical company, Smith, Kline and French, in collaboration with Professor Still (Columbia University), has developed a chemical reaction retrieval program which allows the laboratory scientist to search a database rapidly for reaction citations extracted from literature [22(e)]. SYNLIB uses simple graphic instructions to enter chemical structures questions. This involves drawing a structure on a screen using a light pen. Specific atoms in the query structure may then be selected to direct the library search and are considered to be the key atoms in the structure. The chemist may perform either a narrow search or a generic search. For the *narrow search*, the conditions for a successful match are (a) the keyed atoms in the query structure must be keyed in the product structure of the library entry (b) the keyed atoms in the product structure must be present, but not necessarily keyed, in the query structure. For a broader *generic search*, only condition (a) is required for a successful match. Other options permit the user to search for specific citations, to set additional constraints such as yield, number of steps, reactions conditions. Hardcopy output is possible. The library currently consists of about 17 000 reactions and will be extended on a regular basis.

To be kept alive, a computer-aided synthetic program should also be fed with new data in reactivity (if this is an element of pruning and tactics), and newly available structures (see Chapter 8). Again, the problem of classifying the information to avoid overfeeding is crucial. The foregoing remarks show that much remains to be done in this direction (see also ref. 26(a)), which identifies *it as one of fast progress and dramatic improvements for the next decade.*

5.4   Comparison of CAOS and the abilities of a good chemist

The other problems are less critical unless one strives for a totally non-interactive program. To succinctly describe the situation we have constructed a table which we believe summarizes what, at this point, the chemist does better than the computer and what the computer does better.

Better done by chemist	Better done by computer
Recognizes the *overall target in general terms*: difficult, easy, type of problems, etc.	*Systematically perceives specific structural situations* when taught to do so. Will never forget or overlook them.
Ability to recognize that a reaction usually thought as low yield *could be improved in the case under study.*	Ability to 'remember' *a very large set of reactions.* The chemist with a good library does well also.
Ability *to put marks on the different nodes* because of his knowledge of the constraints associated with a given problem. Even better for an overall judgement of a synthetic tree.	Ability to do repeatedly *the repetitive work* of writing thousands of structures. No overlooked solutions because of fatigue.
Ability to *create new strategies*	Ability to *systematically compare* a given structure (target or precursor) with *thousands of starting materials.*
Experience. The good chemist knows which reactions are reliable which ones are tricky,	Ability to explore the consequences of a new strategy *on hundreds of targets or do*

Better done by chemist	*Better done by computer*
which odd ones could be slightly modified to work on his target. The data-bank of a good chemist has often been fed for more than twenty-five years by oriented reading of the literature.	*the same for a starting commercial compound* whose price suddenly drops or rises.
Mobility. The chemist knows what he does not know and knows where to go in the library to find the relevant data.	Ability to *transport without reluctance a reaction or a strategy* born in one part of chemistry anywhere else even if superficially naive.
Adaptation to new information: the chemist reading a recently published paper is able to extrapolate it in different directions in the space of creativity. The chemist may *caricature a target* to recognize symmetry, by analogy with known material. The chemist busy with one target becomes *one-track-minded*: everything that he reads which could be relevant is at once biased to adapt to his aim.	A data bank *continuously improves if well structured and regularly fed* whereas the human brain forgets. But this is a very long term comparison. What is taught to a good chemist by twenty years of experience would probably take at least forty years for one man to teach the computer (but what about a team of twenty persons)? The computer may *generate thousands of structures and reactions* when it has been taught by a clever program to do so.

We have probably overlooked some points but the reader may complete this table by himself. The overall picture which transpires is that man leads the comparison in matters of judgement, ability to extrapolate, ability to select a few pieces of information from among thousands, ability to go and search for relevant information where it should be, and in recognition of the structure of a problem. In contrast, the computer is very good at comparing a structure, a reaction, a problem to thousands of those already known, at generating the thousands of faceted solutions that a general problem may take, and working without fatigue or prejudice. To caricature, the good chemist is a flyer and the good program is a plodder. Everyone knows that the *best groups result from a balanced composition of such characters*.

5.5   Possible breakthroughs as seen by workers in CAOS

To make easier our work of reviewing the field of CAOS, we sent a questionnaire to the groups involved in this research. Most of whom kindly responded. One of the questions was '*If you were allowed to imagine a possible breakthrough in the field for the next ten years which one would you suggest (even if it is not feasible by your own group)*? The variety of answers illustrates the richness of hope. We quote: 'Centralized Knowledge Base to which a large distributed user community contributes and upon which a large distributed user community draws'; 'A real dream is to get a problem solving system for synthetic organic chemistry, which is able to extract his knowledge base in a completely automatic way from appropriate data-banks by learning mechanisms'; 'In my opinion the directions now being followed were laid out in the first years of the field and the rate of true innovation is very low now.

Our work is directed to examining new approaches since I think the current approaches are limited in many ways'; 'Real synthetic chemists would take it seriously and contribute to logic development'; 'Full success will require both a thoroughgoing reaction library *and* a much higher degree of automation. Current versions requires too much user intervention for numerous trivial tasks'; 'That every chemist has a terminal or small computer at his desk or bench'; 'The field is not conducive to breakthroughs. What is required is hard work over many years'; 'The joining of non-empirical (mathematical in their nature) approaches with the approaches based on coded empirical data of organic chemistry'; 'An expert system based on data written in a natural or a graphic language; a self-teaching CAOS program; automatic extraction of concepts and strategies from literature data'.

Technocrats may wonder if in thirty years synthetic chemists will become 'obsolete': if 'perfect' computer-aided synthesis programs would become available, if reliable reactions could be performed by robots [303] and unreliable ones very systematically improved through planning of experiments [304a,b] what would the chemist's work become? This situation has very little chance to occur even within fifty years. The main reasons for this are: all the laws of chemistry are not yet known and even the best KIS system cannot avoid the need for actual experiments (see table, p. 84). Another factor is that *a much too-small investment in the field has been made up to the present time*. When one sits in a library and looks at the C.A. stacks from the beginning, one wonders what size of group could extract all the relevant information to feed a computer-aided synthesis program. In one way, this is reassuring because one of the *main driving forces of searchers is the feeling of being explorers*: kill this feeling and most will leave the field to search for a more stimulating *terra incognita*. On the other hand, as in alpinism, if the number of undone exploits is still rather large, the equipment of attacking groups is also increasing in quality and has nothing to do with what was available fifty years ago [304c]. It is clear that poor alpinists will not make big exploits just because they are well equipped, but it is clear also that no good team would accept the challenge of a difficult peak using the equipment used by the preceding generation. To further the analogy of an evaluation of the role to be played by computer-aided synthesis, one has to also remember that organic synthesis is at least as multifaceted as geography, which spans the friendly and easy hills as well as the difficult and glamorous Mount Everest. There is certainly room for many useful different programs within such a variety.

I.6   LIST OF ABBREVIATIONS

⇒	Transform
AHMOS	Automated Heuristic Modelling of Organic Synthesis
ALCHEM	A Language for Chemistry
ASI	Atomic Sequence Indice
ASR	Automatic design of Synthetic Routes

ASSOR	Advanced Simulation System of Organic Reactions
BCT	Block Cut point Tree
BE	Bond Electron matrix
BO	Bond Order
CAMEO	Computer-Assisted Mechanistic Evaluation of Organic Reactions
CAOS	Computer-Aided Organic Synthesis
CCR	Current Chemical Reaction
CHIRP	Chemical Engineering Investigation of Reaction Paths
CHEM INF	Chemischer Information
CHMTRN	Chemical Translator
CONGEN	Constrained Generation of isomers
CICLOPS	Chemical Implementation of Computers in the Logic Oriented Planning of Syntheses
DARC	Description, Acquisition, Retrieval, Computer-aided Design
DBONDAT	Doubly Bonded Atom
DCO	Descriptor of Conformation of the chains
DCO′	Descriptor of Conformation of ring substituents
DCOcy	Descriptor of Conformation of rings (cycles in French)
DCY	Descripteur de Cyclisation
DEL	Descripteur d'Environnement Limité
DEX	Descripteur d'Existence
DINASYN	Deoxyribonucleic Synthetiser
DLI	Descripteur des Liaisons (Bonds Descriptor)
DNA	Descripteur de la Nature des Atomes
DSOR	Domain of a Structurally Ordered Population of Compounds
DST	Descriptor of Stereochemistry
DU model	Dugundji–Ugi model
ELCO	Environment which is Limited Concentric and Ordered
EM	Ensemble of Molecules
EROS	Evaluation of Reaction for Organic Synthesis
EWA	Electron Withdrawing Atom
EXTRUS	see ref. 209
FGA	Functional Group Addition
FGI	Functional Group Interchange
FIEM	Family of Isomeric Ensemble of Molecules
FLAMINGOES	Formal-Logical Approach to Molecular Interconversions. Generation, Orientation Evaluation of Syntheses
FMO	Frontier Molecular Orbital
GPAIR	Group Pair Transform
GRACE	Generalized Reactions Analysis for Creation and Estimation
GSINC	Single Group Transform
GSS	General Synthesis Simulation by Synthon Substitution
HEXARR	see ref. 209

IGLOO	Invariant Graph and Logical Ordered Operators
IGOR	Interactive Generation of Organic Reactions
IEM	Isomeric Ensemble of Molecules
ISI	Institute of Scientific Information
KETOREACT	Ketone Reactivity
KIS	Knowledge Interchange System
LCAO	Linear Combination of Atomic Orbitals
LHASA	Logic and Heuristics for Automated Synthesis Analysis
MASSO	Méthode d'Aide en Synthèse Sur Ordinateur
NOON	Number of Outermost Occupied Neighbors
NWV	Neighbor Weight Vector
OCSS	Organic Chemical Synthesis Simulation
ORAC	Organic Reaction Access by Computer
PASCOP	Programme d'Aide à la Synthèse des Composés Organiques et Organophosphorés
PELCO	Perturbation of an ELCO
PFP	Program. Flow Prepn.
PMCD	Principle of Minimum Chemical Distance
PSG	Problem Solving Graph
PSYCHE	Pi-System-Chemie
REACT	see ref. 190 (Powers)
REACT	Reaction Sequence Simulation (Djerassi)
REACS	Reaction Access System
REPAS	Reaction Path Synthesis
SAS	Stimulated Analytical Synthesis
S.B.	Strategic Bonds
SCANMAT	see Chem. Abs. 103, 108597e
SCANSYNTH	see ref. 222
SCI	Science Citation Index
SECS	Simulation and Evaluation of Chemical Synthesis
SEQFGI	SEQuential Functional Group Interchange
SIMUL	see ref. 166
SIMULA	see ref. 160
SLING	Synchem Linear Input Graph
SOS	Stimulated Organic Synthesis
SPRESI	Speicherung und Recherche Strukturchemischer Information
SSSR	Smallest Set of Smallest Rings
SST	Starting Material Selection Strategies (choosed to suggest an analogy with the SST airplane in its ability to rapidly traverse large distances in one jump, non-stop)
STRATOS	see ref. 198a
STRCNT	Stereocenter
SYNTAB	Synthese Ableitung
SYNCHEM	Synthetic Chemistry
SYNGEN	SYNthesis GENerator
SYNLIB	Synthesis Library

SYNOPSYS	Synthesis, Optimization System
SYNPLAN	Synthese Planung
TAMREAC	Transition Metal Reactivity
TSD	Topological Structural Descriptor
WLN	Wisswesser Line Notation

1.7. REFERENCES

[1] E. J. Corey and W. T. Wipke, *Science*, **166**, 178 (1969).

[2] E. J. Corey, *Tetrahedron*, **39**, No 20 (1983).

[3] M. Chastrette, M. Rajzmann, M. Chanon and K. Purcell, *J. Am. Chem. Soc.*, **107**, 1–11 (1985).

[4] A. Eschenmoser and C. E. Wintner, *Science*, **196**, 1410–26 (1977).

[5] E. J. Corey, *Pure Appl. Chem.*, **14**, 19–37 (1967).

[6] R. B. Woodward, in *Perspectives in Organic Chemistry* (A. Todd, ed.), Interscience Pub., NY (1956), pp. 155–160.

[7] R. B. Woodward, *Pointers and Pathways in Research*, G. Hofteizer for Ciba of India, Ltd, Bombay, India (1963), p. 23.

[8] E. J. Corey, *Quart. Rev.*, **25**, 455–482 (1971).

[9] (a) G. E. Vleduts, *Inf. Storage Retr.*, **1**, 101 (1963).

(b) L. H. Sarett, 'Synthetic Organic Chemistry. New Techniques and Targets', presented before the Manufacturers Association, June 9 (1964).

[10] J. E. Dubois and A. Panaye, *Tetrahedron Lett.*, 1501–4 (1969); ibid., 3275–78 (1969).

[11] M. Bersohn and A. Esack, *Chem. Rev.*, **76**, 269–282 (1976).

[12] 'Computer-Assisted Organic Synthesis', (W. T. Wipke and W. J. Howe, eds) *ACS Symposium Series*, **61**, Washington DC (1977).

[13] R. Carlson, C. Albano, L. G. Hammarstöm, E. Johansson, A. Nilsson and R. E. Carter, *J. Molecular Science*, **2**, 1–22 (1983).

[14] A. K. Long, S. D. Rubenstein and L. J. Joncas, *Chem. Eng. News*, **9**, 22–30 (1983).

[15] J. Haggin, *Chem. Eng. News*, **9**, 7–22 (1983).

[16] (a) Y. Kudo, *Kagaku to Kogyo*, **35** (1), 18–20 (1982).

(b) Y. Kudo, *J. Synthetic Org. Chem. Jpn*, **42**, 694–700 (1984).

[17] (a) I. Gutman and M. Milun, *Kemija u Industriji*, **3**, 123–130 (1973).

(b) B. Donova-Jerman—Blazic, *Kemija u Industriji*, **12**, 709–713 (1976), Hedos Program for aromatic chemistry is also described in this report.

[18] (a) A. Weise, *Wiss. Fortsch.*, **30** (7), 268–272 (1980) in German.

(b) H. Bruns, *Naturwissenschaften*, **66**, 197–201 (1979) in German.

(c) J. Gasteiger, *Chim. Ind.*, **64**, 714–720 (1982).

(d) J. H. Winter *Chemische Syntheseplanung* Springer-Verlag, Heidelberg (1982).

[19] V. Nevalainen, E. Pohjala and S. Väisänen, *Kemia-Kemi*, **11** (3), 217–223 (1984).

[20] (a) R. K. Lindsay, B. G. Buchanan, E. A. Feigenbaum and J. Lederberg, *Applications of Artificial Intelligence for Organic Chemistry*, McGraw-Hill Book Company, NY (1980), pp. 28.

(b) D. L. Larsen, *Ann. Rev. Med. Chem.*, **16**, (1981).

[21] R. Carlson, C. Albano, L. G. Hammarström, E. Johansson and A. Nilsson, *Kem. Tidskr.*, **93** (2), 32–41 (1981).

[22] (a) S. Turner, Education in Chemistry, (March), 48–50 (1984)

(b) Anon, *Chem. Brit.*, **19**, 988 (1983).

(c) Anon, *Chem. Ind.*, 905 (1983).

(d) A. P. Johnson, *Chem. Brit.*, **21**, 59–61 (1985).

(e) J. Boother, *Chem. Brit.*, **21**, 36–7 (1985).

[23] P. de Clerq, *Chemie Magazine*, **7** (8), 27–29 (1981).

[24] (a) H. Bockovà and I. Urbankova, *Chem. Listy*, **71**, 715–723 (1977).

(b) J. Benes, *Chem. Prum.*, **31** (7), 379–88 (1981).

[25] A. J. Thakkar, *Topics Curr. Chem.*, **39**, 3–18 (1973).

[26] (a) F. Choplin, *Bulletin Centenaire de l'E.S.P.C.* (1982).

(b) J. Tremoliere, *Electronique Applications*, **17**, 17–22 (1981).

(c) E. Gordon, *Sciences et Techniques*, **78**, 2–6 (1981).

(d) G. Kauffmann, C. Laurenço, *La Recherche*, **123**, 742–44 (1981).

(e) M. Chanon, R. Barone, *Encyclopaedia Universalis, Synt. Ass. Ord.*, Vol. Supplement (1979) pp. 1368–74.

[27] M. Osinga, *Chemie Magazine*, **5**, 283–5 (1982).

[28] H. C. J. Ottenheijm and J. H. Noordik, *Janssen Chimica Acta*, **2**, 3–8 (1984).

[29] (a) M. Marsili, *Chem. Ind.* (Milano), **64**, 722–26 (1982).

(b) G. Sello, *Chim. Ind.* (Milano), **66**, 346 (1984).

[30] S. Y. Zhu and J. P. Zhang, *Hua-Hsueh Tung Pao*, **5** (34), 8–15 (1981).

[31] (a) E. J. Corey, M. Ohno, R. B. Mitra and P. A. Vantakencherry, *J. Am. Chem. Soc*, **86**, 478 (1964).

(b) E. J. Corey, R. B. Mitra and H. Uda, *J. Am. Chem. Soc.*, **86**, 485 (1964).

[32] D. P. G. Hamon and R. N. Young, *Aust. J. Chem.*, **29**, 145–161 (1976).

[33] E. J. Corey, W. T. Wipke, R. D. Cramer and W. J. Howe, *J. Am. Chem. Soc.*, **94**, 421 (1972).

[34] W. J. Howe, PhD. Thesis, Harvard University (1972).

[35] E. J. Corey, W. T. Wipke, R. D. Cramer and W. J. Howe, *J. Am. Chem. Soc.*, **94**, 431 (1972).

[36] E. J. Corey, R. D. Cramer and W. J. Howe, *J. Am. Chem. Soc.*, **94**, 440 (1972).

[37] E. J. Corey, and G. A. Peterson, *J. Am. Chem. Soc.*, **94**, 460 (1972).

[38] D. Pensak, PhD, Harvard University Sept. (1973) University Microfilm International, Ann Arbor, Michigan, USA.

[39] E. J. Corey, and R. D. Balanson, *J. Am. Chem. Soc.*, **96**, 6516 (1974).

[40] E. J. Corey, W. J. Howe and D. A. Pensak, *J. Am. Chem. Soc.*, **96**, 7724 (1974).

[41] E. J. Corey, W. J. Howe, H. W. Orf, D. A. Pensak and G. Peterson, *J. Am. Chem. Soc.*, **97**, 6116 (1975).

[42]　E. J. Corey and W. L. Jorgensen, *J. Am. Chem. Soc.*, **98**, 189 (1976).

[43]　E. J. Corey, and W. L. Jorgensen, *J. Am. Chem. Soc.*, **98**, 203 (1976).

[44]　E. J. Corey, W. W. Orf and D. A. Pensak, *J. Am. Chem. Soc.*, **98**, 210 (1976).

[45]　D. A. Pensak and E. J. Corey, in *Computer Assisted Organic Synthesis, ACS Symposium Series 61*, Washington DC (1977) pp. 1.

[46]　E. J. Corey and A. K. Long, *J. Org. Chem.*, **43**, 2208 (1978).

[47]　E. J. Corey and N. F. Feiner, *J. Org. Chem.*, **45**, 757 (1980) ibid, 765.

[48]　E. J. Corey, A. P. Johnson and A. K. Long, *J. Org. Chem.*, **45**, 2051 (1980).

[49]　(a) E. J. Corey, A. K. Long, J. Mulzer, H. W. Orf, A. P. Johnson and A. P. W. Hewett, *J. Chem. Inf. Comput. Sci.*, **20**, 221, (1980).

　　　(b) E. J. Corey, A. K. Long and S. D. Rubenstein, *Science*, **228**, 408–18 (1985).

　　　(c) E. J. Corey, A. K. Long, T. W. Greene and J. W. Miller, *J. Org. Chem.*, **50**, 1920–27 (1985).

[50]　I. Ugi, *Rec. Chem. Progr.*, **30**, 289 (1969).

[51]　I. Ugi and P. Gillespie, *Angew. Chem. Int. Ed.*, **10**, 914–15 (1971) and 915–919.

[52]　I. Ugi, P. Gillespie and C. Gillespie, *Trans. NY Acad. Sci.*, **34**, 416 (1972).

[53]　J. Dungundji and I. Ugi, *Top. Curr. Chem.*, **39**, 19–64 (1973).

[54]　J. Blair, J. Gasteiger, C. Gillespie, P. Gillespie and I. Ugi, *Tetrahedron*, **30**, 1845 (1974).

[55]　I. Ugi, *IBM-Nachr.*, **24**, 180 (1974).

[56]　I. Ugi, J. Gasteiger, J. Brandt, J. F. Brunnert and W. Schubert, *IBM-Nachr.*, **24**, 185 (1974).

[57]　J. Blair, J. Gasteiger, C. Gillespie, P. Gillespie and I. Ugi, in *Computer Representation and Manipulation of Chemical Information* (W. T. Wipke, S. Heller, R. Feldman and E. Hyde, eds) Wiley, New York (1974) pp. 129.

[58]　J. Brandt, J. Friedich, J. Gasteiger, C. Jochum, W. Schubert and I. Ugi, in *Computers in Chemical Education and Research* (E. V. Ludena, N. H. Sabelli and A. C. Wahl, eds) Plenum Press, NY (1977) p. 401.

[59]　J. Dugundji, D. Marquarding and I. Ugi, *Chemica Scripta*, **11**, 17–24 (1977).

[60]　C. Jochums and J. Gasteiger, *J. Chem. Inf. Comput. Sci.*, **17**, 113 (1977).

[61]　J. Brandt, J. Friedich, J. Gasteiger, C. Jochum, N. Schubert and I. Ugi, in 'Computer Assisted Synthesis', *ACS Symposium Series*, **61** Washington DC (1977). pp. 33–59.

[62]　J. Gasteiger and M. Marsili, *Tetrahedron Lett.*, 3184 (1978).

[63]　J. Brandt, J. Friedich, J. Gasteiger, C. Jochum, P. Lemmen, W. Schuber and I. Ugi, *Pure Appl. Chem.*, **50**, 1303–18 (1978).

[64]　W. Schubert and I. Ugi, *J. Am. Chem. Soc.*, **100**, 37 (1978).

[65]　J. Gasteiger and C. Jochum, *Topics Curr. Chem.*, **74**, 93 (1978).

[66]　J. Gasteiger, *Computer Chem.*, **2**, 85 (1978).

[67] J. Gasteiger and C. Jochum, *J. Chem. Inf. Comp. Sci.*, **19**, 43–48 (1979).

[68] I. Ugi, *Giessener Universitätsbl.*, **11**, 68 (1978).

[69] I. Ugi, J. Bauer, J. Brandt, J. Friedich, J. Gasteiger, C. Jochum and W. Schubert, *Angew. Chem. Int. Ed.*, **18**, 111–123 (1979).

[70] J. Gasteiger, C. Jochum, M. Marsili and J. Thoma, *Informal Commun. Math. Chem.*, **6**, 177 (1979).

[71] W. Schubert and I. Ugi, *Chimia*, **33**, 183–191 (1979).

[72] J. Gasteiger, Z. *Naturforsch*, **34b**, 67–75 (1979).

[73] J. Friedich and I. Ugi, *J. Chem. Res.*, **70** (1980).

[74] I. Ugi, J. Bauer, J. Brandt, J. Dugundji, R. Franck, J. Friedich, A. von Scholley and W. Schubert, *Stud. Phys. Theor. Chem.*, **16**, 219 (1981).

[75] M. Marsili and J. Gasteiger, *Stud. Phys. Theor. Chem.*, **16**, 56–57 (1981).

[76] J. Gasteiger, M. Marsili and B. Paulus, *Stud. Phys. Theor. Chem.*, **16**, 229–246 (1981).

[77] (a) J. Brandt, J. Bauer, R. M. Frank and A. von Scholley, *Chemica Scripta*, **18**, 53–60 (1981).

   (b) J. Bauer, I. Ugi, *J. Chem. Res.*, **11**, 3101 (1982).

[78] C. Jochum, J. Gasteiger, I. Ugi and J. Dugundji, Z. *Naturforsch.*, b, **37B**, 1205–15 (1982).

[79] J. Brandt and A. von Scholley, *Computer Chem.*, **7**, 51–59 (1983).

[80] M. Marsili, J. Gasteiger and R. E. Carter, *Chim. Oggi*, **9**, 11–18 (1984).

[81] J. Gasteiger and Z. Hippe, *Zesz. Nauk Politech. Rezes Zowskiej.*, **3**, 37–48 (1984).

[82] W. Schubert and W. Ellenrieder, *J. Chem. Res. Synop.*, **8**, 258–59 (1984).

[83] J. B. Hendrickson, *J. Am. Chem. Soc.*, **93**, 6847–54 (1971).

[84] J. B. Hendrickson, *J. Am. Chem. Soc.*, **93**, 6854–62 (1971).

[85] J. B. Hendrickson, *Angew. Chem. Int. Edit.*, **13**, 47–76 (1974).

[86] J. B. Hendrickson, *J. Am. Chem. Soc.*, **97**, 5763–84 (1975).

[87] J. B. Hendrickson, *J. Am. Chem. Soc.*, **97**, 5784–5800 (1975).

[88] J. B. Hendrickson, *Topic Curr. Chem.*, **62**, 49 (1976).

[89] J. B. Hendrickson, *J. Am. Chem. Soc.*, **99**, 5439 (1977).

[90] J. B. Hendrickson, *J. Chem. Educ.*, **55**, 216 (1978).

[91] (a) J. B. Hendrickson, *J. Chem. Inf. Comput. Sci.*, **19** (3), 129–136 (1979).

   (b) J. B. Hendrickson, *J. Chem. Educ.*, **62**, 245–9 (1985).

[92] J. B. Hendrickson, and E. Braun-Keller, *J. Comput. Chem.*, **1** (4), 323 (1980).

[93] J. B. Hendrickson, E. Braun-Keller and G. A. Toczko, *Tetrahedron*, **37**, suppl. 1, 359–70 (1981).

[94] J. B. Hendrickson, and A. G. Toczko, *J. Chem. Inf. Comput. Sci.*, **23**, 171–77 (1983).

[95] J. B. Hendrickson, D. L. Grier and A. G. Toczko, *J. Chem. Inf. Comput. Sci.*, **24**, 195–203 (1984).

[96] A. K. Ghose, *J. Scientific Indus. Res.*, **40**, 423–431 (1981).

[97] M. Bersohn, *Bull. Japan Chem. Soc.*, **45**, 1897–1903 (1972) paper submitted in 1970.

[98] M. Bersohn, *JCS Perkin I*, 1239–41 (1973).

[99] A. Esack and M. Bersohn, *JCS Perkin I*, 2463–70 (1974).

[100] A. Esack, *JCS Perkin I*, 1120–24 (1975).

[101] A. Esack and M. Bersohn, *JCS Perkin I*, 1124–29 (1975).

[102] M. Bersohn and A. Esack, *Chimica Scripta*, **9**, 211 (1976).

[103] M. Bersohn and A. Esack, *Comput. Chem.*, **1**, 103 (1976).

[104] M. Bersohn, in 'Computer Assisted Organic Synthesis *ACS Symposium 61*, (eds W. T. Wipke and W. J. Howe) Washington DC (1977) p. 128.

[105] M. Bersohn, A. Esack and J. Luchini, *Computer Chem.*, **2**, 105–111 (1978).

[106] M. Bersohn, *Computer Chem.*, **2**, 113–116 (1978).

[107] M. Bersohn and K. Mackay, *J. Chem. Comput. Sci.*, **19**, 137–41 (1979).

[108] M. Bersohn, in 'Computer Assisted Drug Design', *ACS Symposium, 112* (E. C. Olson and R. E. Christoffersen eds) Washington DC (1979), pp. 341–352.

[109] M. Bersohn, *JCS Perkin I*, 1975–77 (1979).

[110] M. Bersohn, in 'Supercomputers in Chemistry' *ACS Symposium Series*, 173 (P. Lykos, I. Sharitt, eds) (1981). pp. 109–116.

[111] M. Bersohn *JCS Perkin I*, 631–637 (1982).

[112] A. Weise, W. Schäfer, R. Walther and D. Martin, *Z. Chem.*, **12**, 81 (1972).

[113] A. Weise., *Z. Chem.*, **13**, 155–156 (1973).

[114] A. Weise, *Mitteilungsbl. Chem. Ges. DDR*, **20**, 25 (1973).

[115] A. Weise, *Z. Chem.*, **15**, 333–340 (1975).

[116] A. Weise, A. Klebsch and G. Westphal, *Z. Chem.*, **17**, 295 (1977).

[117] A. Weise and H. G. Scharnow, *Z. Chem.*, **19**, 49 (1979).

[118] A. Weise, *Wiss. Fortsch*, **30**, 268 (1980).

[119] A. Weise, *J. Prakt. Chem.*, **322**, 761–68 (1980).

[120] A. Weise, *Z. Chem.*, **21**, 352–53 (1981).

[121] A. Weise and G. Westphal, *Z. Chem.*, **21**, 218–19 (1981).

[122] A. Weise., and H. G. Scharnow, *Z. Chem.*, **21**, 353–54 (1981).

[123] N. S. Sridharan, PhD., State University of New York, Suny at Stony Brook (1971).

[124] H. L. Gelernter, N. S. Sridharan, A. J. Hart, S. C. Yen, F. W. Fowler and H. Shou, *Top. Curr. Chem.*, **41**, 113 (1973).

[125] (a) N. S. Sridharan, Department Computer Science Rutgers, The State University of New Jersey. Technical Report No 43 (1976) 34 pages.

(b) N. S. Sridharan, *Proceedings of International Federation for Information Processing*, Stockholm, August 1974.

[126] H. L. Gelernter, A. F. Sanders, D. L. Larsen, K. K. Argawal, R. H. Boivie, G. A. Spritzer and J. E. Searleman, *Science*, **197**, 1041–49 (1977).

[127] K. K. Agarwal, D. L. Larsen and H. L. Gelernter, *Computers Chemistry*, **2**, 75–84 (1978).

[128] H. Gelernter, G. A. Miller, D. L. Larsen and D. J. Berndt, Technical Report No. 46, Department of Computer Science, State University of New York at Stony Brook (1983) 26 pages.

[129] D. J. Berndt, G. A. Miller and H. Gelernter, Technical Report No 51, Department of Computer Science, State University of New York at Stony Brook (1983) 33 pages.

[130] H. Gelernter, S. S. Bhagwat, D. L. Larsen and G. A. Miller, in *Computer Applications in Chemistry* (S. R. Heller, R. Potenzone, eds) Elsevier Sci. Publ., Amsterdam (1983), pp. 35–59.

[131] R. Barone, Thèse d'Etat Marseille (1976).

[132] R. Barone, M. Chanon and J. Metzger, *Revue de l' I.F.P.*, p. 771 (1973).

[133] R. Barone, M. Chanon and J. Metzger, *Tetrahedron Lett.*, **32**, 2761 (1974).

[134] R. Barone, and M. Chanon, *Nouv. J. Chim.*, **2**, 659 (1978).

[135] R. Barone, M. Chanon and J. Metzger, *Chimia* **32**, 216 (1978).

[136] R. Barone, P. Camps and J. Elguero, *Ann. Quim.*, **75**, 736 (1979).

[137] A. Boch, R. Barone, M. Chanon and J. Metzger, *Comput. Chem.*, **3**, 83 (1979).

[138] (a) R. Barone, M. Chanon and M. L. H. Green, *J. Organomet. Chem.*, **185**, 85 (1980).

      (b) I. Theodosiou, R. Barone, M. Chanon, *J. Molecular Catalysis*, **37**, 1985, 22.

[139] R. Barone, M. Chanon, G. Vernin and J. Metzger, *Rivista Italiana*, **62**, 136 (1980).

[140] R. Barone and M. Chanon, in *The Chemistry of Heterocyclic Flavouring and Aroma Compounds* (G. Vernin Ed.), Horwood, Chichester, J. Wiley NY (1982).

[141] R. Barone and M. Chanon, *Heterocycles*, **16**, 1357 (1981).

[142] R. Barone, M. Chanon, P. Cadiot and J. M. Censé, *Bull. Soc. Chim. Belg.*, **91**, 333 (1982).

[143] (a) R. Barone, M. Chanon and M. L. Contreras, *Nouv. J. Chimie*, **8**, 311–15 (1984).

      (b) M. Istin and J. M. Grognet, Personal Communication (June 1985).

[144] W. T. Wipke and P. Gund, *J. Am. Chem. Soc.*, **96**, 299 (1974).

[145] W. T. Wipke and T. M. Dyott, *J. Am. Chem. Soc.*, **96**, 4825 (1974).

[146] W. T. Wipke, S. R. Heller, R. J. Feldmann and E. Hyde, *Computer Representation and Manipulation of Chemical Information*, Wiley, NY (1974).

[147] W. T. Wipke and T. M. Dyott, *J. Am. Chem. Soc.*, **96**, 4834 (1974).

[148] T. M. Gund, P. V. R. Schleyer, P. Gund and W. T. Wipke, *J. Am. Chem. Soc.*, **97**, 743 (1975).

[149] W. T. Wipke and T. M. Dyott, *J. Chem. Inf. Comput. Sci.*, **15**, 140 (1975).

[150]   W. T. Wipke and P. Gund, *J. Am. Chem. Soc.*, **98**, 8107 (1976).

[151]   W. T. Wipke, in *Computers in Chemical Education and Research*, Plenum Press, NY (1977) p. 381.

[152]   W. T. Wipke, H. Braun, G. Smith, F. Choplin and W. Sieber, 'Computer Assisted Organic Syntheses, *ACS Symposium Series 61*, Washington DC, p. 97 (1977).

[153]   P. Gund, *Annu. Rep. Med. Chem.*, **12**, 288-97 (1977).

[154]   W. T. Wipke, G. I. Ouchi and S. Krishnan, *Artificial Intelligence*, **11**, 173 (1978).

[155]   P. Gund, E. J. J. Grabowski, G. M. Smith, J. D. Endosse, J. B. Rhodes and W. T. Wipke, in 'Computer Assisted Drug Design', *ACS Symposium Series 112*, Washington DC (1979) p. 527.

[156]   P. Gund, E. J. J. Grabowski, D. R. Hoff, G. M. Smith, J. D. Andose, J. B. Rhodes and W. T. Wipke, *J. Chem. Inf. Computer Sci.*, **20**, 88-93 (1980).

[157]   R. E. Carter and W. T. Wipke, *Kem. Tidske.*, **93**, 20 (1981).

[158]   W. T. Wipke, *Cosmet. Toiletries*, **99** (10) 73-82 (1984).

[159]   W. T. Wipke, and D. Rogers, *J. Chem. Inf. Computer Sci.*, **24**, 71-81 (1984).

[160]   W. T. Wipke, and D. Rogers, *J. Chem. Inf. Computer Sci.*, **24**, 255-62 (1984).

[161]   P. Benedek, and M. Mezei, *Magyar Kemikusok Lapja*, **28**, 29 (1973).

[162]   P. Benedek, and P. Vaczi, *Magyar Kemikusok Lapja*, **28**, 623 (1973).

[163]   P. Benedek, *Kemia Közlemények*, **42**, 457-71 (1974).

[164]   A. Nemeth, P. Benedek and P. Vaczi, *Magyar Kemikusok Lapja* **29**, 100 (1974).

[165]   G. Balizs, P. Benedek and P. Vaczi, *Magyar Kemikusok Lapja* **29**, 155 (1974).

[166]   P. Benedek, *Magyar Kemikusok Lapja*, **30**, 48-53 (1975).

[167]   J. E. Dubois, D. Laurent and H. Viellard, *C.R. Acad. Sci.*, **263C**, 764 (1966).

[168]   J. E. Dubois, and H. Viellard, *Bull. Soc. Chim. Fr.*, 900-905, 905-913, 913-921 (1968).

[169]   J. E. Dubois, in *Computer Representation and Manipulation of Chemical Information* (W. T. Wipke, S. R. Heller, R. J. Feldmann and E. Hyde, eds.) J. Wiley, NY (1974) pp. 239-264.

[170]   J. E. Dubois, A. Panaye and M. J. Cojan-Alliot, *C.R. Acad. Sci.*, **280C**, 353 (1975).

[171]   J. E. Dubois, *Isr. J. Chem.*, **14**, 17 (1975).

[172]   J. E. Dubois, Vth International Conference on Computers in Chemical Research and Education, Toyohashi, Japan (1980).

[173]   J. E. Dubois and H. Viellard, *Bull. Soc. Chim. Fr.*, 839-48 (1971).

[174]   J. E. Dubois, *Pure Appl. Chem.*, **53**, 1313-1327 (1981).

[175]   J. E. Dubois, A. Panaye and C. Lion, *Nouveau J. Chimie*, **5**, 371-380 (1981); ibid. 381-91 (1981).

[176]   R. Picchiottino, G. Georgoulis, G. Sicouri, A. Panaye and J. E. Dubois, *J. Chem. Inf. Comput. Sci.*, **24**, 241-249 (1984).

[177] (a) J. E. Dubois, M. J. Alliot, and H. Viellard, *C.R. Acad. Sci.*, **271C**, 1412–15 (1970).

(b) J. E. Dubois, M. J. Alliot and A. Panaye, *C.R. Acad. Sci.*, **273C**, 224–27 (1971).

(c) J. E. Dubois and A. Panaye, *Bull. Soc. Chim. Fr.*, 2100–10 (1975).

(d) J. E. Dubois, and M. J. Cojan-Alliot, *C.R. Acad. Sci.*, **280C**, 13–15 (1975).

(e) P. Cayzergues, A. Panaye and J. E. Dubois, *C.R. Acad. Sci.*, **290C**, 441–44 (1980).

(f) J. E. Dubois, A. Panaye and P. Cayzergues, *C.R. Acad. Sci.*, **290C**, 429–32 (1980).

[178] J. E. Dubois, R. Picchiottino, G. Sicouri and Y. Sobel, *C.R. Hebd. Sceances Acad. Sci. Ser.*, *1*, **294**, 251–56 (1982).

[179] J. E. Dubois, A. Panaye, R. Picchiottino and G. Sicouri, *C.R. Hebd. Sceances Acad. Sci. Ser. 2*, **295**, 1081–86 (1982).

[180] J. E. Dubois, G. Sicouri, Y. Sobel and R. Picchiottino, *C.R. Hebd. Sceances Acad. Sci. Serie B*, **298**, 525–30 (1984).

[181] M. Pfau, *L'Actualité Chimique*, Avril, 45–52 (1985).

[182] J. D. Fellman, G. A. Rupprecht and R. R. Schrock, *J. Am. Chem. Soc.*, **101**, 5099 (1979).

[183] M. J. S. Dewar, *Current Contents Physical, Chemical, Earth Sciences*, **25**, No. 14, 16 (1985).

[184] S. W. Golomb, in *Graph Theory and Computing* (R. C. Read, ed.), Academic Press, New York (1972) pp. 23–37.

[185] P. E. Blower and H. W. Whitlock, *J. Am. Chem. Soc.*, **98**, 1499–1510 (1976).

[186] H. W. Whitlock, *J. Am. Chem. Soc.*, **98**, 3225–3233 (1976).

[187] H. W. Whitlock, in 'Computer-Assisted Organic Synthesis, *ACS Symposium Series*, **61** (W. T. Wipke and W. J. Howe, eds), Washington DC (1977), pp. 60–80.

[188] G. J. Powers and R. L. Jones, *AI Ch EJ*, **19**, 1204 (1973).

[189] G. J. Powers, R. L. Jones, G. A. Randall, M. H. Caruthers, J. H. van de Sande and H. G. Khorana, *J. Am. Chem. Soc.*, **97**, 875 (1975).

[190] R. Govind and G. J. Powers, in Computer Assisted Organic Synthesis, *ACS Symposium Series*, *61* (W. T. Wipke and W. J. Howe, eds), Washington DC (1977) pp. 81–97.

[191] • R. Govind and N. Kedia, *Comput. Chem. Eng.*, **7** (**3**), 157–173 (1983).

[192] And we should warn the reader that for some examples (see 197) the strategy precisely consists in an interactive play with the chemist who corrects overpruning associated with too stringent rules present in a given transform.

[193] P. Gund and J. M. Gund, *J. Am. Chem. Soc.*, **103**, 4458–65 (1981).

[194] P. Gund, X-Ray Crystallogr. 'Drug Action Course Int. Sch. Crystallogr. 9th' (A. S. Horn and C. J. De Ranter, eds) Oxford Univ. Press, Oxford, UK (1984) pp. 495–506.

[195] F. Choplin, R. Marc, C. Laurenço and G. Kaufmann, 'Phosphorus Compounds and their Non-fertilizer Applications', 1st International

Congress on Phosphorus Compounds, Rabat (1977) pp. 87–93 (Pub. 1978).

[196] (a) F. Choplin, C. Laurenço, R. Marc, G. Kaufmann and W. T. Wipke, *Nouveau J. Chim.*, **2**, 285–293 (1978).

(b) F. Choplin, R. Marc, G. Kauffmann and W. T. Wipke, *J. Chem. Inf. Comput. Sci.*, **18**, 110–8 (1978).

[197] F. Choplin, P. Bonnet, M. H. Zimmer and G. Kaufmann, *Nouveau J. Chim.*, **3**, 223–230 (1979).

[198] (a) P. Bonnet, J. C. Derniame, M. H. Zimmer, F. Choplin and G. Kaufmann, *RAIRO Informatique/Comput. Sci.*, **13**, 287–308 (1979).

(b) F. Choplin, R. Dorschner, G. Kaufmann and W. T. Wipke, *J. Organomet. Chem.*, **152**, 101–9 (1978)

[199] M. H. Zimmer, F. Choplin, P. Bonnet and G. Kaufmann, *J. Chem. Inf. Comput. Sci.*, **19**, 235–241 (1979).

[200] (a) C. Laurenço, L. Villien and G. Kaufmann, *Phosphorus and Sulfur*, **18**, 129–32 (1983).

(b) P. Metivier, P. Jauffret, M. P. Heitz, C. Laurenço and G. Kaufmann, Congrès National de la Societé Française de Chimie, Sept. 1984. Résumés 11C4 et 11A17.

[201] F. Choplin, S. Goudiam and G. Kaufmann, *J. Chem. Inf. Comput. Sci.*, **23**, 26–30 (1983).

[202] (a) C. Laurenço, L. Villien and G. Kaufmann, *Tetrahedron*, **40**, 2721–29 (1984) ibid., 2731–40 (1984).

(b) C. Laurenço and G. Kaufmann, *Tetrahedron Lett.*, **21**, 2243 (1980).

[203] G. Moreau, *Nouveau J. Chimie*, **2**, 187–193 (1978).

[204] (a) B. G. Buchanan and D. H. Smith, *Computers in 'Chemical Education and Research'*, Plenum Press, NY (1977), p. 401.

(b) T. H. Varkony, R. E. Cahart, D. H. Smith and C. Djerassi, *J. Chem. Inf. Comput. Sci.*, **18**, 168 (1978).

[205] T. H. Varkony, D. H. Smith, and C. Djerassi, *Tetrahedron*, **34**, 841–52 (1978).

[206] Y. Yoneda, *Saikin No Kagakukôgaku* Edited by the Association for Chemical Engineers of Japan Maruzen, Tokyo (1970), p. 145 (in Japanese).

[207] Y. Yoneda, 'Kemoguramu (Chemogram I), *Maruzen Tokyo* Vol. 1 (1972) (in Japanese) p. 309.

[208] Y. Yoneda, *Bull. Chem. Soc. Japan*, **52**, 8–14 (1979).

[209] T. Brownscombe, PhD (1973) Rice University, Houston, TX, USA.

[210] (a) R. D. Stolow and L. J. Joncas, *J. Chem. Educ.*, **57**, 868–73 (1980).

(b) M. P. Bertrand, H. Monti and R. Barone, *J. Chem., Educ.*, **63**, July (1986).

[211] R. B. Agnihotri and R. L. Motard, in 'Computer Applications to Chemical Engineering,' *A.C.S. Symposium Series*, **124** (R. G. Squires and G. V. Reklaitis, eds) Washington DC (1980), pp. 193–206.

[212] Anonymous, in *Chem. Eng. News*, Sept 24, p. 39–40 (1979).

[213]   T. Salatin and W. L. Jorgensen, *J. Org. Chem.*, **45**, 2043-2051 (1980).

[214]   B. L. Roos-Kozel and W. L. Jorgensen, *J. Chem. Inf. Comput. Sci.*, **21**, 101-111 (1981).

[215]   T. D. Salatin, D. Mc Laughlin and W. L. Jorgensen, *J. Org. Chem.*, **46**, 5284-94 (1981).

[216]   B. L. Roos-Kozel, PhD Thesis Purdue University (1981).

[217]   C. E. Peishoff and W. L. Jorgensen, *J. Org. Chem.*, **48**, 1970-79 (1983).

[218]   J. Schmidt Burnier and W. L. Jorgensen, *J. Org. Chem.*, **48**, 3923-41 (1983).

[219]   W. L. Jorgensen, *Kagaku*, **38**, 483-488 (1983). We thank Professor Jorgensen for giving us the English preprint of this article.

[220]   (a) J. Schmidt Burnier and W. L. Jorgensen, *J. Org. Chem.*, **49**, 3001-20 (1984).

   (b) C. E. Peishoff and W. L. Jorgensen, *J. Org. Chem.*, **50**, 1056-68 (1985).

[221]   Z. Hippe, R. Hippe, M. Dec and W. Szumilo, *Proceedings Intern. Conference Software Reliability*. Ksiaz Techn. Univ. Press, Wroclaw (1979).

[222]   Z. Hippe, *Analytica Chim. Acta*, **133**, 677-83 (1981).

[223]   Z. Hippe, *Stud. Phys. Theor. Chem.*, **16**, 249-258 (1981).

[224]   Z. Hippe, O. Achmatowicz and R. Hippe, *Stud. Phys. Theor. Chem.*, **16**, 207 (1981).

[225]   N. S. Zefirov and S. S. Trach, *Zh. Org. Khim.*, **11**, 225 (1975).

[226]   N. S. Zefirov and S. S. Trach, *Zh. Org. Khim.*, **11**, 1785 (1975).

[227]   N. S. Zefirov and S. S. Trach, *Zh. Org. Khim.*, **12**, 7 (1976).

[228]   N. S. Zefirov and S. S. Trach, *Zh. Org. Khim.*, **12**, 697 (1976).

[229]   N. S. Zefirov and S. S. Trach, *Chem. Scripta*, **15**, 4 (1980).

[230]   N. S. Zefirov and S. S. Trach, *Zh. Org. Khim.*, **17**, 2465-86 (1981).

[231]   S. S. Trach and N. S. Zefirov, *Zh. Org. Khim.*, **18**, 1561-83 (1982).

[232]   N. S. Zefirov and S. S. Trach, *Zh. Org. Khim.*, **20**, 1121-42 (1984).

[233]   (a) X. Zin, M. Chen and M. Song, *Fenzi Kexue Xuebao*, **2**, 113-117 (1982), *CA.* **97**, 71571.

   (b) I. Erdos, 29th IUPAC Congress, 5-10 June 1983, Cologne. Poster P-128.

[234]   T. Uchimaru and S. Hara, *Yuki Gosei Kagaku Kyok*, **42**, 701-707 (1984); *C.A.* **101**, 170219.

[235]   (a) N. Seidel, *Prax Naturwiss.*, **5**, 136-143 (1984).

   (b) P. Cadiot, J. M. Cense and D. Villemin, *L'Actualité Chimique*, Mars 1985, 76-78.

[236]   R. L. Synge, *J. Chem. Inf. Comput. Sci.*, **25 (1)**, 50-55 (1985).

[237]   M. F. Lynch, J. M. Harrison and W. G. Town, *Computer Handling of Chemical Structure Information* McDonald London and American Elsevier Inc. (1971).

[238]   L. J. O'Korn, in 'Algorithms for Chemical Computations' *ACS symposium Series 46* (R. E. Christoffersen ed.). Washington DC (1977), pp. 122-148 (43 re.).

[239]   M. F. Lynch, *Computer-Based Information Services in Science and Technology*, P. Peregrinus Ltd, Herts, England (1974).

[240] P. C. Read, *J. Chem. Inf. Comput. Sci.*, **23**, 135–149 (1983).

[241] J. E. Ash and E. Hyde, *Chemical Information Systems*, E. Horwood, Chichester, England (1975) ch. 11

[242] (a) E. G. Smith and P. A. Becker, *The Wisswesser Line Formula Chemical Notation*, Chemical Information Management Inc., Cherry Hill, NJ (1976).

      (b) J. J. Vollmer, *J. Chem. Ed.*, **60**, 192 (1983).

[243] (a) G. M. Dyson, *A New Notation and Enumeration System for Organic Compounds*, 2nd Ed., Longmans, London (1949).

      (b) *IUPAC Nomenclature of Organic Chemistry* Pergamon Press Oxford (1979).

[244] J. E. Rush, *J. Chem. Inf. Comput. Sci.*, **16**, 202 (1976).

[245] Personal Communication.

[246] (a) C. E. Granito, *J. Chem. Doc.*, **13**, 72–74 (1973).

      (b) A. Zamora and D. L. Dayton, *J. Chem. Inf. Comput. Sci.*, **16**, 219–22 (1976).

      (c) see ref. 238, p. 134.

[247] A. T. Balaban, *Chemical Applications of Graph Theory*, Academic Press, London (1976).

[248] And we do not consider here fragment codes such as Ring Code (249) which play a considerable role in substructure searches.

[249] C. E. Granito, S. Roberts and G. W. Gibson, *J. Chem. Doc.*, **12**, 190–6 (1972).

[250] H. L. Morgan, *J. Chem. Doc.*, **7**, 154–165 (1967).

[251] G. Moreau, *Nouv. J. Chimie*, **4**, 17–22 (1980).

[252] D. M. Bishop, *Group Theory and Chemistry*, Clarendon Press, Oxford (1973).

[253] D. J. Gluck, *J. Chem. Doc.*, **5**, 43 (1965).

[254] (a) T. Nakayama and Y. Fujiwara, *J. Chem. Inf. Comput. Sci.*, **20**, 23–28 (1980).

      (b) Y. Fujiwara and T. Nakayama, *Anal. Chim. Acta*, **133**, 647–656 (1981).

      (c) T. Nakayama and Y. Fujiwara, *J. Chem. Inf. Comput. Sci.*, **23**, 80–7 (1983).

[255] P. C. Jurs, in ref. 146, pp. 265–86.

[256] (a) R. S. Cahn, C. K. Ingold and V. Prelog, *Angew. Chem. Int. Edit.*, **5**, 385 (1966).

      (b) E. L. Eliel, *J. Chem. Educ.*, **62**, 223–224 (1985).

[257] M. F. Lynch, in ref. 146, pp. 31–54.

[258] C. A. Shelley, *J. Chem. Inf. Comput. Sci.*, **23**, 61–65 (1983).

[259] (a) J. F. McOmie *Protective Groups in Organic Chemistry*, Plenum, New York and London (1973).

      (b) T. W. Greene, *Protective Groups in Organic Synthesis*, J. Wiley, New York (1981).

[260] E. Garagnani and J. C. J. Bart, *Z. Naturforsch*, **32b**, 455–464, 465–468 (1977).

[261] J. C. J. Bart and E. Garagnani, *Z. Naturforsch*, **32b**, 678–683 (1977).

[262] J. C. J. Bart and E. Garagnani, *Z. Naturforsch*, **31b**, 1646–53 (1976).

[263] W. P. Jencks, *Catalysis in Chemistry and Enzymology*, McGraw-Hill, New York (1969).

[264] R. F. Heck, *Organotransition Metal Chemistry. A Mechanistic Approach*, Academic press, New York (1974).

[265] J. K. Kochi, *Organometallic Mechanisms and Catalysis*, Academic Press, New York (1978).

[266] J. P. Collman and L. S. Hegedus, *Principles and Applications of Organotransition Metal Chemistry*, University Science Books Mill Valley, CA (1980).

[267] G. Wilkinson, F. G. A. Stone and E. W. Abel, *Comprehensive Organometallic Chemistry* Pergamon Press, Oxford (1982) Vol. 8.

[268] G. Vernin, *The Chemistry of Heterocyclic Flavouring and Aroma Compounds*, E. Horwood, Chichester, J. Wiley, NY (1982).

[269] C. Reichardt, *Solvent Effects in Organic Chemistry*, Verlag Chemie Weinheim, NY (1979).

[270] J. March, *Advanced Organic Chemistry Reactions Mechanisms and Structure*, 2nd Ed., McGraw-Hill International (1977).

[271] Sun Tzu, *The Art of War* (translated by S. B. Griffith, ed.), Clarendon Press, Oxford (1963).

[272] D. H. R. Barton and W. D. Ollis, *Comprehensive Organic Chemistry*, Pergamon Press, Oxford (1979).

[273] R. K. Mackie and D. M. Smith, *Guidebook to Organic Synthesis*, Longman, London (1982) 325 pages.

[274] S. Warren, *Designing Organic Synthesis*, Wiley, NY (1978) 285 pages.

[275] R. P. Lyer A. V. Prabhv, *Synthesis of Drugs. A Synthon Approach*, Sevak Publications Bombay (1985) 122 pages.

[276] T. Lindberg, *Strategies and Tactics in Organic Synthesis*, Academic Press, Orlando, San Diego (1984).

[277] C. A. Payne and L. B. Payne *How to do an Organic Synthesis*, Allyn and Bacon, Boston (1969).

[278] I. Fleming, *Selected Organic Syntheses*, J. Wiley, NY (1978).

[279] J. Mathieu, R. Panico and J. Weill-Reynal, *L'Aménagement Fonctionnel en Synthèse Organique*, Hermann, Paris (1978).

[280] L. Velluz, G. Valls and J. Mathieu, *Angew. Chem. Int. Ed.*, **6**, 778 (1967).

[281] H. W. Whitlock, in 'Computer-Assisted Organic Synthesis' (W. T. Wipke and W. J. Howe, eds) *ACS Symposium Series*, *61*, Washington DC (1977) pp. 60–81.

[282] K. P. C. Vollhardt, in ref. 276, pp. 300–24.

[283] G. S. R. Subba Rao and K. Pramod, *Proceedings of the Indian Acad. Sci.* (Chem. Sci.) **93**, 573–87 (1984).

[284a] R. Barone, and M. Chanon, Chimia, 1986.

[284b] W. Oppolzer and T. Godel, *J. Am. Chem. Soc.*, **100**; 2583–4 (1978).

[285] P. Deslongchamps, *Aldrichimica Acta*, **17**, 59–70 (1984).

[286] B. M. Trost, *Pure Appl. Chem.*, **53**, 2357–70 (1981).

[287] A. G. Fallis, *Canad. J. Chem.*, **62**, 183–234 (1984).

[288]   Ref. 272, Vol. 1, p. 103; Vol. 3, p. 990.

[289]   Ref. 272, Vol. 1, pp. 1049-52.

[290]   (a) W. S. Johnson, *Accts Chem. Res.*, **1**, 1-8 (1968).

   (b) W. S. Johnson, *Angew. Chem. Int. Ed.*, **15**, 9 (1976).

   (c) E. E. Van Tamelen, *Accts Chem. Res.*, **1**, 111-20 (1968).

   (d) P. A. Bartlett, J. I. Brauman, W. S. Johnson and R. A. Volkman, *J. Am. Chem. Soc.*, **95**, 7502-4 (1973).

   (e) K. Ziegler, *Advances Organometal. Chem.*, **6**, 1 (1968).

   (f) P. A. Wender and J. J. Howbert, *Tetrahedron Lett.*, 5325 (1983).

   (g) S. D. Burke and P. A. Grieco, *Organic Reactions*, **26**, 361-475 (1979).

   (h) R. V. Stevens, in *Organic Synthesis-2* (IUPAC), Jerusalem-Haifa (1978) (S. Sarel, eds), Pergamon Press (1979). pp. 1325-35.

   (i) K. B. Sharpless, C. H. Behrens, T. Katsuki, A. W. M. Lee, V. S. Martin, M. Takatani, S. M. Viti, F. J. Walker and S. S. Woodart, *Pure Appl. Chem*, **55**, 589-604 (1983).

[291]   H. Gelernter, *Proceedings of a Symposium on the Mathematical Theory of Automata*, Polytechnic Institute of Brooklyn Press, New York (1962).

[292]   H. Berliner, in *Les Progrès des Mathématiques*, Bibliothèque pour la Science. Librairie, Berlin, Paris (1980).

[293]   (a) A. T. Balaban, D. Farcasiu and R. Banica, *Rev. Roumaine Chim.*, **11**, 1225 (1966).

   (b) J. S. Littler, *J. Org. Chem.*, **44**, 4657-67 (1979).

   (c) D. N. P. Satchell, *Naturwissenschaften*, **64**, 113-21 (1977).

   (d) J. F. Arens, Rcl. *Trav. Chim. Pays-Bas*, **98**, 471-500 (1979); ibid., 395-99 (1979); ibid., 155-61 (1979).

   (e) J. H. Winter, *J. Chem. Inf. Comput. Sci.*, **24**, 263-65 (1984).

[294]   (a) O. Sinanoglu, *J. Am. Chem. Soc.*, **97**, 2309 (1975).

   (b) O. Sinanoglu, *Nippon Kagaku Kaishi*, **9**, 1358-1361 (1976).

[295]   I. Demes, J. Vidoczy, L. Botar and L. Gal, *Theor. Chim. Acta*, **45**, 215, 225 (1977).

[296]   V. Kvasnicka, M. Kratochvil and J. Kocà, *Coll. Czech. Chem. Commun.*, **48**, 2284-2304 (1983).

[297]   J. Kocà, M. Kratochvil, M. Kunz and V. Kvasnicka, *Coll. Czech. Chem. Commun.*, **49**, 1247-1261 (1984).

[298]   D. C. Roberts, *J. Org. Chem.*, **43**, 1473 (1978).

[299]   R. D. Guthrie, *J. Org. Chem.*, **40**, 402 (1975).

[300]   J. Valls, in *Computer Representation and Manipulation of Chemical Information* (W. T. Wipke, S. R. Heller, R. J. Feldmann and E. Hyde, eds) J. Wiley NY (1974) pp. 83-103.

[301]   D. Bawden, T. K. Devon, F. T. Jackson S. I. Wood, M. F. Lynch and P. Willett, *J. Chem. Inf. Comput. Sci.*, **19**, 90-93 (1979).

[302]   (a) Booklet ISI Information Services in the Science, Social Sciences, and Art.

   (b) Anonymous in April 12 issue of *Chem. Eng. News* (1982) p. 26.

[303]   (a) A. R. Frisbee, M. H. Nantz, G. W. Kramer and P. L. Fuchs, *J. Am. Chem. Soc.*, **106**, 7143–7145 (1984).

(b) A. Delacroix, J. N. Veltz and A. Leberre, *Bull. Soc. Chim. Fr.*, **2**, 481 (1978).

(c) C. Porte, R. Borrin, S. Bouma and A. Delacroix, *Bull. Soc. Chim. Fr. I*, 90 (1982).

[304]   (a) D. R. Cox, *Planning of Experiment*, J. Wiley, NY (1958).

(b) G. S. G. Beveridge and R. S. Schechter, *Optimization: Theory and Practice*, McGraw-Hill, New York (1970).

(c) G. Pimentel, *Chemtech*, **16**, 1986, 155.

CHAPTER II

Teaching Chemistry with Microcomputers

Daniel Cabrol and Claude Cachet,

University of Nice, France and

Richard Cornelius,

Lebanon Valley College,
Annville, USA

II.1 INTRODUCTION

The advent of microcomputers has promoted the use of computers in education from a few projects in which researchers tested the capabilities of the machines and explored student reaction to a tidal wave of activity which taxes the ability of instructors to respond to change. Chemical education has played a special role in the changes which have occurred. Chemistry as a discipline is well suited for the effective application of computers in education. The laboratory experimentation that is an integral part of chemical education has a parallel in computer-assisted education modules which allow a student to learn a concept by exploration. Thus the educational technique itself serves in part as an introduction to the discipline. In addition, teachers of chemistry are often accustomed to the use of computers as research tools. For these teachers the use of computers as educational tools is merely an additional application; the technology is already familiar.

The high level of activity with microcomputers in education today has resulted primarily from the new low cost of microcomputers compared with earlier equipment. The change in cost, however, has changed more than simply the number of applications; it has changed the goals of those involved in using computers in education, and it has changed the people that are involved. To understand the changes that have occurred, we must begin by briefly examining the history of computers in education in relation to the way that computers have been used by society in general.

During the sixties computers were an extremely expensive medium for use in education. In business and industry computers were also regarded as very expensive machines. Typically their use could be justified only when the consequence was a reduced workforce. Thus the prevailing attitude in education was that computers could be employed only in those instances where they could replace instructors. The computer manufacturers provided the primary funding for projects using computers in education. Several major implementations were undertaken during this decade. The IBM 1500 was developed to be used as a tool for research into the uses of computers for educational purposes. The PLATO project [1] had specially designed display hardware which included the capability to sense where a user touched the screen. TICCIT [2] concentrated on the overall design of a learning environment in which the student was in control. These projects permitted a significant number of people to gain direct experience in the use of computers in education.

As the prices of computers fell during the seventies, computers began to be used in business and industry in order to make existing personnel more efficient, to eliminate tedious tasks, and to offer new services. Educators reacted against the idea that they should be replaced, and the lower prices of equipment made possible projects which strived to use computers to create new learning environments. The development of LOGO and SOLO occurred during this period. LOGO [3] used computers to foster logical thinking and problem-solving skills in children. SOLO [4] provided opportunities for high school students to use computers in special projects. Most of the uses of computers in education were research-oriented. Large programs such as NDPCAL in the UK [5] and French experiments in secondary schools [6] involved attempts to experiment and evaluate the effectiveness of computer-based instruction in schools without special reference to a particular learning/teaching strategy. NDPCAL was a project in England in the middle 1970s which was aimed at making the use of computers a normal part of the school curriculum. A contemporary project was conducted in France. Both of these projects strongly emphasized the importance of teacher training. Together they involved more than a hundred schools and more than a thousand teachers. Necessarily, these projects were heavily supported by national funds. The costs of implementation were still high enough that computer use had to be justified in terms obtaining new knowledge.

The situation changed abruptly in the eighties with the microcomputer revolution. Hardware costs fell so sharply that the cost/benefit analysis changed completely. Businesses, schools and, for the first time, individuals now buy computers without knowing the tasks for which they will be used. Hardware became one of the least expensive parts of computer-assisted education and the primary factors limiting the introduction of microcomputers into education became teacher training and the availability of software. In addition to their low cost, microcomputers have other features which make them more suitable than mainframe computers for many applications in chemical education. The graphics and sound that are standard features of many microcomputers are powerful capabilities. The independence of microcomputers from a central processor means that they can be moved to anywhere that an electrical outlet can be found. This independence of microcomputers, coupled with their reliability and the absence of special environmental requirements, supports their use in resource rooms, lecture halls, and student laboratories. In fact, they are often placed on carriers in order that they can be readily moved from one location to another.

The impact of microcomputers on chemical education has been widespread, but we observe no parallel changes in the overall method of teaching. Computers have been used in education for more than twenty years. The number of applications has increased greatly since the introduction of microcomputers. The increased activity, however, has not produced many new kinds of applications, nor has it significantly increased our understanding of which factors contribute to the effectiveness of computer-assisted education. Studies have indicated that 'computers can make instruction more effective, and the learning experience can be more exciting, satisfying, and rewarding for the learner and the teacher' [7]. In spite of twenty years of experience, however, relatively little is known about the effects of graphics, animation, sound, humour, individualization, cooperation, or competition on the learning process.

Part of the reason that the microcomputer revolution has not increased our knowledge of the effectiveness of various teaching methods is that today increasingly the teachers using computers are no longer only those who wish to devote their professional lives to exploring the medium. The path is open for all instructors to make use of computers in chemical education. This chapter is written to help all teachers of chemistry understand how computers fit into the curriculum, in the ways they can be applied, and where to seek additional resources.

The subsequent material in this chapter begins with brief introductions to the roles of tools in education and to the characteristics of microcomputers which make them valuable as educational tools. Following this background material is a review of the major uses of microcomputers in chemical education, with attention to areas where future developments are likely or desirable. The last portion of the chapter covers resources (software, and information) available to those wishing to use microcomputers in chemical education.

II.2 MICROCOMPUTERS AS TEACHING AND LEARNING TOOLS

2.1 The educational environment

To understand the value of microcomputers in education, we have to grasp the nature of the learning process itself. Currently, no indisputable model of this complex process is available. We must be partially satisfied with simple descriptions which can serve as pragmatic guidelines for the development of effective uses of computers in chemical education. An introductory reading useful to relate the possible roles of technology to the learning process is offered by the Rushby's book *Introduction to Educational Computing* [8].

One important aspect of education is the communication of information. A student acquires much important scientific knowledge by receiving information. The effectiveness of the communication process is related to:

—the organization of the subject itself,
—the clarity in expressed ideas and
—the use of a communication process which introduces the least distortion.

If education is perceived as only this process of communication, then the teacher is the one who carries out the action. Certainly the teacher has control over whether ideas are expressed clearly and over the means used to communicate the information. In some cases the teacher can also have limited control over the organization of the subject itself. In this description of education, difficulties are attributed to inadequacies in one of the three points mentioned or, more commonly, to the deficiency of students as receivers. The use of technology in education has been focused principally on improvements in the communication process.

Many tools are available to help teachers effectively express the language, observations and theories of chemistry. These tools range from chalk, an old tool in both the anthropological and the geological senses, to the videodisc, one of the most recent technological innovations to be applied to education. Each of these tools, microcomputers included, has its advantages and disadvantages. While it is not appropriate in these pages to catalog and compare all these various techniques and their combinations, a description of the characteristics of microcomputers is presented in later paragraphs.

Considering the teacher as a transmitter and the student as a receiver is an extremely limited view of education. There is a large part of education which is communication, but there is more. The organization of any discipline is held together by a unifying thread of logic, a rational set of relationships. It is not enough for a student to memorize how parts of the discipline are related to one another. In fact, the rational organization of a subject itself may not always be the best way for a student to be introduced to the subject. Usually the teacher has adopted the logic of the discipline, but the learner may not follow the same logic. For instance, studies have shown that students use 'natural' reasoning for solving mechanics problems rather than the logic of physics [9]. Moreover, many pseudo-scientific concepts are already found in the cognitive knowledge of the student, but they are not identical in content with their scientific counterparts and are not related to each other in the same manner. One objective of science education is for the logic of the discipline to become part of the thought processes of the student. Thus *education is a process of transformation*. The necessary changes can occur only as a consequence of a student's learning activity, and not simply as a result of receiving an arbitrary system imposed from the outside. In this view of education, *the student is the one who carries out the action*.

The teacher's role is not restricted to that of transmitter, but also includes the design and control of the learning situation in which the transformation process of education takes place. Learning is a process in which motivation and personal interest are essential. The most stimulating environments are certainly different for different students according to their aptitudes and experiences. The differences demand individualization of students' activities, but the individualization must not be confused with isolation. The importance of a group environment and the significance of communication between students must not be underestimated.

The microcomputer as a tool has characteristics which can be useful in the process of communication, but it also may be used by the student for activities which would be difficult or impossible without the computer. It is more than a deliverer of information; it can provide the learning environment in which students can explore the logic of the discipline, discover for themselves the relationships among concepts, and begin to undergo the transformation process that is called education.

2.2   Characteristics of microcomputers

We shall limit the scope of our discussion to those machines which are either in wide use today or have the likelihood of becoming widely used within the next, say, five years. Microcomputers most commonly used for chemical education include, in approximate number of available machines, the Atari models 400 and 800, the IBM PC, RML 380Z, the TRS-80 Models I, II and III, Commodore models of the 64, 2000 and 3000 series and the Apple II, IIe and IIc. Other machines are in use in isolated locations, but nearly all available software is targeted toward one or more of the machines mentioned. The development of educational software is time-consuming so that it is not surprising to see dominance by the early machines. Educational software in chemistry is just beginning to appear for newcomers like IBM PC, Zenith 100 and Macintosh. Moreover, the large number of machines now in use exerts a stabilizing effect, so that we can expect to see today's machines remain in use for several years even though far more powerful machines may be available at

Table 1

	Commodore 64	Apple IIc	IBM PC	Macintosh
Processor	6510 (Z80)	6502 (Z80)	8088	68000
	8 bits	8 bits	16 bits	32 bits
ROM	20 K	18 K	40 K	64 K
RAM	64 K	128 K	64 K (544)	128 K
Disk	$5\frac{1}{4}$ in 170 K	$5\frac{1}{4}$ in 143 K	$5\frac{1}{4}$ in 160 K	$3\frac{1}{2}$ in 400 K
Display	24 × 40	24 × 40 (or		
		24 × 80)	25 × 80	
Graphics	320 × 200	280 × 192	640 × 200	512 × 342
Colors	16	8	8 320 × 200	1
Remarks	sound			Mouse, sound

comparable prices . The salient features of four of today's microcomputers typical of four price ranges are summarized in Table 1

Technical comparisons are beyond the scope of this chapter, more details have to be found in specialized magazines.

Six important characteristics of these microcomputers as a whole will be described: computational capabilities, data storage and retrieval, decision-making, interaction, graphics, and electrical input/output. For each of these characteristics some indication will be given of its adequecy for educational purposes. Short discussions are presented for the well-known characteristics of computers, while longer descriptions are given for those areas in which microcomputers support new possibilities for education.

2.2.1   *Computational capabilities*

Even the most limited of microcomputers today can accurately perform computations thousands of times more rapidly than any human. Computations such as least-squares analysis of data become trivial exercises with the microcomputer. For such calculations microcomputers have the effect of providing the results more rapidly than mainframe computers. With the larger machines, the amount of time required to access the central computer and provide it with the necessary identification or account numbers can easily exceed the entire elapsed time needed to use a microcomputer for the same purpose.

Microcomputers today are not powerful enough to rapidly perform complex theoretical calculations, but they can be used for simpler molecular orbital problems such as those involving Huckel [10] or extended Huckel calculations [11]. They are especially well-suited for statistical computations of data analysis in which the calculations are simple but extremely tedious to attempt by hand. They can also be very useful for obtaining numerical solutions of simultaneous or polynomial equations as in the exact calculation of pH [12] or for numerical integration of differential equations in chemical kinetics [13].

2.2.2   *Data storage and retrieval*

The very first computers had extremely limited internal storage capacity and no external storage capability at all. Today's microcomputers typically have from 8 000 to 640 000 characters of internal storage. External storage is often in the form of magnetic flexible diskettes, or floppy disks, each of which can record more than 100 000 characters of information. The ease with which these diskettes can be exchanged makes the effective external storage capability almost unlimited.

Many problems involving data storage and retrieval can be handled using commerically available general-purpose programs such as PFS Report [14] which are designed to be flexible in order to be adapted to many different problems. Tasks such as computer handling of literature references, data dealing with physical and chemical properties, chemical stockroom inventory, files of student information (enrollment, grades, completion of laboratory work, etc.) can be accomplished using programs in this category. For specialized problems, specific programs can be written. For example, one program on an Apple II computer has been written to compare mass spectra with a library of 1800 spectra recorded on disk [15]. One thousand compounds in the library can be compared with an unknown compound in roughly one second. The search algorithm is based on a comparison of the six most intense peaks in each spectrum. After completion of a search, the program lists the three most probable compounds with their molecular weights and gives the degree of similarity with the unknown mass spectrum.

2.2.3   Decision-making

One characteristic which has been used to distinguish between calculators and computers is the ability to make decisions based upon information supplied by the user or produced by the program. This ability gives the basis for the user to select among options in a menu, and to have control over the program flow. It is also an essential feature for using microcomputers as instructional tools because it allows programs to provide separate responses for correct and incorrect answers or a series of responses based upon the nature of an incorrect answer.

2.2.4   Interaction

The decision-making power of computers in general, when combined with the rapid computational power and the independence of microcomputers from the effects of other users makes possible direct, rapid interaction between the microcomputer and the student.

Interaction has always been one of the most valuable characteristics of computer use in education, but the microcomputer further enhances this capability. It is a factor which can reinforce the student's motivation and interest. The computer does not even have to wait until the student indicates completion of an answer. One example of how this interaction can be put to use occurs in nomenclature drills in which the computer examines each character as it is typed into the computer [16]. If a student types an incorrect character, the computer can interrupt the drill before the next character is typed, providing a relevant explanatory message. This rapid feedback prevents an incorrect answer from being repeated in a student's mind as the answer is typed.

Another type of interaction is accomplished through *the flexibility of screen design*. In any printed medium, all of the text and graphics must obviously already be on the page before it is given to the student; the publisher cannot change the page as the student reads. The microcomputer screen, however, offers complete flexibility in this area. In the simplest example text and graphics appear on the screen in order that they should be examined, rather than simply from the top to the bottom. Words that deserve special attention may flash or change colors. Errors in answers that students type may be marked to show errors. This technique has been used in nomenclature drills to indicate where problems exist [17]. Inverted pairs of letters, missing characters, and incorrect capitalization have been pointed out to students in this way. The flexibility of screen design permits student input to be planned to occur at any position on the screen. Thus, for example, the numbers needed to balance an equation may be entered directly as coefficients in the equation.

2.2.5   Graphics

The limited graphics of most computers ten years ago was a serious impediment to their use in chemical education. Often the proper representation of symbols as fundamental to chemistry as subscripts and superscripts was impossible on remote computer terminals which were restricted to a scrolling teletype format. The possibility of using graphics and controlling the screen design is one of the obvious advantages of microcomputers over the earlier machines. Many microcomputers support high resolution color graphics with the potential for animation. Thus the computer screen can be used as an electronic, animated chalkboard with the advantage over conventional audiovisual devices such as slides, films, and television that interaction with the user is possible.

The static nature of the graphics in this book prevents an accurate representation of interactive microcomputer graphics. In spite of the limitation of the printed medium, we will give some examples of animated graphics from commercially available programs with the goal of describing the advantages over other media.

(*a*)   *Representation of molecules in three dimensions*   Chemists have developed systems and conventions to represent molecules. The level of abstraction increases from solid molecular models to formal representation such as Newman or Fisher projections. Students in chemistry must learn to think using these symbolic representations. Computer programs such as Molecular Animator by Howbert [18] or MOLEC by Owen and Curie [19] can be used to construct and display symbolic molecular structures. Under keyboard control the structures can be rotated, enlarged or reduced, or stopped in any orientation. The programs include the means for the user to modify or create new structures using atomic coordinates. Perspective is achieved by ball and stick representations using hidden lines and movement. The continuous rotation from one orientation to another in these programs, when used in conjunction with molecular models, serves as an aid to help students grasp the connection between each representation.

(*b*)   *Representation of reaction mechanism*   The dynamic representation of reaction mechanisms is a typical case where animation is essential for a complete understanding of molecular processes. *Introduction to Organic Chemistry* by Smith [17] makes use of animation to illustrate reaction mechanisms in organic chemistry. Figure 2 is extracted from a sequence

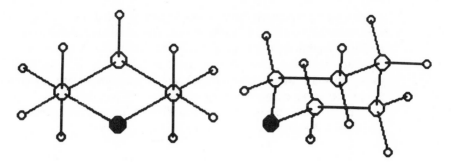

Fig. 1   Representation of tetrahydropyran by the 'Molecular Animator' by J. J. Howbert, published by COMPress, a division of Wadsworth, Inc. (Reproduced by permission of COMPress.)

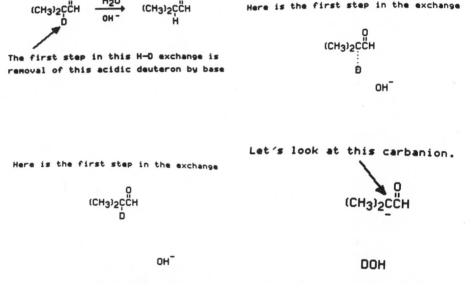

Fig. 2   Pictures from mechanisms in *Introduction to Organic Chemistry* by S. G. Smith, published by COMPress, a division of Wadsworth, Inc. (Reproduced by permission of COMPress.)

illustrating the mechanism of acidic hydrogen–deuterium exchange in ketones. Both the direct and reverse mechanisms can be repeated at the discretion of the user. The progress of the animated reaction may be placed under user control through the keyboard or accessory game paddles.

Interactive computer graphics indiscutably have a positive effect on student interest. Whether the animation makes the learning process more effective is a different question. *Introduction to Organic Chemistry* was chosen as the subject of a case study which was carried out by *Conduit* on this particular question [20]. The study concluded overall that the package was instructionally effective, that the students enjoyed using the material, and that the students felt they learned some chemistry by using the software. Interestingly, the study found in this case 'no evidence that the animation, *per se*, directly contributed to the learning'. Graphics remain necessary, but creating animation can be tedious and time-consuming. The author must be careful to closely relate the animation to the learning objectives. At the molecular level of representation, the author should also convey to the students that no one has ever seen a molecule reacting with another and that *what is displayed on the screen is just a model which represents current theory*.

(*c*)   *Drawing graphs*   Physical chemistry makes extensive use of graphs to help students visualize the relationships between variables. Any type of graph can be displayed on the computer screen, subject to the limitation of the resolution of the system in use. Even complex three-dimensional graphs such as orbital functions or energy functions of transition states can be drawn in isometric projection. Nearly any kind of computer can be used to create such figures off-line on a digital plotter for reproduction and distribution. The additional value of microcomputers arises from the control that they can give to the user the values of parameters and scaling for real-time plotting. In one application [21] rapid switching between two slightly different curves has been used to emphasize the small differences between weak and strong acid titration curves. The identical portions of the two curves remain unchanged, while the portions that are different flicker. This technique can be used in many other cases. For instance, comparison of rigorous and approximate solutions of a given chemical system is a challenging task using mathematical derivations. The delimitation of the applicability of classical approximations such as the steady state treatment in kinetics or simplified pH formulae in dilute solutions through mathematical discussions can be replaced by an empirical approach. For a given set of parameters, both the rigorous and approximate curves can be computed, plotted, and compared as just described. The student can then change the value of some parameters (concentrations, rate constants, acidity constants, etc.) and observe the consequences of the new choices on the coincidence and differences of the two curves.

This association of computer graphics, interactivity and computational power of micros can change the way physical chemistry is taught. This approach has been systematically explored by Barrow [22].

2.2.6   *Electrical input/output*

Microcomputers typically communicate with students by means of text, graphics, and sound, but *microcomputers are also capable of communicating with other machines. Most microcomputers can be easily interfaced with almost any laboratory apparatus.* A brief guide for interfacing equipment with widely available microcomputers (S-100 machines, TRS-80, Commodore, Apple II) can be found in reference 23. Commercial interface modules are available but in some cases direct input to the computers is possible without additional hardware. Examples of coupling microcomputers with pH meters [24, 25], gas chromatographs [26, 27], liquid chromatographs [27], UV-visible spectrophotometers [24, 28], gamma spectrophotometer [29], potentiostats [30], conductimeters [31] and mass spectrometers [32] have been described. *This coupling capability allows rapid and reliable data acquisition and control of laboratory instruments.*

The capability of microcomputers to communicate with other apparatus has also been used, to a lesser extent, to provide control over audiovisual equipment and create a computer-controlled multimedia learning environment. A number of audiovisual devices can be connected as peripheral to a microcomputer: random-access slide projectors, audio or video tape players, and video disc players [33]. For some purposes, these devices have a number of advantages over microcomputers. These advantages include high quality photographs, storage capacity for a great number of pictures, fast access, and accelerated or reverse motion. The interconnection of these devices with microcomputers permits the advantages of interactivity, computational power, and memory to be applied to the audiovisual equipment. There are no examples available from chemistry to discuss at the present time, but it is possible to imagine a learning situation which would closely approximate on-the-job training. For example, the student could be put in a decision-making situation such as being in charge of a water treatment plant. The student could use the computer to get the necessary information to make decisions. Information would be provided through computer displays, and such a program is under development by the ChemCom project [34]. It would also be possible to give information through real-life pictures or filmed sequences. The consequences of the choices and decisions could be commented and illustrated by short sequences. A major problem for this approach, however, lies in the high cost of professional photography for the amount of filming that would be required.

II.3 APPLICATIONS SOFTWARE

3.1 Software styles

This section will describe software currently in use in chemical education and will identify areas where progress needs to be made. Wherever possible examples are taken from software which has been published commercially or which has been described by published articles. A piece of software may serve as an excellent example of a particular point while having deficiencies in other areas. We have made no attempt to provide an overall evaluation of any software package. Those interested in formal evaluations of available software should refer to Section 4.2 and consult such sources as the *Journal of Chemical Education.*

No particular style of software is necessarily better than another. The particular type of software that can be most useful depends upon the subject matter, the student, the instructor, and the learning environment. It is useful, however, to realize that the computer supports a great variety of approaches for education. *The medium has the potential for a greater variety of instructional materials than are available as printed matter.* The various styles are identified in the following paragraphs before specific applications are discussed.

One way to classify software is to distinguish between *supplementary* and *self-contained* software. *Supplementary software is designed to complement other resources used in teaching.* It is incomplete by itself and relies upon the other resources in order to be pedagogically useful. This software may be supplementary in the same sense as a handbook of chemical data. No matter how well organized such a handbook may be, students never read it from beginning to end. Instead, they use it as adjunct to other instructional materials. Supplementary software may require, for example, that students receive direction from the instructor regarding what can be learned before it is used. A program which permits students to work with a physical system, changing parameters to see the resulting changes in the system, may be supplementary software. The student must understand something about the physical system and what can be learned from its study before using the program.

At the other extreme, *self-contained software can be used even in the absence of a formal course.* In fact, a software package may be intended as a complete course in itself. Such packages include basic information on the subject, and they also ask the student questions on the material that is described. No other resources are required to use self-contained software.

In an intermediate case, software may be supplementary in the same sense as a book of problems, which is not useful in the absence of a basic source of information such as a textbook or teacher. *Many software packages* which generate practice problems for students *are intermediate between supplementary and self-contained.*

3.1.1   The interaction plane

One of the features of microcomputers which was identified in Section 2 is their ability to interact with students. The interaction of a piece of software can be viewed as having two components: (1) the action required from the student and (2) the variety of responses by the software. If the student is challenged, has a wide variety of possible inputs, and has control over parts of the program, then the first component contributes to the interaction of the software. If the program has many possible reponses and a variety of branches in response to student needs, then the second component contributes to the interaction of the software. These two components can be combined to form the 'Interaction Plane.' This combination is shown schematically in Fig. 3.

Of the various factors contributing to the level of activity for the student, the easiest factor to understand is the nature of student input. The kind of input can be divided into seven categories which will be discussed in approximate order of increased demands upon the student and therefore in order of increased interaction: Multiple Choice, Numerical, Textual, Mathematical Formula, Chemical Formula, Analog, and Graphic.

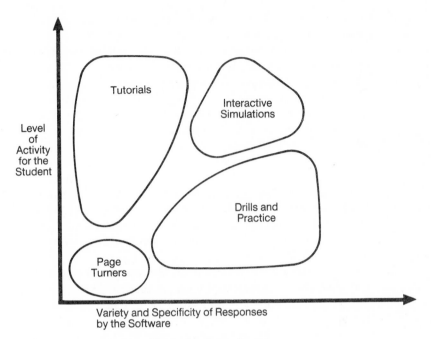

Fig. 3   The interaction plane.

Multiple choice is *the simplest of input styles* and is the easiest for a computer program to process. Two or more possibilities are available to the student, and the student must pick one. By this definition questions which must be answered with 'Yes' or 'No' are included. Usually in multiple choice input only a single character is required, sometimes with the return key also used. Typing out a complete answer, such as the word 'Yes' instead of simply the letter 'Y' changes the appearance of the input, but does not increase the level of interaction. *The multiple choice* style of input may be the most desirable when the student is merely selecting from among a series of options for the program, but *provides little challenge to the student when used for answering questions* over the subject being studied. Yes/No questions may be answered at random with a 50% chance of the answer being correct, and multiple choice questions may be nearly as easy to get correct unless the distractors are carefully chosen.

Numerical answers can have up to three components: the symbol for the quantity, the numerical value, and the unit. The first programs for microcomputers usually requested only the numerical value, but now often the unit is also required. When only the numerical value is required as input, the analysis of the answer is reduced to simple arithmetic tests. Provisions for entering exponential notation can make the problem of analysis by the program more difficult as can checks for the number of significant figures. If more than one component is required, then the program must distinguish and analyze separately the different parts of the answer.

Textual answers can be handled by programs in two different ways. Where the answer is restricted to one of several short possibilities, each character can be analyzed as it is typed in order that the student may make a correction before continuing. This approach is useful with a problem such as naming a class of compounds. It is a much more difficult task for a program to understand an answer to an open-ended question such as 'What are the conditions required to obtain a singlet in NMR spectroscopy?' The analysis of textual answers is based upon pattern matching. Each pattern is defined by a Boolean expression built on keywords. Orthographic tolerances may be introduced to deal with misspelled words. Synonyms and negatively worded answers must also be taken into account. For open-ended questions careful analysis is essential to avoid incorrect interpretations of answers. The process of software development for this type of input should include recording student inputs and the response of the computer so that the matching procedure can be adjusted to reduce the number of misinterpreted and unrecognized answers to an acceptable level.

Some questions require students to enter **mathematical formulae**. Unfortunately, the keyboard may be a limiting factor in this form of input. Special symbols such as Greek letters or the integral sign can be easily produced on the screen, but they are not available on microcomputer keyboards. In addition, handling formulae which cannot be written on a single line can present a challenge to the software author. Some specialized authoring languages offer facilities which reduce mathematical formulae into a canonical form, but these languages are not widely available on microcomputers.

The input of chemical formulae presents many of the same difficulties as for mathematical formulae, but fortunately the necessary symbols are found on the keyboard. Early applications of computers to chemical education used computers for which the display of input was restricted to a linear format. Sometimes authors attempted to represent chemical formulae in a linear format, but students then had to learn a formalism in order to read the chemical formulae. Most microcomputers are flexible enough that can display subscripts and superscripts, but the problem of accepting input from the student remains. If both subscripts and superscripts are accepted as input, then some formalism must be introduced for indicating where on the screen a number should appear. In one approach the SHIFT key has been used successfully to indicate that a number should be a superscript. Otherwise numbers appear as subscripts or coefficients, and the program can distinguish between the latter two choices from the context [16].

Analog input is the label given here to input in which the student is given control over some continuous simulated event that is usually shown graphically. Often one key is used to initiate the event and a different one is used to stop it. Examples of this approach include turning on the heater under an oil bath, releasing liquid from a buret, and moving the piston to change the pressure of a gas. The most difficult aspect of the programming for the input in this case is creating the graphics which move in response to the student's actions. By comparison, analyzing the meaning of the student's input requires little effort.

Graphical input into a microcomputer is also possible. In fact the input, whether through the keyboard, light-pen, game paddles, joystick, mouse or graphic tablets, is easy to accomplish. The difficulty, however, is writing the software to interpret the input. Simple lines and curves can be analyzed by curve-fitting procedures, but analysis of more complex figures becomes extremely difficult. Another obstacle to implementing this type of input is that the hardware for entering graphic input (except for the keyboard) is not widely available.

3.1.2   Challenge and control for the student

The extent to which a student is challenged is a more abstract concept than the nature of the input, but it is an important factor contributing to the feeling of interaction between the computer and the student. To see the difference that a challenge can make, we will consider an example of very simple input: answering 'Yes' or 'No' to a question. If the question is 'Does a reaction occur when these two reagents are mixed ?' and it is designed merely to test the student's recall of chemical data, then the challenge is very low; an incorrect response is inconsequential. The situation is very different, however, if the student is using a simulation of a chemical plant and the question is 'Can the temperature now be raised safely in order to increase the yield ?' An incorrect response in this instance could potentially damage the plant and end the simulation. If the consequences of a student's action are minor, the student has no incentive to be certain that an answer is correct. If the consequences are great, then all of the student's attention will focus on the answer.

The feeling of direct control over a situation is also something that *can hold the attention of a student*. In the case of increasing the temperature in the chemical plant, two possible options would be having the student type in a new temperature and letting the student adjust the level of a burner. The direct control of the burner will likely increase the student's feeling of interaction with software.

3.1.3   Responses by the software

The position of a piece of software on the horizontal axis in the Plane of Interaction is determined by the variety and specificity of responses that the software can make. 'Response' here is meant to include two related actions. One is the immediate visual (and/or audible) response to a student's input, and the other is the consequential branching that occurs within the program as a result of the student's input. There is no absolute distinction between these two actions, because providing different visual responses may well involve branching within the program. The distinction, however, is useful for discussion, and the two actions in general call for different strategies upon the part of the software author.

An example of an immediate response is the response that a student receives after entering a correct answer. One program may say 'Correct' after every correct answer. Another one may attempt to be encouraging and include such variety as 'You answered that one correctly' and 'That's three in a row now. Very good!'. The last response requires the program to store information about the immediate past answers in order to compose a sentence that will let the student know the program is 'paying attention' to what the student is doing. The second response provides variety, while the third one also provides specificity. Of the three statements given, the first is characteristic of software to the left on the horizontal axis of the interaction plane, and the last is characteristic of software to the right on the axis.

The simplest kind of branching that a program can do involves different approaches for correct and incorrect answers. A program may provide an explanation of the problem if an answer is incorrect. The specificity of the response can be increased if the program can analyze the incorrect answer to determine the source of the error. If the problem is complex, remediation may be appropriate. At a higher level, the program may change the nature of the information that it presents to the student on the basis of how well the student is doing. For example, the program may alter the difficulty of problems that are presented after judging the ability of a student.

Most media can be placed in the lower left corner of the interaction plane. In order for software to move away from the corner, it must differ from other media in some of the ways that have been mentioned. Often the best software lies far from each axis, but software which is near the origin on one of the axes is not necessarily poor. The extent of interaction desirable depends upon the nature of the subject, the student, and the learning environment.

3.2   Software for student use

3.2.1   Page turners

'Page Turner' is a label for educational software which lies close to the intersection of the axes in the interaction plane. In its simplest form, a page turner displays a new page of information each time that the user presses a key. This category of software also includes programs in which the user has control over the order in which pages are displayed. This format may use the power of microcomputer to place the appearance of text at different locations on the screen and may include spectacular graphics, but the same material is presented each time that the software is used. In its simplest form, this presentation differs only slightly from a movie of the same material. The differences from a movie are that the graphics cannot be nearly as good, but the student does have control over the speed at which the material appears. A piece of software in which the user has control over the sequence in which pages are displayed is a slight improvement, but offers few advantages of a book containing the same material. The one important advantage that computer displays hold over movies or books is that the displays have the potential to be altered. If the software is not protected and the user has the requisite technical skills, the presentation of an individual frame of material can be changed without affecting the remainder of the software. A piece of software can, in fact, include the techniques for changing it as an added feature. Thus a handbook of chemical data could be kept current without the requirement for reprinting if it were published as software instead of as a printed volume.

Page turners may have their place in chemical education, but they are not recognized as being effective teaching tools when compared with software which provides greater interaction. No major work of educational software in chemistry is entirely a page turner. Most software packages do, however, contain portions which are simply page turning.

3.2.2   Drill and practice

Homework, or written exercises, have long been an accepted part of chemistry courses. Learning chemistry involves considerable memorization and practice in solving exercises and problems. No matter how well a student seems to understand a subject from reading a textbook or listening to a lecture, there seems to be no substitute for practice problems. Most textbooks include exercises at the end of each chapter. The label 'Drill and Practice' applies to such exercises given to a student interactively by a computer program.

Having homework problems on a computer offers a number of advantages over exercises printed in a textbook or given to a class on printed sheets. Probably *the greatest single advantage arises from the immediate feedback that students receive for right or wrong answers*. Rarely are written homework assignments returned to students within twenty-four hours. Under unfavorable circumstances students may have to wait a week or more to know whether their work was correct. Microcomputers often reduce the time to less than a second. At the same time that the microcomputer is providing more rapid feedback to the student, it is lightening the workload of the teacher. Typically the grading process for written homework is so time-consuming that problems are marked only as correct or incorrect with little or no explanation of where the error lies. Depending upon the nature of the problem, the program may be able to provide analysis to suggest to the student what kind of error was made. Even if the program cannot determine the point of error, it can always supply the correct answer for the student to compare with the answer that was entered. It can also offer a different question on the same subject so that the student may try again immediately. *The computer can serve as an infinitely patient supplier of different problems*. It may also be able to *adjust the level of difficulty of the problems* in order to match the needs of the student using the computer. The recording of scores for use by the instructor is an element of some drill and practice programs.

Drill and practice programs usually are intended as supplementary material in a course. They do not begin by explaining material to the student, but rely upon the textbook and lecturer for that purpose. In the ideal case the programs and textbook would be written to complement each other. This possibility makes software publishing attractive to textbook publishers. The first major package of drill and practice programs written to complement a specific textbook was *Concentrated Chemical Concepts*, published by John Wiley and Sons and written by Cornelius [16]. In this software package individual drills may refer to specific pages, tables and figures in an introductory chemistry textbook written by Holum and also published by Wiley [35]. The flexibility of the computer is demonstrated by the option for the instructor to include or omit the textbook references.

Two examples will be given from *Concentrated Chemical Concepts* to demonstrate how the power of the computer may be applied to produce effective exercises. In the drill on electron configuration the student places

```
3. Show the electron configuration for
oxygen (O) and press 'RETURN.'

1s  2s  2px  2py  2pz  3s  3px  3py  3pz  4s
↑↓  ↑↓  ↑↓   ↑    ↓
__  __  __   __   __   __  __   __   __   __

Unpaired electrons in p orbitals must
have parallel spins.

Do you want to try again? (Y=yes, N=no)


Type 'U' for an 'up' arrow and 'D' for a
'down' arrow. The space bar removes an
arrow, and the '←' and '→' keys move the
cursor. Type '?' for a periodic table.
```

Fig. 4 Drill on electron configuration from *Concentrated Chemical Concepts* by R. Cornelius, published by John Wiley & Sons. Copyright 1982 (Reproduced by permission of John Wiley & Sons, Inc.)

'up' and 'down' arrows in the correct locations to represent electrons in orbitals as shown in Fig. 4. The program checks the number of electrons, their locations, and their spins. If the answer is incorrect it responds with one of six different messages to specifically identify which error was made. This problem is an ideal one for analysis by computer because the set of possible answers is closed. The analysis of incorrect answers can be complete.

Another drill in the series asks students to determine whether two substituted methane structures are the same, optical isomers, or unrelated (see Fig. 5). The program randomly generates structures from nearly 15 000 possible combinations. The student may rotate one of the structures about any bond in order to better compare the two structures. *The program thus bridges the gap between the two-dimensional structures found in most textbooks and the three-dimensional models students can manipulate with their hands.*

An alternative format for drill and practice is the 'educational game.' In this area we have much to learn from the incredible attraction of video games. In many video games the player does not actually win. *The goal of those who play a video game is to improve their performance.* Improving the performance

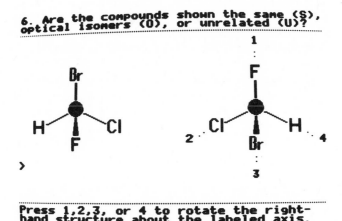

Fig. 5 Drill on optical isomers from *Concentrated Chemical Concepts* by R. Cornelius, published by John Wiley & Sons. Copyright 1982. (Reproduced by permission of John Wiley & Sons, Inc.)

of students in chemistry is precisely the goal of educational drill programs. A study of popular video games has concluded that *interaction, competition, the absence of an upper limit to the score, and fantasy are the factors that make the games appealing.* We should consider to what extent these factors can be and should be introduced into educational software.

Smith has used many of the elements of two popular video games, Space Invaders and Pac-Man, in a pair of educational games called Chem Rain [36] and Chem Maze [37]. The competitive factor introduced in these programs holds the student's interest while progress depends upon the recall of chemical information. Figure 6 shows an example from *Chem Rain.* Reagents fall steadily from above while the student manipulates compounds along the bottom of the screen to find ones that will react with the reagents. The student then 'shoots' the compound at the reagent. If the combination is correct, the product of the reaction replaces the original compound and another reagent appears at the top of the screen. The speed of the rainfall increases as the score increases. An interesting option at the end allows students with good scores to record them on the disk for others to see.

It would be interesting to see a thorough analysis and evaluation of this style of approach and to understand how students learn by using it. Today *we cannot be certain whether the game format might introduce unexpected effects in the way that students think about chemical compounds and reactions.*

3.2.3   Tutorial dialogs

Tutorial programs are those in which the computer engages the student in an interactive dialog. Such programs present course material in successive elements, ask questions to check whether the student properly understands

Your score = 30   Best score = 390

CrO_3

NaBr   HBr

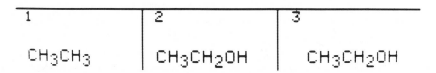

1	2	3
CH_3CH_3	CH_3CH_2OH	CH_3CH_2OH

Fig. 6 *Chemical Rain* by S. Smith. Published by COMPress, a division of Wadsworth, Inc. (Reproduced by permission of COMPress.)

the material, and analyze and comment on the student's answers. In this type of dialog, the computer can be programmed not only to evaluate student's answers but also to respond to various requests for such items as special 'Help' pages, a calculator for more complex calculations, or access to a dictionary or to specialized databases.

Most tutorial programs have a linear structure of the type proposed by Skinner. The subject matter is divided into small concepts, and displays for each concept appear in a predefined order. Each display is followed by a question and a message for reinforcement. Failure to answer a question properly results in a brief explanation or hint, and the question is asked again.

An example of a tutorial program is given by the software entitled *Introduction to Polymer Chemistry* developed for the American Chemical Society by Gibson, Pochan and Smith [38]. This self-contained package serves as a course covering the major concepts applying to polymer structure, nomenclature, molecular weight, step-growth versus chain-growth, etc. The user can determine the best pace for the material and can return to review any section at any time. The course is principally intended for in-service training of adults. Extensive field evaluation of this course has been undertaken by Chapman [39]. Although in a strict sense the conclusions of the evaluation apply only to the software examined, they provide a useful indication of the value of using (well-written) tutorial programs in general. The major conclusions of the evaluation were that:

—*Users expressed highly positive attitudes toward computer-assisted instruction* both before and after their exposure to the course and were impressed by the capabilities and quality of this type of instruction.

—*Supplementary course materials* (texts, study guides, etc.) *are important* in providing enough information to satisfy the broad range of user background typical in continuing education clientele.

—*The course produced a levelling effect* as measured by much greater uniformity of post-course test results when compared with pre-course test results.

—*The course topic is probably the principal factor determining user participation*; computer interaction may also be a strong factor in attracting course users.

Other tutorial programs have a branched structure of the type described by Crowder. The major units of such dialogs include three elements: the scene, the input-evaluation, and the linking subunits.

The scene subunit displays information in various forms. The input-evaluation subunit accepts user inputs, evaluates answers, and gives instant feedback. The linking subunit branches to another unit of the dialog depending upon the last and previous answers. *Thus the dialog is structured as a network.* A student will use different branches of this network depending upon the nature of the answers and commands that are given.

Construction of highly branched interactive tutorial programs is a tedious task using standard computer languages. Specialized languages and systems have been developed to help authors (see Section 4.1). *One of the major difficulties in designing such branched dialogs lies in the evaluation of students' answers.* This evaluation can be broken down into two stages. First the author of the dialog must define the various classes of answers. Each class is associated with specific actions of the program. For each wrong answer, the possible reasons for the failure must be ascertained in order to provide appropriate complementary information or comments or to branch to a remedial section of the dialog. The second step is for the author to describe the procedures for identifying the students' inputs with one of the predefined classes. This simplifying categorization is indispensable for accepting various equivalent forms of answers which have the same didactic meaning. As discussed in Section 3.1, the difficulty of analyzing the answers varies with the kind of input that is supplied. The first step is based on didactic hypothesis, while the second raises both technical and linguistic problems.

In spite of all the limitations and constraints mentioned above, *it is possible to design rather complex interactive tutorial dialogs which are efficient.* The advantage of the branched structure is that *the fraction of students able to reach a specified educational goal is larger than with the linear methods or that the mean completion time is reduced.*

An example of a program with a branched structure is one designed to teach the basic principles of interpreting NMR spectra [40]. The program is totally under the control of the user who can propose a solution at any time or choose to be guided to the solution through a logical sequence of questions. Various aids are at the student's disposal: tables of chemical shifts, detailed explanations, and short overviews of basic concepts. In the event of incorrect responses to key questions, the program branches to remedial sections. Depending upon the information available to the user at a given time, the order and formulation of questions are tailored to the hypothesis stated by the user or inferred by the program. The program ends with a comment on the student's performance and a review of the student work which stresses methodical aspects of spectra analysis.

Tutorial software tends to be self-contained, thus it can be used outside educational institutions in programs such as in-service training or as a complement to curricula in high schools and universities.

3.2.4   Simulation, model-building

(a) *Definition of simulation   Simulation is the use of a model in order to describe or predict the behavior of a system.* Simulations are used in chemistry for many reasons. Perhaps the most important reason is that the science of chemistry is one built upon models as a means of correlating, understanding, and predicting all aspects of chemistry. In chemical education, simulations are also used for systems that are difficult to examine experimentally. The difficulty may be simply the absence of a visual image of the system. The molecular behavior of a gas cannot be seen with the human eye, but we can create a model of a gas behavior which can be seen. For other systems direct investigation may be difficult because a system may:

—require too long or short time period for their examination (the formation of coal or molecular vibrations),
—be too difficult technically to use with a large class (gas chromatography-mass spectrometer instrumentation),
—present technical difficulties (experiments in the absence of gravity),
—be too expensive (analysis of diamonds for impurities),
—pose ethical problems (testing of potential drugs in humans), or
—be too dangerous (nuclear fission).

Thus simulations are a part of chemistry education and they are also a part of chemistry itself.

(b) *Simulation by computer*   The model does not necessarily have to be mathematical in form in order to be used in computer simulation, but nearly all chemical simulations on microcomputers are expressed by an equation or set of equations. For such a model:

—each relevant property of the system is identified as a variable (given a mathematical symbol), and the behavior of the system is written using equations containing these variables,
—these equations include a set of parameters whose values must be fixed before the model can be used, and
—a set of conditions defines the limits within which the model approximates the behavior of the original system.

A computer program can be written to compute the values of each observable property for a given set of constrained properties. The set of constraints constitutes the input to the program and the observables are the

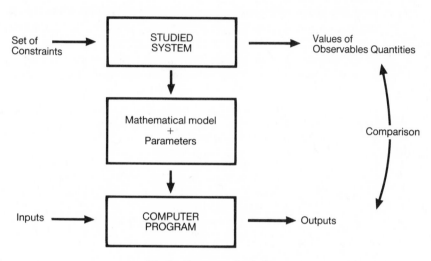

Fig. 7   Computer simulation.

output. Figure 7 shows the relationships between the original studied system, the mathematical model, and the computer program.

(c)   *The construction of simulations*   Models fall roughly into two broad categories: empirical models and theoretical models. Empirical models are of a descriptive nature; they represent the observations on a system. Such models can be built empirically by curve-fitting or by using more elaborate multivariate statistical analysis. By its very nature an empirical model can be safely used to simulate the behavior of the original system only for the limited range of constraints within which the fitting was done. In general, the predictive capability of such a model is not so reliable because the extrapolation is based solely on the behavior of the mathematical functions used and not on the actual behavior of the system. Theoretical models, on the other hand, are built through an analysis of the system in an attempt to identify the underlying causes. This analysis implies:

—simplification of the observed phenomenon to eliminate aspects which are considered to be irrelevant for the purposes at hand,
—the formation of hypotheses regarding the physical nature of the system, and
—application of scientific theories to the considered phenomenon.

In chemistry the explanatory nature of theoretical models is usually based either upon events at the atomic and molecular level or upon thermodynamic principles. If the simplifications and hypotheses used in building the model do not distort the behavior of the system, then the model acquires predictive capabilities outside the domain in which it was first built.

Once the model is constructed, *the parameters of the model must be adjusted so that the simulation program gives results close to the measured values of the original system.* This adjustment is achieved after varying the constraints and measuring the values of typical quantities. An optimization procedure can then be applied to minimize the differences between the behavior of the model and the original system. When this process converges to an optimum it is necessary to carefully compare the final results in order to determine whether the model is acceptable. At this stage, alteration or modification of the model itself may be necessary. After the optimization process is complete, *the limits of accuracy for the model within the domain of applicability can be specified.*

(*d*) *Applications* An inspection of published reports of microcomputer programs for use in chemical education [41] reveals that more than half are simulations. *One reason for the wide use of computer simulations is that the computer may be placed under the control of either individual students or the instructor.* Because simulations with microcomputers are useful in the hands of the instructor, they can be used even when only one computer is available for teaching purposes.

Computer simulations differ according to the educational objective and the knowledge of the student. If the model and the parameters are known, then *the simulation can be used to illustrate the model and demonstrate the effects of changing the values of the parameters.* If the model is known and if the parameters must be determined by the student, then the simulation can be used to *train students in data collection and treatment.* If neither the model nor the parameters are known, then the simulation can be used to *engage the student in an exploratory activity to discover the model.* Such a program can be designed in a realistic format to mimic actual laborartory experiments. Students can also use simulation programs to gain experience without emphasis on the analysis of the model.

(*e*) *Illustration of a model* Students have usually not developed their abstract thinking to a level allowing them to understand models through equations, and they need to rely upon a concrete, visual representation to comprehend models. Thus *it is easier to introduce fundamental models by means of animated pictures than by the conceptual or mathematical representation* of the model or of the theory.

One of the first well written simulations for a microcomputer provides an excellent example of a pictorial representation of a model. The program is one on ideal gases written by Gelder [42]. One part of this program permits the dynamic representation of molecules during the expansion of an ideal gas. Static photographs of the simulation are shown in Fig. 8. The molecules are shown as small moving dots on the screen, which is divided into two equal compartments. At the beginning of the simulation, all the molecules are confined in one compartment. With one key-stroke the user can open a small aperture in the wall separating the two compartments and watch the

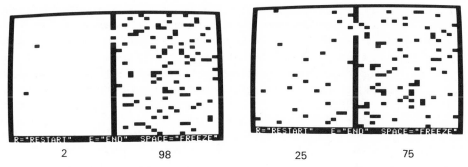

Fig. 8   Expansion of ideal gases from *Chem Lab Simulations 2* by J. Gelder. Published by High Technology Software Products Inc. Copyright 1979. (Reproduced by permission of High Technology Software Inc. Chem Lab Simulations is a trademark of High Technology Software Products Inc.)

molecules going through this aperture. The repartitioning of molecules between the two compartments can be continuously observed, and the dynamic nature of gas equilibrium is illustrated as the molecules move from one compartment to the other. This part of the program acts as a motion picture; the user can change the number of molecules and start and stop the experiment.

An example from electrochemistry [43] serves to show how the power of a microcomputer for interactive operation, display, and computation can be combined. A set of programs is used to illustrate the concepts of diffusion and convection near an electrode and the electron-transfer processes at its surface. The programs use one-half of the screen to display the theoretical interpretation while the other half gives experimental results. For example, profiles of concentration versus distance from the electrode appear next to the corresponding polarization curve (i versus E). The game paddles can be used to control the electrode potential and the thickness of the Nernst layer.

(f) Finding the values of parameters   Another use of simulations arises when *the student knows the model but must determine the values of the parameters*. In the absence of a computer, data analysis is often the major part of chemical problems given to students, but the necessary data must be given with the problem. Thus usually only one type of analysis is appropriate for the given set of data. *A simulation program can be used by instructors to generate the data which are needed to determine the values of the parameters*. The program can produce problems sheets with new data or the program can be placed under student control. When students use the program, they may choose which data they want and apply different methods to solve the problem.

Chemical kinetics give us many examples of using simulations for students to determine parameters. For a given kinetic model many approaches can be taken to determine the values of the rate constants and thermodynamic quantities. For example, one can use the concentration versus time curves with the integrated forms of rate laws, initial rates for various values of initial concentrations, or rate versus time curves. For each method given, the necessary data are different. Simulation programs which generate kinetic data are not difficult to write, and many are available [44, 45]. It is much more difficult to write interactive programs in which students are provided with options for choosing the initial concentrations, the temperature, the time interval between each point, and even which quantities are to be observed.

(g) *Discovering the model* *The presentation of theories and models in lectures carries the potential danger that students may learn them as if they were facts.* To avoid such a misinterpretation and give students a clear understanding of the nature of theories and models, it is important to engage students in learning activities in which they can collect experimental data, develop hypotheses, make predictions, determine the validity of the predictions by collecting more data, and refine the hypotheses. *Students should be able to discover that a model as a conceptual construction can be modified to account for new results, and that a given phenomenon can be described by different models.* Simulation programs can help to reduce the time and effort needed for the observations required during the formulation and refinement of a model.

Two contrasting styles of simulation programs exist: free exploration programs, and programs which lead the student to a solution.

One part of the ideal gas simulation by Gelder [42] serves as an example of the first type of program. The user can watch an illustration of a gas confined in a cylinder in which the volume may be adjusted by means of a piston. The user fixes the values of two of the four quantities: pressure, volume, temperature, and number of moles of gas. When the third quantity is varied by the user, the position of the piston and the speed of the gas molecules change accordingly, and the values of the four quantities are continuously displayed on the screen as shown in Fig. 9.

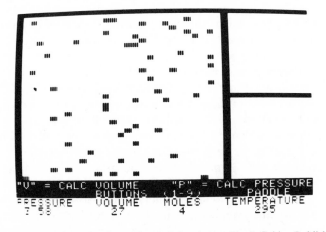

Fig. 9   Ideal gas simulation from *Chem Lab Simulations 2* by J. Gelder. Published by High Technology Software Products Inc. Copyright 1979. (Reproduced by permission of High Technology Software Products Inc.)

This program does not impose any particular pattern for learning on the user. It does not lead the student in any manner, it does not control the student's activity, and it does not check the level of understanding and/or the validity of conclusions which may be drawn. The program is complete in the sense that the student (or instructor) has complete control, but it is not self-contained. Most students would need the aid of a lecture or separate study material about gases in order to benefit from using the program.

In contrast, some parts of *Introduction to General Chemistry* by Smith [37] cover closely related material in such a way that the student is led to discover relationships between quantities. As an example, a simulation of the behavior of an ideal gas is incorporated into a tutorial dialog in which the concept of an absolute temperature scale is introduced. The student has some control

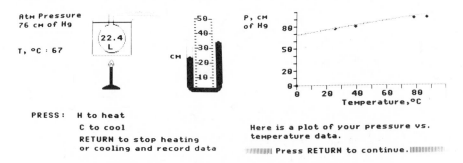

Fig. 10   Pressure temperature measurement from *Introduction to General Chemistry* by S. Smith, R. Chabay and E. Kean, published by COMPress, a division of Wadsworth Inc. (Reproduced by permission of COMPress.)

over the program in choosing the temperatures at which to record data (Fig. 10). The program does, however, demand that an adequate range of temperatures be covered. After data collection is complete, the program presents the conclusion which is to be drawn from an examination of the results. The linear structure of these programs prevents students from exploring aspects of the model that they find interesting, but it also prevents them from wasting time in activities which the author deems unproductive for the purpose at hand.

Neither style should be regarded as superior to the other. The choice may vary with the topic, the teaching conditions, the level of students, and the educational objectives. Another possibility would be a free exploration format in which the students' activities are monitored. The program could give comments or provide guidance upon request or when it becomes apparent (to the program) that the student is examining only a limited portion of the model. Writing a program which would decide when to help a student would entail developing methods for characterizing student activity to determine when help should be given. At this time no examples of microcomputer programs in chemistry with this capability can be given.

Simulations on microcomputers provide experiences which are otherwise time-consuming or which are difficult or impossible for chemistry teachers to use and to write, and the frequency of their use is likely to continue to increase.

3.2.5   Microcomputers in the teaching laboratory

'Chemistry is an experimental science.' Although this statement is commonly accepted as relevant for research or applied chemistry, its importance in chemical education is often underestimated. Personal laboratory work not only contributes to the development of practical skills but also helps students to relate theoretical relationships and concepts to their observations and operations. Moreover, the role of practical work in increasing student's interest and motivation for chemistry cannot be neglected. Unfortunately, in most curricula the time devoted to laboratory work is drastically limited. The increasing cost of chemicals and apparatus makes unlikely an increase in laboratory time in the near future.

Low-cost, stand-alone *microcomputers can be conveniently used in the teaching laboratory to improve the effectiveness of the time that students do spend in the laboratory*. Various approaches can be used independently or in combination: pre-lab. simulations, pre-lab. testing of a student's preparation, in-lab. data acquisition and reduction, in-lab. checking of student's results, post-lab. simulation, etc. While the general principles and methods of simulation have been described in the previous section, specific examples of simulated laboratory experiments will be presented here.

Instructions for laboratory work are usually presented to students using printed laboratory manuals and/or various audiovisual devices. These types of teaching materials do not require active student participation. Computer programs can be used, as complements to other materials *to display the apparatus to be used and present the successive operations to be performed later in the laboratory*. Simple page-turner programs can be sufficient to present the glassware and its use to the student. The graphic capabilities of micro-computers today enable the drawing of animated pictures of apparatus. An example taken from a series by Olmsted and Olmsted entitled *Pre-Lab. Studies for General, Organic and Biology Chemistry* [46] is given in Fig. 11.

```
         Volumetric Transfer Pipet
Step 1 - Using the pipet
bulb, draw liquid up into
the pipet. (There is no
need to fill it.) Remove
the bulb and use this small
amount of liquid to wet the
entire inside of the pipet.
Drain the liquid out and
discard it.
```

```
Step 2 - Again using the bulb,
draw liquid up until it is
above the level of the mark
on the neck of the pipet.

Quickly slip the bulb off
and close the end with
your finger.
```

```
Press the SPACE BAR to continue.
```
```
Press the SPACE BAR to continue.
```

Fig. 11   Using glassware from *Pre-Lab Studies* by S. L. Olmsted and R. D. Olmsted, published by John Wiley & Sons. Copyright 1982. (Reproduced by permission of John Wiley & Sons, Inc.)

Training students to operate more complex pieces of laboratory equipment can also be achieved through interactive pre-lab. simulations. The available time in the laboratory is too short to let students gain experience with instruments at their own pace before using them for measurements. Simulation may help to overcome the time problem. The functional role of each part of the instrument can be explained and demonstrated by simulations. The user can be instructed step by step how to operate the equipment, and the results of each action can be illustrated in real time by animation. Each necessary operation (such as standardization, calibration, choice of measurement range, reading, etc.) can be simulated in a few seconds and repeated until mastery is achieved. This approach can replace the reading of operating manuals for equipment which are often boring or too complex for students. The same package of programs quoted above [46] includes an example of training students to use a single-pan analytical balance as shown in Fig. 12.

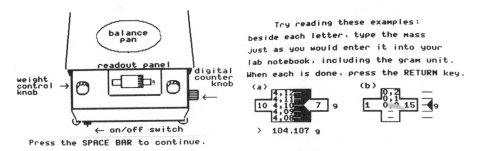

Fig. 12   Reading an analytical balance from *Pre-Lab Studies* by S. L. Olmsted and R. D. Olmsted, published by John Wiley & Sons. Copyright 1982. (Reproduced by permission of John Wiley & Sons, Inc.)

In another approach, *the complete sequence of operations to be performed can be simulated before students enter the laboratory.* As an example, a pre-laboratory simulation by Bendall covers the determination of molecular weight using freezing point depression measurements [47]. The program first illustrates laboratory techniques (weighing the sample, filling the capillary tube, measuring the melting point) using animated graphics. Next the student must read and record the weight of the empty tube and of the tube filled with pure camphor. The student may then control the rate of heating and read the temperature when the sample melts or solidifies. The process is repeated with an unknown sample in the tube. After all the necessary measurements have been completed, the student is trained to use the results to compute the molecular weight of the unknown and identify it from a list of compounds.

Simulated experiments can also let students gain experience by using an empirical approach. The simulated system functions as a black box. The user obtains results but is not principally concerned with the internal operations of the simulation. One example of such approach is a gas chromatography simulation program [48] which is designed to familiarize introductory students with some aspects of GLC analysis. With the time required for equilibration following the change of a column or temperature, it is difficult for a student in the laboratory to gain extensive experience in choosing a column, adjusting the temperature or setting the attenuation of a chromatograph. The program shows the paper of a chart recorder moving across the screen. Diagnostic comments appear on the screen if the peak is off-scale or if peaks are not resolved. The user is allowed to choose a column, temperature, and attenuation and obtain a chromatogram for his choices within 30 seconds. In a subsequent portion of the program, the user is asked to separate a mixture of three compounds randomly selected by the program. A similar program allowing the user to 'create' the column has been reported [49].

During simulated experiments the methodological aspects of experiments may be emphasized without interference from practical problems. This dissociation of logical and practical difficulties seems to be a very fruitful approach. Students are required to select options, make decisions, observe consequences, and draw conclusions. Each student's action can be evaluated and commented upon immediately by the program which supports the simulation.

Qualitative analysis of inorganic solutions is a good example of such pre-lab simulations. Obviously, labotatory techniques (precipitation, centrifugation, filtration, etc.) and the observation of colors and characteristics of precipitates and solutions cannot be practiced outside the laboratory, but the comprehension of the analytical method itself is often hidden by the practical difficulties encountered by students. To avoid the loss of time and to help students to obtain good results, laboratory experiments are often turned into cookbook exercises, leaving very little space for initiative on the part of the student. A simulation of the entire qualitative analysis scheme has been programmed on mainframe computers and tested with large groups of students [50, 51]. The adaptation of the program for microcomputers is now possible. In such a simulation students can focus their attention on the logic of the method without being disturbed by practical aspects. It is also possible to draw students' attention to the chemical concepts involved in each labotarory operation (solution equilibrium, solubility product, effect of pH on solubility and complexation, etc.). This added instruction is difficult during laboratory work since it requires the constant assistance of an instructor and would slow down the experiments which are to be performed in a limited amount of time. Gaps in the student's knowledge can be identified during the simulation so that remedial teaching materials can be assigned.

Another example is given by programs in the ESSOR Series [52] which simulate experiments in chemical kinetics, equilibrium measurements, and Compton scattering. The simulations are not intended to replace laboratory work, but to permit students *to focus their attention on the planning of successive experiments and on data analysis.* The programs are highly interactive and are intentionally designed not to suggest particular paths of investigation to the student. All of the chemical examples included in the series have in fact been investigated in detail, but the repetition of the work by the students would require lengthly sessions in the laboratory. During this type of simulated experiment, *it is essential to allow the student a maximum degree of freedom in the formulation and realization of a plan for experimentation.* For example, one program of the series allows the kinetic study of heterogeneous catalytic dehydrogenation of secondary alcohols [53]. The user may vary almost all of the factors which could potentially influence the reaction (temperature, initial concentrations, presence of catalysis, pressure, duration of the experiments, time of measurements, trapping of hydrogen, etc.). Each input by the student is checked by the program and appropriate comments are provided where appropriate. The results of the simulated experiments are given in a format very close to that which would be obtained if the experiment were performed in the laboratory, including simulated errors. These simulations are used in connection with genuine experiments to extend the field of investigation. In addition to the benefits in methodological training, *this side-by-side comparison of real and simulated measurements helps the student to understand the nature of simulation and modelling.*

Microcomputers are frequency used in the teaching laboratory for *data acquisition and reduction.* Interfacing micros with most laboratory instruments is not very complex (see Section 2.2.6) but requires a significant increase in the investment for each working station. In contrast, only one machine is needed for processing the results of a whole group of students. Data reduction programs are easy to write and do not present pedagogical questions. Examples of data reduction programs are numerous (see references 54 and 55 as two possibilities). *This is certainly one good area in which to start using microcomputers in chemical education.* Combining on-line data acquisition and data reduction is very efficient for improving students' work in the lab. Several different experiments have been described [56]. Two examples are computer-assisted generation of calibration curves and data logging for the use of ion-selective electrodes.

Two other inexpensive ways to use microcomputers in connexion with experimental work remain: using computers for pre-laboratory quizzes and for checking students' results during the lab period. Students may endanger themselves and their classmates by being inadequately prepared for laboratory work. Using microcomputer generated quizzes for safety [57] and for reviewing techniques of the manipulation and related concepts [58] have proved to be effective and enjoyable means for ensuring thorough laboratory preparation. A single machine may be sufficient for a large group. *Each student càn check critical results as they are obtained.* These results can be evaluated by the program and compared with standard results expected. Comments delivered by the program can be adjusted by instructors. One of the strong points of this approach is that a large amount of time is not wasted when a single result is not good. Students' confidence is also reinforced. Statistics can be maintained by the program and systematic deviations from expected results may be detected.

3.2.6   *Problem-solving approach*

The fundamental format of problem-solving is one in which the learner must find a way to produce a solution to a given problem [59–61]. Situations in which the solution can be reached directly by memory or by applying known, fixed procedures are not problem-solving situations.

Examples of problem-solving can be characterized both by the nature of the solution to be reached and by the variety of information required to reach the solution [62]. A problem may have a unique solution (a closed problem), or it may have several solutions (an open problem). In the latter case one solution may not be as good as another for a particular purpose. The information required to solve a problem may all be present in the statement of the problem, or some of it may have to be found elsewhere. The necessary supplemental information may come from a student's memory, it may be found in tangible sources such as books, papers, experts, retrieval systems, or it may be obtainable from observation, experimentation, or simulation.

The way in which students solve problems has been the object of numerous studies (see citations given in reference 62). The process apparently involves at least four stages:

—recognition of the existence and nature of the problem,
—selection of appropriate pieces of information,
—combination of the separate pieces of information, and
—evaluation of the solutions.

Microcomputers can help students during the last three stages of *problem-solving* identified above. For example, *computerized retrieval systems* can be made available to students to help them locate the necessary information. *Simulation programs* can reduce the time required to obtain information which might otherwise have to come from experimentation. Simulations can also be used to permit students to attempt to build solutions. Combining separate pieces of information can involve complex calculations and manipulation of data. These operations can be facilitated by *dedicated programs*. Once a solution has been reached, it can be checked by a program under the control of the student. In the case of open problems, solutions proposed by students must be evaluated. Here again, a dedicated program incorporating evaluation can be used by the student. Microcomputers can also help teachers to check or evaluate students' solutions and activity.

For the moment there are no examples of programs using all of the possibilities mentioned above. The programs described below do include some of the features that characterize problem-solving on microcomputers.

In the first example the problem is to find a synthetic route to an organic product starting from available chemicals. The usual ways to solve this problem involve either an intuitive approach founded on long chemical experience or a considerable amount of data retrieval. Obviously students lack the chemical experience required and must therefore face the difficulty of obtaining the pertinent information to build the synthetic route. Classical documentation systems (books, paper files, and even standard computerized retrieval systems) are not well suited for this task, because the long access times remove the student from the problem. Since general programs written for organic synthesis design need large computers, they are difficult to use for teaching purposes. The program Micro-Synthese [63] uses a standard Apple

II computer. The program is used with a file of elementary organic synthesis steps. For a given target molecule the program searches through the file of possible reactions to form the target and indicates all of the possible precursors. The user decides to retain or discard each proposed step. Each selected step opens the way to one synthetic route. The process of examining possible precursors is repeated recursively starting with each selected precursor as a new target. A synthetic tree summarizes in a graphical way all of the choices. The program includes the facilities to incorporate new reactions in the file, store target molecules and precursors, display the synthetic tree in totality or in parts, and give the structural formulas of each compound in the tree.

Although Micro-synthese was originally written for research purposes, it can also be fruitfully used for teaching. While in the first case there is only one user (the researcher), in the second case there are at least two users (the teacher and the student). Thus the teacher clearly states the task to be performed and the way in which the program can help. The student must examine each step proposed by the program to select a few leading to an acceptable solution. The contribution of this program to problem-solving for students lies in its help in the retrieval of selected information and in combining various pieces of information. For the instructor, it is also useful in the record it maintains of the synthetic tree built by the students. The tree can serve as the basis for discussion, and can be modified or completed after consultation between the student and the instructor.

The second example involves the design and optimization of a continuous-flow distillation tower. The problem requires collecting and combining information from

— the thermodynamic quantities for the reactions involved,
— the technical aspects of column design, and
— economic considerations.

The problem is complex and requires simplification for students outside of chemical engineering. A combination of the above programs for microcomputers (without the economic factors) has been in use for several years at the University of Nice [64]:

— Selected data on the thermodynamic and ecomomic aspects are placed in a database which is at the disposal of the student.
— Specific programs to draw the vapor-liquid equilibrium diagrams can help to determine the number of theoretical plates necessary to achieve a given separation.
— A simulation program can be used to study the effects of various parameters on the column state.
— The same simulation program can be used by students to check the validity of their solutions. An evaluation function taking into account the economical aspects could be incorporated into the the simulation.

Table 2   List of adjustable parameters

Real number of plates in the column above the feeding level
Real number of plates in the column below the feeding level
Composition of the input mixture
Flow of input
Flow of vapor
Flow of recycling liquid
Flow of distillate
Reflux ratio
Mean efficiency of the plates

The design and operation of the column can be defined interactively by the student. For each input parameter (see Table 2) the program knows feasible values and provides explanatory comments for rejected values. Additional information can be provided upon request.

After all the necessary parameters are entered, the program checks the consistency of the entire set before beginning the simulation. Comments are provided for the case of malfunction (such as when the light product cannot rise to the top of the column), and a report is generated indicating the composition and the temperature of each plate and of the output ports. All the attempts by students are recorded onto a disk for use by the instructor to analyze and use to provide computer-generated comments to individual students.

Programs which support a problem–solving environment are difficult to write because of the number of elements which must be included in order to provide freedom to the student to pursue many possible paths of action. Working out suitable examples can be quite difficult. The solutions to simpler problems may already be known to students, so that the objectives may be circumvented. More complex cases may pose a difficulty with the amount of chemical background required. Yet the examples should appear solvable to students in order not to drive them to try to find a solution by means other than problem-solving [65].

These programs provide exercises in which the student must pull together many pieces of information in order to solve a problem or reach a conclusion. The time constraints under which chemistry must be taught usually prevent students from having to solve problems in this way. Although the computer may represent a step away from laboratory experience, it may be able to give students better training in solving problems than is possible in the laboratory.

3.2.7   *Artificial intelligence approach*

Applications of artificial intelligence techniques open new lines of development to computer-assisted instruction. A survey [66] analyzed prototype systems available in 1980. Recent advances in this fast moving field are to be found in research papers in specialized journals such as *Artificial Intelligence* or in the proceedings of the IJCAI (International Joint Conference on Artificial Intelligence).

One of the first practical applications of artificial intelligence has been the development of 'expert systems.' One of the aims of *expert systems* is to *simulate the steps followed by a human expert in resolving a problem* or carrying out a complex task. Expert systems are particularly well suited in fields in which no algorithmic solution is known *a priori*, for they use a heuristic approach to search for a solution. Contrary to conventional computer programs which are designed to deal repeatedly with the same kind of data and to process the data in the same way, *expert systems deal with a class of problems and can find different solutions.*

They comprise two separate parts: a database of knowledge and an 'inference engine'. The database incorporates in a declarative language the knowledge of experts in a limited field. Usually it is composed of two parts: a database of facts and a database of rules. The inference engine must select the rules in the database and apply them to the facts to produce new information in order to answer the stated problem.

In a general sense, expert systems are able to find facts which are not present in the database but which may be deduced from the database. Expert systems are particularly well suited for situations in which the database of facts and rules is very large and can evolve. These systems are built to allow easy addition, deletion, or modification of facts or rules, and the systems are able to check the consistency of the new information with the existing data base. Fields in which the knowledge is of qualitative nature rather than quantitative lend themselves very well to this approach since algorithmic solutions are usually not feasible. Most operational expert systems have been developed in domains in which the result is a diagnosis: medicine, geological prospection, trouble-shooting. In chemistry the DENDRAL [67] system is able to analyze the mass spectrum of an unknown compound and propose a structural formula by application of fragmentation rules rather than pattern matching through a library of spectra. Most of the operational expert systems run on mainframe computers, but applications on micros are for the near future.

Heuristic problem-solver programs are closely related to expert system. Such programs for microcomputers capable of solving systems of equations [68] and other mathematical problems [69] are available. One example in chemical education is under development by the authors [70] for a limited area of introductory chemistry. This program runs on an Apple II computer. It can solve most quantitative problems dealing with chemical solutions: concentrations, dilutions, mixing, titrations, etc. The user must supply to the system the quantity to be computed and the values of available data. Relations between quantities may also be indicated. If the necessary data are available, *the system will solve the problem and explain to the user how the solution was obtained.* It uses diagrams to show the network relating the answer and the pieces of information supplied. If the problem cannot be solved, the system will indicate to the user which additional data are needed.

The extent to which such systems will be valuable in chemical education is uncertain. Certainly this program will be valuable to students as an aid to organizing their thinking about problems. The planned extension will help students to find a solution by suggesting logical steps.

The potential impact upon chemical education by expert systems and heuristic problem solvers is unknown and will remain unknown until experimentation reveals the benefits and difficulties associated with the use of such programs. Almost all existing programs used today in computer-assisted instruction can present questions to students but are unable to answer themselves similar questions. Expert systems not only can find an answer to questions and solve problems for the student, but more importantly, they can explain how they reach the solution.

3.3   Software used by the instructor

3.3.1   Instructional aids

The discussion to this point has centered around software which is designed to be used by students. Often instructors use the same software for related purposes. As was pointed out previously, learning cannot be accomplished merely by a communication process, but nevertheless communication is an important part of the teacher's action. Thus the interactive graphic capability of microcomputers can be put in action during lectures using existing video equipment or large screens. This approach is especially useful in order to demonstrate the behavior of modelled systems without struggling with mathematical treatment.

On several occasions in the previous pages, we have mentioned the possiblity of keeping track of students activities using educational software. The potential impact of this possibility is commonly underestimated. Under the usual teaching conditions, only the final answer or result by a student is available to provide feedback to the instructor. Thus it may be difficult to identify the causes of failure. During student interaction with a piece of software, all the information provided to the student is known at each stage. Not only is it possible for the software to adjust its behavior to individual needs but the 'history' of each student's activity can be recorded. Analysis of the resulting records might help to understand why students fail, to identify individuals needing personal attention, and to improve educational software.

3.3.2   Microcomputers as personal tools

Another class of software used in education is designed for the personal benefit of the instructor. Programs in this category will likely benefit students indirectly by making the teacher more aware of the needs of individual students, but the student may not even be aware that the programs are being used.

Word processing is certainly the most popular application. In connexion with short programs written to generate data and answers for problems and exercises, word processing is very helpful for preparing problem sets for distribution. Minor modifications are so easy that updating and diversification of problem sets for different groups is no longer a problem.

Personal databases are very convenient for organizing files of students, even for fairly large enrolments. Producing lists, searching for students with characteristics, and preparing statistical reports become easy tasks. The same software can be used for other purposes: personal bibliographies, stock-room inventories, files of experiments or demonstrations, etc.

Calculation programs, (VisiCalc [71] being the archetype) can help instructors records and process quantitative information. One application is the use of these programs to record and process grades. Dedicated grade management programs are also available [72, 73].

3.3.3 Computer-assisted test construction

Computer-Assisted Test Construction Systems (CATCS) can save time and produce individualized tests of equivalent difficulty on a given topic or subject matter. Some systems can directly print exams ready for use. Two types of CATC Systems exist: question retrieval systems and question generator systems. Question retrieval systems make use of large questions banks in which questions are identified by subject matter, topic, difficulty, and discrimination index. A teacher can select the questions or the computer can select them on the basis of a set of characteristics. The teacher can eliminate some questions and obtain alternative questions with similar characteristics. The most extensive question retrieval system is SOCRATES [74] which holds more than 7 000 questions in chemistry and runs on a large CDC Cyber 174 computer. Question generator systems use a bank of programs, one for each question format. Each program varies words and numerical values in the specific format. For multiple choice questions the programs also produce the distractors. An example of question generator CATCS is given by a system in use at the University of Pittsburgh [75]. It comprises 450 subroutines written in Fortran for a DEC System 1099.

Computer-assisted test contruction (CATC) greatly facilitates the work of assessment. As a result, instructors can use tests more frequently. For large enrollments, or with several groups, tests of equivalent difficulty and discrimination can be prepared. Moreover, students can be given multiple opportunities to pass the tests. CATC systems also open the way to objective assesment. The very process of setting up the system requires evaluation of the questions.

Evaluation and discussion of advantages for CATCS which are in extensive use, are given in a survey paper [75]. Although existing systems run on large computers, their adaptation to microcomputers is being considered.

3.3.4 Computer-managed instruction

The use of CATCS to generate tests contains the seed of a further use of computers. Just as the process of generating tests is tedious, so are the tasks of correcting many copies, and providing comments in response to errors or well written answers. The statistical treatment of data from exams is rarely carried out without the use of computers. The goals of computer-managed instructions (CMI) are

— automated scoring of tests.
— recording of scores and other statistics.
— processing of recorded information.

Automated scoring is accomplished most easily for multiple choice questions. Numerical answers are not difficult to analyze, but pose difficulties of having the answers in a computer readable format. Short textual answers pose the same problem, and answers which include graphics or long text are beyond the capabilities of today's microcomputers. The advantages of automated scoring are in the elimination of effort for the instructor but also in the extent to which specific comments may be generated, and the speed with which feedback is provided to students. Even if the scores are not saved directly by means of automated scoring, teachers can use microcomputers to store information about each exam. The analysis of discrimination values for questions permits the identification of questions which may have been poorly worded or which have more than one possible correct response. An example of a package of programs for CMI in given in reference 75.

Both CATCS and CMI have proved to save a substantial amount of time in preparing exam questions, scoring, record-keeping, and grading. Their implementation required a considerable investment which was possible because major projects were heavily supported. Using mainframe computers was a necessity for these collective efforts. Microcomputers of the first generation could not support these systems. The availability of inexpensive hard disks and the development of networks of microcomputers have changed this situation. In addition to the changes in the costs, microcomputers offer capabilities for working with subscripts, superscripts, special symbols, and graphics, features which were not available on standard line-printers.

II.4 AUTHORING LANGUAGES AND AVAILABLE RESOURCES

A major problem in the development of effective educational software is that the development is very time consuming. Efforts to reduce this problem can be undertaken in two complementary directions: (1) to try to cut the time necessary to produce software and (2) to encourage the dissemination of available software.

4.1 Authoring languages

Programming the computer is, unfortunately, a significant obstacle in the development of educational software. Every hour spent by a teacher struggling with the computer writing programs is an hour lost from the application of that teacher's primary professional skills. The careful definition of educational objectives, the choice of an instructional strategy, and the elaboration of the content should come first in time and in importance. A great deal of time is also needed for revision on the basis of experience gained as students interact with the software. But for non-expert programmers, writing 'user-friendly' programs sufficiently crash proof, even with very simple content, is a real challenge. In principle any general purpose programming language could be used to code educational software. Most of the available programs are written in BASIC, some in PASCAL, and a very few in other languages [76, 77]. The reason is simply that BASIC is the standard language on microcomputers. Neither BASIC nor PASCAL are well suited to compare students' responses with expected answers or to handle even minor spelling mistakes. Standard programming statements for transferring information from the keyboard to the program, such as INPUT or READLN, have weak points, and serious authors should avoid using them [78]. Special routines for input should be used. In order to facilitate the programming of educational software, specialized authoring languages have been developed since the early age of CAI. Moore has made a comparison of these languages that have been used to create chemistry courseware [79]. Very few of them have survived to the microcomputer revolution.

Two different approaches have been pursued: (1) adding special features useful for CAI to a standard language and (2) designing a dedicated language. EnBASIC [80] and PILOT [81] are typical examples of these approaches available on microcomputers.

EnBASIC is a package of editors, subroutines and other tools added to BASIC on the Apple computer to facilitate the creation of user-oriented software while keeping access to the full power of the computer. Its design is based on the experience of Smith, who was heavily involved in the development of PLATO for twenty years.

The package adds to Basic special routines for accepting and judging user responses constructed in a semi-natural language. In order to 'permit discourse within a limited context by automaticaly handling misspelled words and alternative phrasing without requiring that the designer specify every alternative or erroneous form' [80]. Software authors may choose to have errors automatically marked on user input by the judging routine. Because the routines are in machine language, they can make the necessary comparisons almost instantaneously. EnBASIC also include a character generator for character-based graphics routines and supports features such a subscripts and superscripts which are important in software for chemistry. EnBASIC has been used by Smith to produce complete series of software in organic [17] and general [37] chemistry.

Pilot (for Programmed Inquiry, Learning Or Teaching) is a very simple language which uses single commands to perform the most common tasks encountered in CAI. It was developed to enable instructors with little computer background to write educational software and to insure its transportability. As a result Pilot is available on almost all microcomputers. The simplicity of this authoring language has been achieved by reducing the number of commands to a minimum. Six commands constitute the core of the language. They are: *Type* to input information on the screen, *Accept* to receive the user response, *Match* to compare this response to expected answers, *Compute* to perform arithmetic calculations or string modifications, *Jump* to make conditional or unconditional branching and *Use* to call a subroutine. To take advantage of characteristics of microcomputers on which they run, most of the version of Pilot have several additional commands. Thus the transportability of software, is not complete, especially for graphics and sound. A Common Pilot standard has been defined to overcome this difficulty [81] the most powerful versions such as Apple Pilot [82] are not fully compatible. Apple Pilot includes four very useful editors: a text editor, a character editor, a graphic editor and a sound editor. Development of tutorial software with Pilot is simpler and more rapid than with general purpose languages due to the flexibility of the *Match* command which allows successful analyzing of free form responses. Unfortunately, other features of Pilot are so limited that authors may not be able to employ the best teaching strategies [83]. For example, calculations of moderate complexity are difficult to implement, thus Pilot is poorly suited for simulation, one of the most attractive approaches for computer use in chemical education.

Too often the philosophy behind an authoring system is to make the language simple by limiting its capabilities. Serious authors, however, require a powerful language. Novice authors should be able to begin to write programs using a subset of the complete authoring system. They should be helped by templates in which the instructional content is placed, but these templates should be optional. The language should be modular so that higher order commands for instructional applications can be defined so as to expand the language's capability. Communication with programs written in general purpose languages should also be made possible.

4.2 Available software

A few years ago the amount of software available for chemical education was very small and its dissemination was based almost exclusively on exchange between individuals. The large number of machines now available coupled with the ease of producing and shipping floppy disks and cassettes has made commercial software distribution possible. At the same time, non-profit projects have been funded to encourage production, evaluation, and dissemination of high quality educational software.

Presently, the distribution of software is in the hands of major traditional publishers, new specialized firms, and non-profit organizations. The Education Division of American Chemical Society played a pioneer role by producing and distributing the first microcomputer based courses. Table 3 lists distributors having chemistry modules in their catalogs. The list is increasing rapidly and will be incomplete by the time it is printed.

The number of programs available from the distributors listed in Table 3 ranges from a very few to a large number covering entire courses. A current list of programs available in the USA can be obtained from the Seraphim Project [76]. Complete catalogs must be obtained from the distributors. Lists of programs in French are available from CDCIEC [77] for the university level and from CNDP [84] for the secondary level. The total number is growing rapidly, so that more than 500 programs were available at the end of 1984.

Prices range from the cost of duplication and mailing to $250 and more for one diskette. The quality and support provided vary also but are not necessarily correlated with the price. In fact, it is very difficult to ascertain the pedagogical value of a program from a simple reading of catalogs. Some publishers offer preview diskettes free or for a nominal fee, but a more effective procedure for evaluation is needed. First of all, a precise description of the goals or objectives for the program should be available, and the prerequisites on the part of the student shoud be identified. The software style and instructional strategy adopted should also be indicated. On the basis of this information the potential user can determine whether a piece of software has a chance to fit the particular educational settings, but there is no assurance that the program will be effective or even function properly. Several guidelines for reviewing educational software have been published [85–88], but it remains a tedious task for the potential user to get through the process. Thus, the initiative of SERAPHIM to promote review and user tests of major software packages is most welcome. Reports of the evaluation by two independent reviewers using the same criteria [88] are available and some of them have been published in the *Journal of Chemical Education*. The first review of commercial materials appeared in June 1983 [89]. A similar procedure of evaluation exists for programs in French [90].

4.3   Sources of information

Indisputably the most complete source of information on computer applications in chemical education is the *Journal of Chemical Education*. A computer series began in 1979 [91] and nearly every issue of the journal contains a contribution to the series. The first sixteen articles of the series together with a dozen more full length descriptions of applications have been reprinted in a volume *Iterations, Computing in the Journal of Chemical Education* edited by Moore [92]. The *Computers in Chemical Education Newsletter* published quarterly by the ACS Division of Chemical Education [93] is another valuable resource. It contains anouncements of workshops, meetings, conferences, courses, short news from various universities in the USA, book reviews, hardware and software queries and regular news from the SERAPHIM project. This project, which has been frequently cited in this chapter, can be very helpful to anyone who wants to start using computers in chemical education. Project Seraphim not ony maintains a list of available software and distributes programs for mostwidely available microcomputers, it also distributes information modules, author modules, review modules, and instructional text modules. Information modules are written documents that provide information on a single specific topic. Author modules are designed to help authors to write programs that are well suited for student use, Review modules present reviews of instructional computer programs both commercial and non-commercial, and instructional text modules provide instruction for users of some of the software modules.

Table 3   Distributors of software for chemical education

Commercial distributors

Avant Garde Creations
  PO Box 30160, Eugene, OR 97403

Cambridge Development Lab
  Pleasant Street, Watertown, MA 02172

COMPress Science
  PO Box 102, Wentworth, MH 03282

Control Data Corp
  Education Division, Minneapolis, MN 55440

Creative Publications
  PO Box 10328, Palo Alto, CA 94303

Digipac Computer Consulting
  907 River Street East, PO Box 0 HQB-02G, Prince Albert Sask., Canada S6V 0B3

Dr Daley's Software
  Water Street, Darby, MT 59829

Educational Materials & Equip.
  PO Box 17, Pelham, NY 10803

Educational Software & Design
  PO Box 2801, Flagstaff, AZ 86003

Edutech. Inc.
  303 Lamartine St., Jamaica Plain, MA 02130

Edu-Soft
  4639 Spruce St., Philadelphia, PA 19139

Edward Arnold Marketing Div. Ltd
  41 Bedford Square, London WC1B 3DQ, UK

Elsevier Scientific Software
  PO Box 330, 1000 AH Amsterdam, NL

GSN Educational Software
  49 Coverhill Road, Grotton, Oldham, Lancashire OL4 5RE, UK

Heinemann Computers Educ. Ltd
  Bedford Square, London WC1B 3HH, UK

High Technology
  PO Box 14665, Oklahoma City, OK 73113

Houghton Mifflin Company
  1900 S. Batavia Avenue, Geneva, IL 60134

John Wiley & Sons, Inc
  605 3rd Avenue, New York, NY 10158

J & S Software
  140 Reid Avenue, Port Washington, NY 11050

Macmillan Publishing Company
  866 Third Avenue, New York, NY 10022

McGraw-Hill Book Company
  PO Box 445, Highstown, NJ 08520

Micro Learningware
  PO Box 2134, Mankato, MN 56002

Microphys Programs
  2048 Ford Street, Brooklyn, NY 11229

Micropi
  Box 5524, Bellingham, WA 98226

NELCAL Thomas Nelson & Sons Ltd
  Nelson House, Walton on Thames, Surrey KT12 4BR, UK

Programs for Learning Inc.
  30 Elm St., PO Box 954, New Milford, CT 06776

Radio Shack Education Division
  1400 One Tandy Center, Forth Worth, TX 76102

W B Saunders Co
  West Washington Square, Philadelphia, PE 19105

Other distributors

American Chemical Society
  Education Division, 1155 Sixteen Street NW, Washington DC 20036

BP Educational Service
  PO Box 5 Wetherby, W Yorks. LS23 7EH, UK

CALCHEM, Dept of Phys. Chem.
  Leeds University, Leeds LS2 9JT, UK

Central Program Exchange
  The Polytechnic, Wolverhampton WV1 1LY, UK

CONDUIT
  PO Box 388, Iowa City, IA 52240

Instructional Media Center
  Marketing Division, Michigan State University, East Lansing, MI 48824

SERAPHIM Project NSF DISE
  Eastern Michigan University, Ypsilanti, MI 48197

Journals more specifically devoted to the educational use of computers such as *Computers in Education, Educational Technology, Journal of Computer Assisted Learning* and *Journal of Computer Based Instruction* can be valuable to those interested in research in this field.

II.5 DEVELOPMENT OF EDUCATIONAL SOFTWARE

Writing *supplementary materials to be used by the teacher* does not raise more problems than writing other computer programs. Knowing what the program should do, choosing an appropriate method, coding the program into computer language, and testing and debugging the programs are common tasks for anyone using computers in chemistry.

Writing *supplementary materials to be used by students* is more difficult because the program must be easy to use and completely protected against any unexpected input. Few pedagogical questions arise since the supplementary material is intended to help students engaged in more or less classical learning activities.

Moving from supplementary to *self-contained materials* raises other issues because the software must embody the educational objectives. Educational software which is designed to be used by students without the control of an instructor is apparently intended to help students master precise educational goals. Thus the *development of such software is not merely a problem of computers*; *it is principally a problem of education.*

Prospective authors should realize that preparing high quality educational software is a long and arduous task. *It is commony accepted that over two hundred hours are needed to prepare one hour of student-software interaction.* Individuals unable to devote this amount of time to a project would be ill-advised to write educational software.

The following recommendations should be considered as general guidelines for beginners and not as imperative rules. The computer is a new medium, and innovative approaches and creativity are most desirable. We shall limit our scope to the first steps of designing a project.

The first step is to choose a topic. Expertise in the subject matter is necessary, and it is also important to have a personal experience with teaching students in this topic. Without this direct experience, an author cannot have a thorough understanding of the obstacles that students encounter in trying to master a subject. In addition to the content, it is highly desirable to *specify educational objectives.* The earlier that the objectives can be defined, the less effort will be wasted in design and development. For an authoritative introduction to educational objectives see Mager [94].

The second step is to imagine students' activities which can help them reach the objectives. Important qualities here are interaction and individualization. Reading this chapter may be valuable, but personal experience is indispensable. It is necessary to know what software has been written and to have used existing programs. The major point in computer-assisted learning is not what the teacher should do, nor what the computer does, but what the student will be able to do with the software. The best approach in examining existing software is to mimic an average student, not to give unexpected and silly answers. The challenge is to think of ways in which the computer can help create a motivating and effective learning situation in which the actor is the student.

The third step is to look for cooperation. A good innovative and creative idea (rather than merely one more program to draw a titration curve) deserves a team to realize it. As indicated above, even for a relativey small project, going from this design stage to a useful piece of software worthy of distribution is a long process. It will require different skills: subject matter expertise, teaching experience, familiarity with instructional design principles, awareness of educational computing, and computer programming skills. It is unusual to find all these proficiencies in a single individual. Moreover, creativity and perseverance are reinforced by a team effort. The programming may be done by someone other than the designer, but a minimal knowledge of computers and their capabilities is indispensable for effective communication. In addition some awareness of the field of educational psychology is certainly helpful.

The next goal is to detail the first ideas in a precise reference document. This document shoud describe the proposed software in such a way that it can be used to:

— solicit support from institutions or publishers,

— reinforce the design and development team,

— establish the external references for future evaluation.

Lieblum [95] suggests the use of a screening procedure as an aid in the decision whether to initiate a CAI project. While some points he quotes are not relevant here, his approach may help an author to interest others in the implementation of the project.

The particular form of the reference document is unimportant, but it should contain information on the following points:

— Identify the educational objectives.
  Can a test be defined to verify whether the goals are reached?
  Will this post-test be included in the software?

— What is the content? Give the organization of the content.
  If several theories exist why has a particular approach been adopted?
  At what level will the content be treated?

— What are the characteristics of the potential audience? (remedial, enrichment, supplementary, age, etc.)

— What are the prerequisites for use of the program?
  Can a test be defined to verify whether these prerequisites are completed?
  Will this pre-test be included in the software?

— What is the expected mode of use? (supplementary, by the instructor, by students in class, at home, in laboratories, in library, etc.)

— What methods are used? (tutorial, drill and practice, simulation, game, problem-solving, testing, etc.)

— What is the program personality (see ref. 87)? (humor, neutral, warmth, praise, criticism, patience, impatience, competition, etc.)

— Are there related existing materials?

— What will the program do?

— What will the instructor do?

— What will the student do?

What is the next phase of development? The Author's Guide from CONDUIT [86] covers the entire process of production of educational software (design, development, style, packaging and review). This guide contains many useful references and we strongly encourage its reading to anyone seriously interested in creating educational software. One area which might deserve special attention from the beginner is the 'user–computer interface'. Studying the results of previous research can help an author avoid mistakes that others have made. Selected articles dealing with user–computer interface are identified here for those interested in further reading.

The TICCIT design gives useful ideas of how to develop effective *learner control* over the software which are applicable to many instructional situations [96]. Asking effective questions appropriate for the content, structure (i.e., association, concepts, problem-solving), and level of learning (i.e., recall, recognition, comprehension, application, analysis, synthesis, evaluation) is a delicate problem. Hall [97] suggests guidelines for this task. The user-computer interface encompasses both the student's input and the screen design. The development of a 'user-friendly' interface for input has been discussed in detail [98, 99]. Screen design, presentation of data, use of colours to stimulate, sustain motivation, and facilitate comprehension and retention are analyzed in references 100 and 101. Although these articles do not deal with chemistry, they are useful for any subject area, and they provide a good introduction to more general literature on man and machine interaction. There are no rules that one must follow absolutely but rather hints that might suggest further experiments and innovations.

II.6 CONCLUSION

Innovations are always slow to be accepted in education. Computers have been used in education for twenty years, but until recently their use was principally the focus of research projects with limited impact on real practice. The microcomputer revolution has sharply changed the situation. Perhaps for the first time since Gutenberg we find a new learning tool for students (as opposed to teaching tools for instructors such as audiovisual devices).

During the last five years the number of chemistry teachers coming in contact with computers has changed from a very few to a majority. The number and variety of educational software packages available continue to grow rapidly. As authors gain experience and users become more knowledgeable and demanding, the quality of software increases. As prices continue to fall, more and more students have their own computers. Some schools, following the lead of Clarkson College and Drexel University, require every freshmen to own a computer. Other schools make them availabe to students in classes, laboratories, library, resource rooms etc.

With the increased opportunities students have to interact with computers for an extended length of time, the question of how to best use computers for improving education remains. Two complementary directions are open: to consider the computer and associated software as a new instructional medium and to view the computer as a new learning tool.

As a new instructional medium, computers have unique characteristics. The flexibility of microcomputers to store and recall information, perform otherwise difficult calculations, display text and graphics, accept and interpret student input, and respond instantaneously provides a powerful, interactive instructional medium. In spite of recent developments we have just begun to explore the possibilities of this medium. Most applications may be labelled conservative in that they reproduce on the computer existing instructional methods.

The application of *microcomputers as personal tools for students* is virtually unexplored but may have the greatest potential. The most obvious uses include word processing, electronic spreadsheet analysis, and database management. Examples of possible student use as planned at Drexel University are given in reference 102. In addition mathematical problem-solver programs may free students from tedious calculations.

Evolving expert systems on personal computers owned by the students could change the way many specialities are taught. If an expert system can be enriched by its user by the addition of new rules, facts, and definitions, then in a certain sense the system becomes a product of the user's own efforts. We are at the point that microcomputer programs are able to solve most of the problems in an introductory chemistry course. Shoud students be permitted to use such programs during exams? If not, the question arises why the students should be required to do the tedious tasks that a computer can do. The same question arose in the minds of many educators when hand-held calculators were first introduced. Surely students should be permitted to use tools which they would be expected to use in their jobs. Their education should help to prepare them for jobs in chemistry. If students are permitted to use expert systems during exams, then major changes may be required in both the type of material which is presented and the manner in which it is taught. As microcomputers become more powerful and application software more flexible, so will the forces of change for chemical education.

II.7 REFERENCES

[1]   D. Alpert and D. L. Bitzer, *Science*, **167**, 1582–90 (1970).

[2]   C. V. Bunderson, *International Journal of Man-Machine Studies*, **6**, 479–91 (1974).

[3]   S. Papert, *Mindstorms*, Basic Book, New York (1980).

[4]   T. A. Dwyer, *International Journal of Man-Machine Studies*, **6**, 137–154 (1974).

[5]   J. Fielden and P. K. Pearson, 'The cost of learning with computers' Council for Educational Technology, London (1974).

[6]   'Dix ans d'informatique dans l'enseignement secondaire 1970–1980', Rapport I.N.R.P. 113, in *Recherches Pédagogiques*, Paris (1982).

[7]   G. Kearsley, B. Hunter and R. J. Deidel, *T.H.E. Journal*, **1**, 90–94 (1983); ibid., **2**, 88–96 (1983)

[8]   N. J. Rushby, *An Introduction to Educational Computing*, Croom Helm, London (1979).

[9]   L. Viennot, *Eur. J. Sci. Educ.*, **1**(2), 205–221 (1979).

[10]   B. M. Peake and R. Grauwmeijer, *J. Chem. Educ.*, **58**(9), 692 (1981).

[11]   J. P. Chesick, *J. Chem. Educ.*, **59**(6), 517–9 (1982).

[12]   G. M. Muha, *J. Chem. Educ.*, **60**(1), 49 (1983).

[13]   R. J. Field, *J. Chem. Educ.*, **58**(5), 408 (1981).

[14]   PFS Software Publishing Corporation, Palo Alto CA., USA.

[15]   J. C. Traeger, *J. Chem. Educ.*, **59**(9), 779–80 (1982).

[16]   R. D. Cornelius, *Concentrated Chemical Concepts. General, Organic and Biological Chemistry.* John Wiley and Sons, New York NY (1983).

[17]   S. Smith, *Introduction to Organic Chemistry.*, COMpress, Wentworth NH (1981).

[18]   J. J. Howbert, *The Molecular Animator.*, COMpress, Wentworth NH (1983).

[19]   S. Owen and J. Curie, 'MOLEC', Cambridge Development Lab., Watertown MA (1983).

[20]   M. J. Peters and K. C. Daiker, *Pipeline*, Spring, 11–13 (1982).

[21]   G. L. Breneman, *J. Chem. Educ.*, **58**(12), 987–8 (1981).

[22]   G. M. J. Barrow, *J. Chem. Educ.*, **57**(10), 697–702 (1980).

[23]   K. L. Ratzlaff, *J. Chem. Educ.*, **58**(6), 470–5 (1981).

[24]   R. L. Johnson, *J. Chem. Educ.*, **59**(9) 784–6 (1982).

[25]   R. D. Cornelius and P. R. Norman, *J. Chem. Educ.*, **60**(2), 98–9 (1983).

[26]   K. J. Dien, R. S. Bell and M. D. Morris, *J. Chem. Educ.*, **58**(3), 243–4 (1981).

[27]   C. B. Pate and H. B. Herman, *J. Chem. Educ.*, **58**(3), 244–5 (1981).

[28]   G. S. Owen, D. Travis and T. Green, *J. Chem. Educ.*, **58**(9), 690–1 (1981).

[29]   J. W. Long, *J. Chem. Educ.*, **58**(7), 550–1 (1981).

[30]   R: von Wandruszka, *J. Chem. Educ.*, **59**(9), 781–2 (1982).

[31]   G. F. Pollnow, *J. Chem. Educ.*, **59**(2), 134–5 (1982).

[32]   J. T. Burt, J. E. Byrd, G. L. Helm, M. J. Perona, A. R. Ristow and D. E. Wilkinson, *J. Chem. Educ.*, **58**(7), 549–50 (1981).

[33]   C. F. A. Bryce and A. M. Stewart, *CALNews*, **22**, 1–3 (1983).

[34]   J. K. Estell and J. W. Moore, Program under development for the 'SERAPHIM/ChemCom Interface Project'.

[35]   J. P. Holum, *Fundamentals of General, Organic, and Biological Chemistry*, 2nd ed., John Wiley and Sons, New York NY (1982).

[36]   S. Smith, *Chemical Rain*, COMPress, Wentworth NH (1983).

[37]   S. Smith, R. Chabay and E. Kean, *Introduction to General Chemistry*, Compress, Wentworth NH (1983).

[38]   H. Gibson, J. Pochan and S. Smith, *Introduction to Polymer Chemistry*, American Chemical Society, Education Division, Washington DC (1981).

[39]   N. Chapman and J. Fleming, *J. Chem. Educ.*, **58**(11), 904–7 (1981).

[40]   J. P. Rabine, M. Rouillard, and D. Cabrol, REMANO, Disks AP713 & AP719 From Project SERAPHIM, Department of Chemistry, Eastern Michigan University. Ypsilanti, MI 48197 USA.

[41]   D. Cabrol, 'Bibliographie commentée 1978–1983', Centre Documentaire Coopératif Informatique-Enseignement Chimie, Université de Nice F 06034 Nice Cedex, Franc.e

[42]   J. Gelder, *Chem Lab Simulation* 2, High Technology Inc. Tulsa OK USA (1979).

[43]   O. R. Brown, *J. Chem. Educ.*, **59**(5), 409–413 (1982).

[44]   S. W. Orchard and M. B. Moolman, *J. Chem. Educ.*, **58**(5), 409 (1982).

[45]   C. H. Suelter and Don Hill, *J. Chem. Educ.*, **59**(12), 984 (1982).

[46]   S. L. Olmsted and R. D. Olmsted, *Pre-Lab Studies*, John Wiley and Sons, New York NY (1983).

[47]   V. Bendall, *Rast Formula* Apple Disk 501 available from Project SERAPHIM see complete address in ref. 76.

[48]   D. M. Whisnant, *J. Chem. Educ.*, **60**(1), 46 (1983).

[49]   J. K. Hardy and D. H. O'Keeffe, *J. Chem. Educ.*, **60**(12), 1061–62 (1983).

[50]   J. J. Kessis, J. Martin and J. P. Cambrini, *L'Actualité Chimique*, 23–27 (1975).

[51]   L. D. Francis, *J. Chem. Educ.*, **50**(8), 556–8 (1973).

[52]   D. Cabrol and C. Cachet, *Eur. J. Sci. Educ.*, **3**(3), 303–12 (1981).

[53]   D. Cabrol and C. Cachet, *L'Actualité Chimique*, (10), 35–6 (1980).

[54]   B. J. Pankuch, *J. Chem. Educ.*, **59**(9), 774 (1982).

[55]   J. G. Roff, *J. Chem. Educ.*, **60**(2), 100–1 (1983).

[56]   S. L. Burden, *J. Chem. Educ.*, **61**(1), 29–30 (1984).

[57]   J. S. Levkov and U. Thaker, *J. Chem. Educ.*, **59**(7), 599 (1982).

[58]   N. H. Kolodny and R. Bayly, *J. Chem. Educ.*, **60**(10), 896–7 (1983).

[59]   K. F. Jackson, *The Art of Solving Problems*, Heinemann, London (1975).

[60]   R. M. Gagné, *The Conditions of Learning*, Holt, Rinehart and Winston, New York NY (1970).

[61]   D. P. Ausubel, *Educational Psychology-a Cognitive View*, Holt, Rinehart and Winston, New York NY (1970).

[62]   A. D. Ashmore, M. J. Frazer, and R. J. Casey, *J. Chem. Educ.*, **56**(6), 377–9 (1979).

[63]   R. Barone, M. Chanon, P. Cadiot and J. M. Cense, *Bull. Soc. Chim. Belg.*, **91**(4), 333–6 (1982).

[64]   D. Cabrol, S. Pastour, R. Luft, and J. P. Rabine, *L'Actualité Chimique*, (2), 40–43 (1980).

[65]   A. Kornhauser, M. Vrtacnik and G. Djokic, 'Chemical Research Results for Teaching Problem Solving', *Proceedings of the Sixth International Conference on Chemical Education*, Maryland, USA (1981).

[66]   A. Gable and C. V. Page, *Int. J. Man-Machine Studies*, **12**, 259–282 (1980).

[67]   R. K. Lindsay, B. G. Buchanam, E. A. Feigenbaum and J. Lederberg, *Application of Artificial Intelligence to Organic Chemistry: The Dendral Project*, McGraw-Hill, NY (1981).

[68]   TK! Solver, Software Arts, Cambridge, MA.

[69]   muSIMp/muMATH, Microsoft, Bellevue WA.

[70]   R. D. Cornelius, D. Cabrol and C. Cachet, *George a Problem-Solver for Chemists and Chemistry Students* COMpress, Wentworth NH (1984).

[71] 'VisiCalc' Personal Software Inc., Visicorp, San Jose CA.

[72] K. M. Wellman, *J. Chem. Educ.*, **59**(7), 602–603 (1982).

[73] R. Cornelius, *GRADISK*, John Wiley and Sons, New York NY (1983).

[74] K. J. Johnson, *Comput. Educ.*, **5**, 147–62 (1981).

[75] K. J. Johnson, W. van Willis, O. Seely Jr. and J. W. Moore, *J. Chem. Educ.*, **58**(2), 117–21 (1981).

[76] J. W. Moore, 'List of available software', Project SERAPHIM, Department of Chemistry, Eastern Michigan University, Ypsilanti, MI 48197 USA

[77] Répertoire de programmes en langue francaise, CDIEC see complete address in ref 41.

[78] R. Cornelius, *COMPUTE!*, **28**, 56—62 (1982).

[79] J. W. Moore, 'Calculators and Computers in Chemistry', in *Source Book for Chemistry Teachers*, Chapter 10, 125–144, 6th International Conference on Chemical Education, Maryland, 1981

[80] *Enhanced BASIC For Inproved Human-Computer Interaction*, COM-Press, Wentworth NH (1982).

[81] 'Common Pilot' developed at Western Washington Univ.

[82] 'Apple Pilot' Apple Computer Inc. Cupertino CA. for a short review see M. Smith, *Creative Computing*, **8**(7), 62–68 (1982)

[83] P. F. Merill, *Creative Computing*, **8**(7), 70–77 (1982)

[84] Centre National de Documentation Pédagogique, Service Logiciels, 29 Rue de Vannes F 92120 MONTROUGE France.

[85] C. McPherson-Turner, *J. of Computer Based Instruction*, **6**(2), 47–49 (1979).

[86] H. J. Peters and J. W. Johson, *CONDUIT's Authors Guide*. CONDUIT University of Iowa, Iowa City IA (1978).

[87] H. Burkhardt, R. Frase and C. Wells, *Computer & Education*, **6**(1), 77–84 (1982).

[88] SERAPHIM, 'Criteria for Reviews of Instructional Software' See complete address in ref. 76.

[89] *J. Chem. Educ.*, **60**(6), A119–83 (1983).

[90] D. Cabrol, 'Procédure de certification des logiciels d'enseignement' CDIEC. See complete address in ref. 41.

[91] J. W. Moore and R. W. Collins, *J. Chem. Educ.*, **56**(3), 140–7 (1979).

[92] *Iterations: Computing in the Journal of Chemical Education*, J. W. Moore, (ed.), American Chemical Society, Education Division, Washington DC (1981).

[93] Chemical Education Newsletter, D. Rosenthal (ed.), Clarkson College of Technology, Postdam NY 13676 USA.

[94] R. F. Mager, *Preparing Instructional Objectives*, Fearon Publishers, Belmont CA (1962).

[95] M. D. Leiblum, *Comput. Educ.*, **3**(4), 313–323 (1979).

[96] M. D. Merill, *Comput. Educ.*, **4**(2), 77–95 (1980).

[97] K. A. Hall, *Journal of Comput. Based Educ.*, **10**(1 & 2), 1–7 (1983).

[98]   P. H. McCann, *Comput. Educ.*, **7**(4), 189–196 (1983).
[99]   S. R. Seidel, *Comput. Educ.*, **7**(3), 153–165 (1983).
[100] M. E. Kidd and G. Holmes, *Comput. Educ.*, **6**(3), 299–303 (1982).
[101] J. M. Jenkin, *Comput. Educ.*, **6**(1), 25–31 (1982).
[102] A. L. Smith, *Comput. Educ.*, **61**(1), 28–9 (1984).

CHAPTER III

Computer Graphics: A New Tool in Chemical Education

Jacques Weber

University of Geneva, Switzerland

III.1 INTRODUCTION

The enormous advantages in using computers for scientific purposes are well known: virtually unlimited facilities for information storage, impressive performances in large-scale computation, maximum degree of flexibility in laboratory data acquisition and analysis, etc. Both fields of fundamental and applied research in chemistry have been using to a large extent such facilities during the last ten years, as exemplified by the development of expert systems such as DARC [1, 2] or by the growing importance of chemometrics [3]. However, it is only recently that *computer graphics* has been recognized as an authentic tool in chemistry. Computer graphics may be defined as '*the creation, storage and manipulation of models of objects and their pictures via computer*' [4] and as such, it is of prime utility in chemistry. Indeed, since the pioneering development of spectacular applications by well-known laboratories such as those of R. Langridge, R. J. Feldmann, P. Gund, W. T. Wipke and others, graphics has known a rapid progress in chemical research. Among the most popular applications, let us mention the following:

— interactive molecular modelling [5];
— drug design [6, 7];
— three-dimensional visualization of physico-chemical properties [8];
— quantitative structure–activity relationship [QSAR, 9].

All these applications, which take place in the current trend of *computer-aided molecular design*, are based on the three-dimensional (3D) representation of molecular structures and functions such as electron density, hydrophilicity, intermolecular interactions, etc.

Even for the inexperienced user, the advantages of dealing with molecular modelling on the computer are enormous: there is an exact one-to-one correspondence between the displayed structure and the data and it is simple to build, manipulate, modify or store any type of compound, even large ones like proteins. In addition, the high degree of man–machine interaction offered by interactive modelling allows [6–9] one to add or suppress atoms or bonds, to make stereochemical predictions, to visualize in three-dimensions any structure entered as two-dimensional and to perform many other applications.

One may therefore naïvely ask whether chemical education has similarly benefited from computer graphics, since the potentialities of graphics for teaching chemistry are tremendous. Unfortunately, the answer to this question is no and we ascribe three reasons for this:

(1) Chemical education did not benefit from the same investments as research and development for exploring the possible applications of graphics.

(2) Both the disparity of teaching materials and the large degree of personalization which prevail in education do not help promote the use of standard graphics hardware and software. Except for the PLATO system, this leads to myriads of small scale applications developed on microcomputers.

(3) The conservatism of many educators does not favor the use of new materials such as computer graphics.

The slower growing rate in the use of computers for teaching, as compared with research, is best exemplified by the Proceedings of the recent International Conference on Computers in Chemical Research and Education [10a], held in Washington in 1982: among 16 plenary lectures and 63 posters, 7 presentations (of which only one was a plenary lecture) were devoted to chemical education.

The purpose of this chapter is to help fill this gap. Indeed, computer graphics applications to both fundamental and applied chemical research have been recently received [10b]. After having summarized the main features of computer graphics technology and programming, we will review the major applications of graphics to chemical education. It will be shown that, whereas most applications are presently devoted to simulation, molecular modelling and quantum chemistry, the real-time animated representation of dynamic processes such as rearrangements and (both organic and inorganic) reactions is a promising field which has considerable potential for teaching these elaborate mechanisms. Recently, reviews on applications of molecular graphics for the medicinal chemist have shown the widespread interest of this field [58, 59].

III.2 BASIC CONCEPTS OF COMPUTER GRAPHICS

2.1 Introduction to graphics

Through the multiple advantages of picture generation and object manipulation, computer graphics has significantly improved the processes of *man–machine communication*: it is indeed much easier for the user to interact with the computer by means of *pictures* (displaying two- or three-dimensional models, with or without animation), *diagrams*, *photographs*, etc. The use of computer graphics devices can be made both for *input* (e.g. a two-dimensional chemical structure introduced as a stick model on a data tablet) and *output* (e.g. the same structure displayed in three-dimensions as a ball-and-stick model on a screen).

Basically, computer graphics equipment allows pictorial input/output (I/O) communication. Among the very large range of such equipment, it is important to make the distinction between those allowing some *interactivity*, and those where the user has a *passive* role. A good example of the latter type is the well-known plotter: during the line drawing generation, the user has little, if any, possibility of controlling and modifying the picture. Interactive graphics, however, allows the user to dynamically control a picture generated on a display and to modify in real-time its content. Some *transformations* of the object, such as translations, rotations, zooming, etc., or the use of multiple windows on the screen [4, 11] are facilities generally offered by such equipment. From the user point of view, interactive graphics presents important advantages, but the cost of hardware rapidly increases as a function of the degree of interactivity. It is therefore not surprising to find a great diversity in computer graphics applications to chemical education, depending on the degree of sophistication of the system on which they have been developed. In order to enable the reader to estimate at best the capabilities and limitations as well of these applications, it is necessary to briefly summarize the basic principles of computer graphics hardware and software.

2.2   Graphics hardware

Without going into too many details (for further information, the reader should refer to refs 4 and 11), it is useful to mention here the characteristics of the three types of graphics output technologies. All of them are based on the cathode-ray tube (CRT), but they utilize display devices which fundamentally differ by the techniques used for drawing pictures.

2.2.1   *The calligraphic (or vector) system*

The image is made of vectors generated on the screen by a random displacement of the electron beam of the CRT. This equipment consists of a display processor, a refresh buffer memory and a CRT which may be seen as a sophisticated oscilloscope.

Advantages   Very good resolution (i.e. pictures of high quality), capability to generate in real-time animated images, high level of interactivity.

Drawbacks   High cost, pictures made of a limited number of lines, no possibility to display solid areas (i.e. no shading), elaborate programming.

2.2.2   The raster-scan system

The image is formed from a raster of points (pixels) on a screen (typically 512×512), the technology of this system being that of television. In addition to this screen, the system consists of a refresh buffer which stores the whole image in a bit map mode. As with television, it is very easy to produce gray scale or color pictures by using additional memory planes.

Advantages   Cheap system, the cost being much smaller than for calligraphic systems (at least for low- to middle-class raster equipment), possibility to display solid areas, easy programming.

Drawbacks   For low-cost systems, low resolution (1024×1024 pixels is a minimum for obtaining pictures of good quality), no real-time motion dynamics, low level of interactivity.

2.2.3   The direct-view storage tube (DVST)

The image is stored by writing it by means of an electron beam as a pattern of positive charges on a grid mounted just behind the screen.

Advantages   Low-cost system of simple conception, absence of flickering (it is not necessary to refresh the phosphor screen).

Drawbacks   Very slow writing mode, difficulty of selectively erasing parts of the picture, low level of interactivity, no real-time motion dynamics.

From the point of view of chemical applications, the calligraphic system is mainly used by the industry for drug design and QSAR purposes [5, 8, 9, 10]. To this end, it is generally connected to a supermini or a large mainframe as a host-computer. The raster system, and to a lesser extent the DVST tube, is an ideal equipment to be used as a graphics terminal of a micro or minicomputer, and as such raster systems are extremely popular for low-cost graphics applications. It is therefore not surprising that most graphics applications to chemical education have been developed on these types of equipment, our laboratory in Geneva being one of the few exceptions. Figure 1 shows a diagram of our configuration, which is based on a PDP-11/60 minicomputer and a Vector General 3400 calligraphic system.

2.3   Graphics software

In order to display pictures, the designer of a graphics program must define an object using primitives which are collected in a *graphics library*. To this end, he generally uses a high-level language, such as PASCAL, FORTRAN or BASIC, the primitives being graphics commands or macro-commands which are directly callable from program statements. The role of the primitives is to define the elements of the object (points, lines, polygons, character strings, etc.) or the display mode (type of line, color, etc.). Examples follow:

(1)   CALL MOVE (X1, Y1): move the electron beam of the CRT, without drawing, from the current position (CP) to the point X1, Y1.

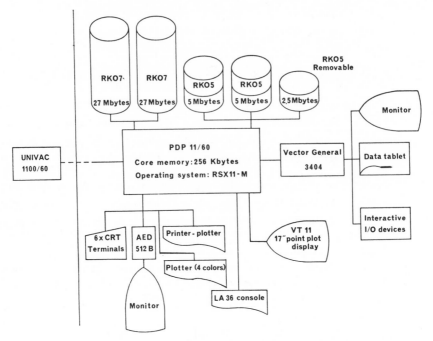

Fig. 1   Block diagram of the hardware configuration of our equipment.

CALL DRAW (X2, X2): draw a line from the CP to the point X2, Y2. The result of these two consecutive commands is to draw a vector with endpoints X1, Y1 and X2, Y2.

(2)   CALL COLOR (5): switch to color number 5 (for example red) for the next part of the picture.

Generally, the possibilities offered by the graphics library are directly related to the sophistication of the equipment: simple raster systems offer only a few primitives, which are analogous to those of plotters, and their programming is thus very easy. Calligraphic or sophisticated raster systems, however, operate with graphics libraries made of several hundreds of primitives.

In order to produce *dynamic images*, such as those illustrating a chemical reaction path, a connection between the host-computer and the graphics system allowing a very high transfer rate (or the order of 100 kbits/sec) is needed. In the case of our equipment, this is achieved by a direct access from the calligraphic system into the core memory of the minicomputer. The data base made of the atomic coordinates $x(t)$, $y(t)$, $z(t)$, where t is an arbitrary time scale corresponding to the reaction coordinate, is stored on the host and used to define vectors on the display system which are refreshed as a function of t. The refresh rate must be high enough so as to produce smoothly animated pictures. In addition to the definition of the objects to be displayed on the screen (three-dimensional model of the structure, potential energy profile, cursor), the application program must also take care of refreshing adequately the vectors built from the database.

III.3   RECENT APPLICATIONS OF COMPUTER GRAPHICS TO CHEMICAL EDUCATION

Among the recent applications (i.e. published since 1979) of computer graphics to chemical education, it is convenient to distinguish those developed on low-cost systems (micro with plotter or low-performance raster) from those using sophisticated equipment (calligraphic or high-performance raster). The applications earlier than 1979 have been already reviewed [12].

It is often difficult to make a clear distinction between typical computer graphics and computer-assisted instruction (CAI) applications using graphics output, which is not surprising because graphics is indeed a privileged tool for student–machine communication. To solve this problem, we have included in the present review all the applications with graphics as the central theme only, but the choice is in some cases arbitrary. A recent review of most CAI applications to chemistry has been presented by Cabrol [13].

3.1   Low-cost graphics systems

Because of their flexibility and easy programming, micro- or minicomputers using low-cost graphics systems as output devices have been widely used in chemical education [14–45]. A large variety of subjects have been treated in these applications, ranging from simulation to molecular modelling, quantum chemistry, etc.

Simulation   As computers are very efficient for simulation tasks, several CAI applications centered on graphics have appeared recently in this field [17, 18, 24]: acid-base titration curves, first- and second-order kinetic mechanisms, electron-transfer processes at the surface of an electrode, radioactive decay, etc. However, these applications are very simple and in most cases non-interactive.

Molecular modelling   Depending on the graphics system used, the display of molecular structures may be carried out at different levels of sophistication. In our opinion, a 512×512 raster system with 64 colors simultaneously displayed is a prerequisite for an efficient and aesthetic modelling, which should exhibit the following features: three-dimensional or pseudo-three-dimensional representation of molecules containing up to several hundreds of atoms in the sticks only, ball-and-sticks and space-filling modes; treatment of hidden lines and surfaces, shading, perspective; interactive manipulation such as rotations, translations, zooming. Among the application packages which have appeared so far [16, 19, 23, 34, 35, 39], only a few, if any, present these characteristics mainly because of limitations of current graphics systems. However, with the promising performance of future personal work-stations (PWS), there is no doubt that a high-quality molecular modelling, which is now only available on sophisticated systems, will soon be possible on low-cost graphics terminals.

Quantum chemistry   Most of the results of quantum chemical calculations can be advantageously visualized on graphics displays: atomic and molecular orbitals, electron densities, potential energy surfaces, reaction pathways, etc. Some education oriented developments have been recently achieved: theory of harmonic oscillator [20, 36], group theory [31], molecular vibrations [33] and orbitals [38].

Chemical engineering   Some applications dealing with industrial processes of simulation and control have been developed [41, 43]. They are related to plant operation procedures and may be of great interest for CAI purposes in chemical engineering.

Miscellaneous   Chemistry computer games have recently appeared [29, 41], with themes centered on drug design and organic reactions. Finally, though PLATO is not, properly speaking, a graphics package, a microcomputer version of it is now available with graphics output on a 512×512 raster display [46].

Systems used   The low-cost systems most commonly used in these applications are of two types: (i) 16 bits minicomputers (typically of the PDP-11 series) with DVST (Tektronix) or low-resolution raster display; (ii) Apple II, TRS-80 or IBM-PC microcomputers with raster display. The programming languages used are generally FORTRAN in the first case and BASIC or APPLESOFT in the second one. Some applications have been developed on plotters, but they could be easily modified for graphic CRTs. However, taken as a whole, the low-cost graphics equipment used in these applications does not allow production of high-quality pictures, nor does it offer a satisfactory level of interactivity. This is disappointing, since there is no reason whatsoever why the needs of chemical education should be satisfied with lower standards than research. The future high-performance work-station based graphics will undoubtedly help solve this problem.

3.2   Sophisticated graphics systems

Calligraphic or high-performance raster systems have been rarely used in chemical education, because they are expensive tools devoted mostly to computer-assisted applications in chemical industries. The only applications developed on these systems are those of Meyer [47], Calo [48] and our own [49–56]. Taking into account the capability of these systems to display dynamic pictures, it is not surprising that these few applications represent the behavior of objects as they vary over time, namely three-dimensional interactive modelling, including the visualization of dynamic processes in molecules, and quantum chemistry.

Fig. 2   The Vector General calligraphic system representing an intermediate of the rearranging C_8H_9 polycyclic cation.

Molecular modelling   The three-dimensional representation and real-time manipulation of molecular models is of considerable interest in chemical education. Application programs allowing high-quality modelling of static structures are now operational [47, 49, 55], and quite recently the animated representation of dynamic structures has been made possible: inversions (ammonia), interconversions (cyclohexane), pseudo-rotations (S-methyl thiepanium), rearrangements (cation C_8H_9, Fig. 2), electrocyclic reactions (dis- and conrotatory ring opening of cyclobutene), sigmatropic rearrangements (1,3 allylic migration of the methyl group), etc. [49, 50, 52, 56]. In these dynamic applications, the following elements are simultaneously represented: (i) the stereochemistry (i.e. the three-dimensional structural deformation) and (ii) the energetic and/or orbital aspects, namely the corresponding potential energy profile and/or the transformation and correlation of the molecular orbitals involved in the process. Among other things, this allows one to visualize in detail the very nature of stereoisomers (stable forms, reaction intermediates), their relative stabilities and the stereospecificity of the reactions by using selection rules of the Woodward–Hoffmann type. In addition, these applications allow much interactivity: the reaction can be stopped by a key, which permits manipulation of the objects displayed. It is possible to translate, rotate and enlarge any part of the picture, look at structure and energy profile independently. However, the development of these dynamic modelling applications requires in general much more effort than static modelling: an adequate data base, made of optimum geometries and corresponding energies, must first be calculated at several points along the reaction pathway using a reliable quantum chemical model. Then, one has to define a strategy for the optimal illustration of the process, a scenario allowing an adequate arrangement of the objects on the screen and a technique of representation leading to aesthetic images.

Quantum chemistry   To our knowledge, the only graphics applications which have appeared so far for teaching quantum chemistry using high-performance systems have been carried out by our Geneva group [51–54]. The first of them consists of a new version of the well-known Hückel model using only graphics I/Os. It is indeed important to present graphically the electronic structure calculated for any compound, together with a schematic representation of the molecular orbitals, atomic charges, etc., and extension of this application to other quantum mechanical models is under way. We have noticed that this graphics approach greatly helps students to understand the content of a quantum chemical calculation. Another application deals with the representation of electron densities (Fig. 3), electrostatic potentials and related properties in the form of color filled contour maps or surface mesh maps. These types of properties modelling may be extremely useful for teaching basic concepts of chemical bonding or intermolecular interaction.

Systems used   The typical configuration of the high-performance systems used in computer-assisted chemistry is based on a 32 bits supermini (most often a VAX) and a calligraphic equipment of the Evans and Sutherland PS 300 type. Our configuration is somewhat different in this respect, but it certainly allows an easier representation of dynamic processes.

III.4   FUTURE DEVELOPMENTS IN COMPUTER GRAPHICS APPLIED TO CHEMICAL EDUCATION

Only a small part of the very large potential of computer graphics in chemical education has been exploited so far. The field is thus open to both new applications and new distribution media of the existing applications among potential users. Before reviewing the main themes of possible applications yet to be developed, let us briefly summarize the main features of the future personal work-station, based on third generation microcomputers, which will undoubtedly allow one to carry out high-performance graphics at a very low cost. This *future PWS* will have 32-bit architecture, 2 Mbytes of main memory, a 10 Mbyte hard disk drive and a 1 000 × 1 000 raster display with

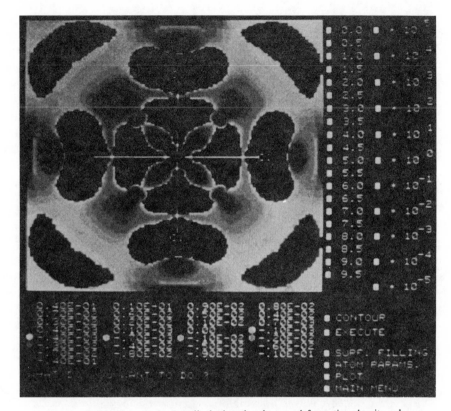

Fig. 3   The AED raster system displaying the electron deformation density calculated for the VCl_6^{2-} complex in a plane containing the metal and four ligands. Dark regions in the color-filled contour map indicate charge accumulation whereas pale ones reflect depletion of charge.

64 color graphics. It is announced for 1986 at a price of 3 000 dollars [57]. If in addition it includes dedicated graphics processors, with hardware implemented simple transformations, it might be used as a high-performance graphics work-station allowing much interactivity. The distinction between the different levels of graphics will thus hardly make sense, since virtually everybody will be able to afford this equipment and to start high-level applications. The only problem could be to perform real-time dynamic graphics, such as the representation of the molecular rearrangements described above, since they require important computational and refresh capabilities; but a solution could be to design the application programs differently so as to perform a nearly real-time animation taking full advantage of the new hardware capabilities associated with these PWS.

Turning now to future applications, there is no doubt that they are going to offer higher quality graphics in terms of user-friendliness, transportability, interactivity, animation, resolution, colors, etc. As it is essential in chemistry to visualize three-dimensional objects and their behavior as they vary over time, we present here a list of applications mainly oriented towards dynamic graphics and which are still needed in chemical education.

Molecular modelling It is important to make a distinction between the modelling of static, rigid structures and the animated representation of dynamic structures. In the first case, many applications have appeared so far, but a general software offering a high-level of interactivity is still to be developed on a high-performance PWS. This program should exhibit the following features: graphic input as two-dimensional drawings with software generation of three-dimensional corresponding structures; display of molecules containing up to 1 000 atoms in the stick only, ball-and-stick and space-filling models with real-time manipulation; full hidden-surface removal and light source shadow casting. The interest of this project lies in the development of a high-performance PWS as an independent molecular modelling station, which would be of great value both in education and research. However, it is in the field of dynamic modelling that much remains to be done. Computer graphics is indeed an ideal tool for visualizing the changes of a system as a function of time, in particular the relative displacements of atoms in a molecule. Therefore, it should be used as a priority for the representation of some basic reaction mechanisms, such as those listed below.

Organic chemistry
— mechanism of enzymatic catalysis
— Cope and Walk sigmatropic rearrangements
— electrocyclic ring opening of cyclopropyl chloride
— S_{N_2} reaction
— rearrangement of cyclobutyl cation into cyclopropyl carbinyle
— 'ene' reaction
— rearrangement of adamantane
— Fischer–Tropf reaction

Organometallic chemistry
— synthesis of dopa-amine (Halpern mechanism)
— oxidizing addition, reducing elimination
— oxidizing cyclization, reducing fragmentation
— α (or β) insertion, α (or β) elimination
— cyclo-insertion, cyclo-elimination

Inorganic chemistry
— construction of polyhedra as a function of the number of vertices (borane chemistry)
— fluxionality of molecules: PF_5 (Berry pseudo-rotation), $Fe(CO)_5$, $PtCl_3(C_2H_4)^-$, ferrocene, allyl-metal complexes, etc.
— infrared spectroscopy: representation of normal modes of vibration (ex: $Cr(CO)_6$); influence of the interaction between carbonyl groups on the infrared spectrum
— reaction mechanisms such as $I_2 + I^- \rightarrow I^- + I_2$.

In most cases, it is necessary first to perform quantum chemical or molecular mechanics calculations so as to have an adequate data base. This means that the development of each these applications represents an investment of several man-months.

Quantum chemistry   Among the great variety of quantum chemical concepts which would undoubtedly benefit from graphics applications, let us quote:

— formation of chemical bond from isolated atoms
— harmonic oscillator: time-dependent solution (classical and quantum theory)
— Walsh rules and the geometry of triatomics
— bonding, non-bonding and anti-bonding orbitals: energy and charge distribution
— expansion and contraction of d and f orbitals in coordination compounds
— concept of hybridization
— dynamic correlation diagram (as a function of distance) for the formation of a coordination compound
— localization of molecular orbitals
— formation of van der Waals' complexes.

Simulation
— evolution of an nmr or Mössbauer spectrum as a function of time (e.g. the ^{19}F spectrum of PF_5 as a function of rearrangement rate)
— variation of an nmr spectrum as a function of the difference between chemical shifts (e.g. $A \rightarrow AB \rightarrow AX$)
— evolution of the intermediates of a multistep reaction as a function of the relative rates of the different steps.

III.5   CONCLUSION

A closer collaboration among all the educators in chemistry contributing to graphics applications appears desirable so as to avoid duplication of efforts. This could be obtained by the creation of specialized groups such as RECODIC 'Infographie didactique de la chimie' and by the publication of a periodical reviewing the latest applications in this field. Finally, standards should be found both for graphics hardware and software so as to enable educators to take advantage of the fascinating potentialities of computer graphics for teaching chemistry.

ACKNOWLEDGEMENTS

The author would like to thank J. J. Combremont, M. Roch, D. Mottier and all the students who participated in the graphics applications developed in Geneva. He is also grateful to Professors P. Vogel and A. Williams for helpful discussions and to Digital Equipment Co. (Europe) for a generous grant of computer time on their VAX-11/780. This work is part of project Nr 2.615-0.82 of the Swiss National Science Foundation.

III.6   REFERENCES

[1]   J. E. Dubois, in *Computer Representation and Manipulation of Chemical Information* (W. T. Wipke, S. R. Heller, R. J. Feldmann and E. Hyde, eds), John Wiley, New York, pp. 239 (1974).

[2]   R. Attias, *J. Chem. Inf. Comp. Sci.*, **23**, 102 (1983).

[3]   I. E. Frank and B. R. Kowalski, *Anal. Chem.*, **54**, 232R (1982).

[4]   J. D. Foley and A. Van Dam, *Fundamentals of Interactive Computer Graphics*, Addison-Wesley, Reading (1982).

[5]   R. Langridge, T. E. Ferrin, I. D. Kuntz and M. L. Connolly, *Science*, **211**, 4483 (1981).

[6]   P. Gund, J. D. Andose, J. B. Rhodes, and G. M. Smith, *Science*, **208**, 1425 (1980).

[7]   P. Gund, E. J. Grabowski, G. M. Smith, J. D. Andose and W. T. Wipke, in *Computer-Assisted Drug Design* (E. C. Olson and R. E. Christoffersen, eds), ACS Symposium Series, 112, 527 (1979).

[8]   P. K. Weiner, R. Langridge, J. M. Blaney, R. Schaefer and P. A. Kollman, *Proc. Natl. Acad. Sci. USA*, **79**, 3734 (1982).

[9]   C. Hansch, R. Li, J. M. Blaney and R. Langridge, *J. Med. Chem.*, **25**, 777 (1982).

[10a] 'Computer Applications in Chemistry' (S. R. Heller and R. Potenzone, eds), *Anal. Chem. Symp. Ser. 15*, Elsevier, Amsterdam (1983).

[10b] J. E. Dubois, D. Laureut and J. Weber, The Visual Computer, **1**, 49 (1985).

[11]   W. M. Newman and R. F. Sproull, *Principles on Interactive Computer Graphics*, 2nd Edition, McGraw-Hill, New York (1979).

[12]   L. J. Soltzberg, *J. Chem. Educ.*, **56**, 644 (1979).

[13]   D. Cabrol, 'Informatique-Enseignement Chemie: Bibliographie Commentée, Centre Documentaire Coopératif Informatique-Enseignement Chimie, Université de Nice (1983).

[14]   G. F. Pollnow, in *Personal Computers in Chemistry* (P. Lykos, ed.), Wiley-Interscience, New York, pp. 153 (1981).

[15]   P. C. Kahn and S. A. Bailey, in *Personal Computers in Chemistry* (P. Lykos, ed.), Wiley-Interscience, New York, pp. 130 (1981).

[16]   V. I. Bendall, *J. Chem. Educ.*, **57**, 250 (1980).

[17]   S. V. Orchard and M. B. Mooiman, *J. Chem. Educ.*, **58**, 409 (1981).

[18]   G. L. Breneman, *J. Chem. Educ.*, **58**, 410, 987 (1981).

[19]   M. L. Deu and S. Perez, *J. Chem. Educ.*, **58**, 552 (1981).

[20]   F. Rioux, *J. Chem. Educ.*, **58**, 553 (1981).

[21]   P. A. Dobosh, *J. Chem. Educ.*, **58**, 693 (1981).

[22] R. H. Batt, *J. Chem. Educ.*, **58**, 990 (1981).

[23] R. Keat, *J. Chem. Educ.*, **59**, 128 (1982).

[24] O. R. Brown, *J. Chem. Educ.*, **59**, 409 (1982).

[25] V. Kumar, J. I. McAndrews and J. W. Mauch, *J. Chem. Educ.*, **59**, 519 (1982).

[26] R. G. Ford, *J. Chem. Educ.*, **59**, 520 (1982).

[27] M. J. Miller and J. F. Johnson, *J. Chem. Educ.*, **59**, 521 (1982).

[28] E. D. Cavin and C. S. Cavin, *J. Chem. Educ.*, **59**, 589 (1982).

[29] J. L. Meisenheimer, V. I. Bendall and G. T. Road, *J. Chem. Educ.*, **59**, 600 (1982).

[30] D. D. Gilbert, T. T. Mounts and A. A. Frost, *J. Chem. Educ.*, **59**, 661 (1982).

[31] B. R. Penfold and R. S. Temple, *J. Chem. Educ.*, **59**, 774 (1982).

[32] C. W. Eaker and E. L. Jacobs, *J. Chem. Educ.*, **59**, 939 (1982).

[33] E. L. Varetti, *J. Chem. Educ.*, **60**, 44 (1983).

[34] L. A. Hull, *J. Chem. Educ.*, **60**, 96 (1983).

[35] H. Nakano, O. Sangen and Y. Yamamoto, *J. Chem. Educ.*, **60**, 98 (1983).

[36] C. Kubach, *J. Chem. Educ.*, **60**, 212 (1983).

[37] G. A. Gerhardt, *J. Chem. Educ.*, **60**, 568 (1983).

[38] J. J. Houser, *J. Chem. Educ.*, **60**, 731 (1983).

[39] G. A. Dauphinee and T. P. Forrest, *J. Chem. Educ.*, **60**, 732 (1983).

[40] R. G. Meyer and R. Barone, *J. Chem. Educ.*, **60**, 736 (1983).

[41] COMPress, Catalog of Educational Software, PO Box 102, Wentworth, N.H. 03282, USA.

[42] M. Ben-Zion and S. Hoz, *Educ. Chem.*, **17**, 101 (1980).

[43] D. M. Himmelblau, *Comput. Chem. Eng.*, **5**, 187 (1981).

[44] H. J. Peters and K. C. Daiker, *Pipeline 1982*, Spring Issue, 11.

[45] R. J. Beynon and G. A. Place, *Biochem. Educ.*, **10**, 104 (1982).

[46] S. Smith, in *Personal Computers in Chemistry* (P. Lykos, ed.), Wiley-Interscience, New York, 177 (1981).

[47] W. G. White, S. M. Swanson and E. F. Meyer, *J. Chem. Educ.*, **59**, 515 (1982).

[48] J. M. Calo and R. P. Andress, *Comput. Chem. Eng.*, **5**, 197 (1981).

[49] J. Weber, G. Bernardinelli, J. J. Combremont and M. Roch, in *Proceedings of Eurographics 80* (C. E. Vandoni, ed.), North-Holland, Amsterdam, 139 (1980).

[50] J. Weber, G. Bernardinelli, J. J. Combremont and M. Roch, *Computer Graphics World*, **4**, 37 (1981).

[51] M. Roch, J. J. Combremont and J. Weber, *Chimia*, **36**, 154 (1982).

[52] J. Weber, M. Roch, J. J. Combremont, P. Vogel and P. A. Carrupt, *J. Mol. Struct. Theochem.*, **93**, 189 (1983).

[53] J. Weber, J. J. Combremont and M. Roch, in *Comptes Rendus des Quatrièmes Journées Internationales sur l'Education Scientifique: l'Informatisation dans l'Education Scientifique* (A. Giordan and J. L. Martinand, eds), Edition Université Paris VII, 49 (1983).

[54]   J. J. Combremont, M. Roch and J. Weber, *Computer Graphics Forum*, **2**, 89 (1983).

[55]   G. Bernardinelli, J. J. Combremont, M. Roch, D. Mottier, J. P. Barras, Y. Mentha and J. Weber, *Acta Cryst.*, **A40**, Suppl. C439 (1984).

[56]   C. W. Jefford, J. Mareda, J. J. Combremont and J. Weber, *Chimia*, **38**, 354 (1984).

[57]   G. S. Owen, *J. Chem. Educ.*, **61**, 440 (1984).

[58]   J. G. Vinter, *Chem. Brit.*, **21**, 1–5 (1985).

[59]   M. Ishiguro and S. Imajo, *J. Synthetic Org. Chem. Japn*, **44**, 722–731 (1984).

CHAPTER IV

Kinetics

J. L. Larice
University of Avignon, France

IV.1 INTRODUCTION

The goal of this chapter is an overview of the methods used in the treatment of kinetic data or the simulation of kinetic reactions. We treat the applications available on medium or small size calculators. Thus, this chapter is limited to the study of the three spheres most frequently encountered in kinetic reactions. In each case, the numerical treatment has been explained so that the novice computer chemist will understand the principles and limitations of methods chosen, thereby avoiding, if possible, the use of a program as a black box.

The first part concerns stoichiometric calculations. We intend to discuss independent reactions, which are very important in the study of sets of possible simultaneous reactions and in chemical engineering. The numerical method used to solve linear equation systems is explained in detail because of its use in other domains of computational chemistry, such as in linear regression and derived methods.

The second part discusses the case in which the chemist needs to estimate the parameters of a theoretical or empirical law extrapolated from experimentally measured values. These methods are allied with regression analysis and curve fitting. We have applied such techniques to concentration vs. time measurements in order to estimate kinetic constants or the reaction order. Obviously, the treatment is the same when a relation is known between two kinds of data. Thus, the study of reaction rate-temperature pairs so as to estimate activation parameters has not been specifically developed.

The final section is dedicated to those numerical integration methods used to simulate a set of elementary reactions, starting with a postulated kinetic law. The principal methods used are described and references to available programs are noted.

This chapter is restricted to the macroscopic treatment of kinetic reactions. The reader interested in chemical kinetic theory will find a good introduction to the topic in the G. G. Hammes book [1] and in the publication by R. K. Boyd [2]. An interesting tentative kinetic equation classification from operators representing species interactions has also recently been published [3]. The study of microscopic chemical kinetics uses more complicated numerical techniques to estimate molecular interactions by statistical methods or quantum mechanics, and is not within the scope of this chapter.

IV.2 STOICHIOMETRY AND REACTION SCHEMES

2.1 Introduction

The establishment of a possible reaction scheme is the starting point of most kinetic studies. When the number of species involved is small, this step is easy. The difficulty increases, however, with the number of elementary steps involved. One rapidly reaches the point where the question to be answered is 'Besides the obvious elementary steps, is it possible to find other steps which agree with the overall stoichiometry of the reaction?' The goal of this section, therefore, is to show how one may inventory all the possible elementary steps agreeing with a set of identified species.

2.2 Matter conservation in chemical reactions

The first constraint to be treated is that of Lavoisier's law. This is essentially the need to solve a set of linear equations. This problem was solved a long time ago [4, 5]. To correctly establish the system of linear equations describing a chemical system, one needs to express the system in the form of a matrix product [6, 7]

$$[A] \cdot [S] = [0] \qquad\qquad (IV.1)$$

In the rectangular matrix $[A]$, every row describes an atom and every column describes a molecule (see Scheme 1). In the column vector $[S]$, the number of elements equals the number of columns in the $[A]$ matrix and in which every element represents a stoichiometric coefficient (unknown) associated with a molecule present in the equation describing the overall transformation. By convention, these elements are positive for products and negative for reactants. The matrix $[A]$ is called the 'reaction matrix' in the following discussion.

For example, if the species: HI, HCl, Cl_2 and I_2 have been identified in an experimental sample, what set of elementary reactions may have been

Numbering of rows	(atoms)	Numbering of columns (molecules)			
		1	2	3	4
		HI	Cl_2	HCl	I_2
1	H	1	0	1	0
2	Cl	0	2	1	0
3	I	1	0	0	2

Scheme 1

involved? The reaction matrix is as shown in Scheme 1, and the column vector is

$$[s] = \begin{bmatrix} \sigma_1 \\ \sigma_2 \\ \sigma_3 \\ \sigma_4 \end{bmatrix}$$

in which the σ's are the stoichiometric coefficients to be found (σ_1 for HI, σ_2 for Cl_2, etc.)

To determine $[S]$, one uses the relationship (IV.1):

$$\left. \begin{array}{c} \sigma_1 + \sigma_3 = 0 \\ 2\sigma_2 + \sigma_3 = 0 \\ \sigma_1 + 2\sigma_4 = 0 \end{array} \right\} \qquad \text{(IV.2)}$$

With three equations and four unknowns, one variable must be chosen arbitrarily if a solution is to be found. Taking $\sigma_4 = 1$ yields $\sigma_1 = -2$ $\sigma_2 = -1$, $\sigma_3 = 2$; thus the final reaction becomes

$$2IH + Cl_2 = 2HCl + I_2$$

We must stress the point that in choosing σ as a positive number, we implicitly ascribe I_2 to the products. To find other sets of σ's agreeing with the identified species one has only to multiply the foregoing set of σ's by either a positive or a negative number.

An interesting special case is found when all $\sigma = 0$. This corresponds to the case in which the species identified do not correspond to an overall reaction. The reader may check this proposition for the system composed of $(NH_4)_2CO_3$, NH_3, H_2O and CO. In order to create a reactive system, one needs only to replace CO by CO_2.

2.3   The case of non-independent equations

The foregoing examples correspond to ideal cases: the degrees of freedom are obtained as the difference between the number of molecules involved (number of unknowns) and the number of atoms involved (number of equations). This difference is found easily only when the equations are independent. Independent equations are not always obvious. Let us consider the system composed of CH_3OH, HCl, CH_3Cl and H_2O. The reaction matrix may be written

$$
\begin{array}{c} \\ C \\ H \\ O \\ Cl \end{array}
\begin{array}{cccc}
CH_3OH & HCl & CH_3Cl & H_2O \\
\left[\begin{array}{cccc}
1 & 0 & 1 & 0 \\
4 & 1 & 3 & 2 \\
1 & 0 & 0 & 1 \\
0 & 1 & 1 & 0
\end{array}\right]
\end{array}
$$

which yields the following set of equations

$$
\left.\begin{array}{l}
\sigma_1 \quad\quad + \sigma_3 \quad\quad = 0 \\
4\sigma_1 + \sigma_2 + 3\sigma_3 + 2\sigma_4 = 0 \\
\sigma_1 \quad\quad\quad\quad + \sigma_4 = 0 \\
\sigma_2 + \sigma_3 \quad\quad = 0
\end{array}\right\} \qquad (IV.3)
$$

In this case, the first equation is simply a linear combination of the second, third and fourth equations. Arbitrarily taking $\sigma_1 = 1$ gives the overall reaction:

$$CH_3OH + HCl = CH_3Cl + H_2O$$

2.4   Finding the number of degrees of freedom

The preceding sub-section has shown that the main difficulty lies in the determination of the number of independent equations in a reaction scheme. Comes [8] recently provided a solution to this problem based on the examination of the rank of the reaction matrix. We can show that the application of the Gauss–Jordan algorithm simultaneously provides the number of degrees of freedom and the possible elementary steps associated with a system of experimentally identified species (see ref. [9] for a more detailed treatment).

The Gauss–Jordan method is a general-purpose method used to invert square matrices or to solve systems of linear equations. Since we find such cases in many aspects of computational chemistry (see multilinear regression below), it is interesting here to describe the method, which is very easily translated into programming languages. The method begins with an examination of each equation. We assume that equation i is currently examined. The terms of this equation are multiplied by a constant and the equivalent equation thus obtained is added to the $n - 1$ remaining equations. The constant is chosen such that the term number i vanishes in all of the remaining $n - 1$ equations thus obtained. A new equivalent set of equations is, therefore, obtained in which each equation presents only one unknown and a constant term.

In order to minimize roundoff errors, it is convenient to modify the order in the set of equations such that the equation with the highest coefficient examined becomes the first in the set. We will treat the example dealt with earlier in this section to illustrate this approach. The system obtained by assigning $\sigma_4 = 1$ in (IV.2) is

$$\left.\begin{array}{r} \sigma_1 + \sigma_3 = 0 \\ 2\sigma_2 + \sigma_3 = 0 \\ \sigma_1 = -2 \end{array}\right\} \qquad\qquad (\text{IV.4})$$

and is associated with the matrix

$$\begin{bmatrix} 1 & 0 & 1 & 0 \\ 0 & 2 & 1 & 0 \\ 1 & 0 & 0 & -2 \end{bmatrix} \qquad\qquad \begin{array}{l} (\text{IV.4.1}) \\ (\text{IV.4.2}) \\ (\text{IV.4.3}) \end{array}$$

The new equation is obtained when (IV.4.1) is multiplied by a coefficient (-1). This equation is then added to (IV.4.3), yielding a new set equivalent to the preceding one:

$$\begin{bmatrix} 1 & 0 & 1 & 0 \\ 0 & 2 & 1 & 0 \\ 0 & 0 & -1 & -2 \end{bmatrix} \qquad\qquad \begin{array}{l} (\text{IV.5.1}) \\ (\text{IV.5.2}) \\ (\text{IV.5.3}) \end{array}$$

The next step consists in applying the same succession of operations to (IV.5.2) but, in our example, it is unnecessary because the second column terms for equations (IV.5.1) and (IV.5.3) are zero. Finally, equation (IV.5.3) is added to equations (IV.5.1) and (IV.5.2) such that the third column terms vanish. We now obtain the matrix

$$\begin{bmatrix} 1 & 0 & 0 & -2 \\ 0 & 2 & 0 & -2 \\ 0 & 0 & -1 & -2 \end{bmatrix} \qquad\qquad (\text{IV.6})$$

which is also equivalent to

$$\begin{bmatrix} 1 & 0 & 0 & -2 \\ 0 & 1 & 0 & -1 \\ 0 & 0 & 1 & 2 \end{bmatrix} \qquad\qquad (\text{IV.7})$$

The solutions are

$$\sigma_1 = -2, \qquad \sigma_2 = -1, \qquad \sigma_3 = +2$$

The Gauss–Jordan approach may be easily extended to matrix equation containing as many or more rows than columns. The calculation in such a case yields a matrix containing as many or fewer non-cancelling rows than columns. The rank (i.e., the number of independent equations) of the matrix equals the difference between the number of columns and the number of rows.

We can consider the second example dealt with in section 4.2.2: The initial reaction matrix

$$
\begin{array}{c}
 \\
C \\ H \\ O \\ N
\end{array}
\begin{array}{cccc}
(NH_4)_2CO_3 & NH_3 & H_2O & CO \\
\left[\begin{array}{cccc}
1 & 0 & 0 & 1 \\
8 & 3 & 2 & 0 \\
3 & 0 & 1 & 1 \\
2 & 1 & 0 & 0
\end{array}\right]
\end{array}
\tag{IV.8}
$$

transforms into

$$
\begin{bmatrix}
1 & 0 & 0 & 0 \\
0 & 1 & 0 & 0 \\
0 & 0 & 1 & 0 \\
0 & 0 & 0 & 1
\end{bmatrix}
\tag{IV.9}
$$

This new matrix describes a system in which no possible reaction connects the molecules involved. If, however, CO is replaced by CO_2, the matrix becomes

$$
\begin{array}{c}
 \\
C \\ H \\ O \\ N
\end{array}
\begin{array}{cccc}
(NH_4)_2CO_3 & NH_3 & H_2O & CO_2 \\
\left[\begin{array}{cccc}
1 & 0 & 0 & 1 \\
8 & 3 & 2 & 0 \\
3 & 0 & 1 & 2 \\
2 & 1 & 0 & 0
\end{array}\right]
\end{array}
\tag{IV.10}
$$

which transforms into

$$
\begin{bmatrix}
1 & 0 & 0 & 1 \\
0 & 1 & 0 & -2 \\
0 & 0 & 1 & -1 \\
0 & 0 & 0 & 0
\end{bmatrix}
\tag{IV.11}
$$

Setting $\sigma_4 = \sigma_{CO_2} = -1$, the coefficients of the other compounds are given, respectively, by the first 3 elements of the last column in the second matrix, i.e.

$$
\sigma_{(NH_4)_2CO_3} = 1, \qquad \sigma_{NH_3} = -2, \qquad \sigma_{H_2O} = -1
$$

2.5 Choice of solution sets: slave and master compounds

In most cases considered, the Gauss–Jordan procedure, when applied to the reaction matrix, transforms it into a matrix counting fewer non–zero rows than the number of columns -1. Photochlorination of methane illustrates this situation. The initial reaction matrix:

$$
\begin{array}{c}
 \\
C \\ H \\ Cl
\end{array}
\begin{array}{ccccccc}
CH_4 & Cl_2 & HCl & CH_3Cl & CH_2Cl_2 & CHCl_3 & CCl_4 \\
\left[\begin{array}{ccccccc}
1 & 0 & 0 & 1 & 1 & 1 & 1 \\
4 & 0 & 1 & 3 & 2 & 1 & 0 \\
0 & 2 & 1 & 1 & 2 & 3 & 4
\end{array}\right]
\end{array}
\tag{IV.12}
$$

transforms into

$$\begin{bmatrix} 1 & 0 & 0 & 1 & 1 & 1 & 1 \\ 0 & 1 & 0 & 1 & 2 & 3 & 4 \\ 0 & 0 & 1 & -1 & -2 & -3 & -4 \end{bmatrix} \qquad \text{(IV.13)}$$

The foregoing system indeed contains $7 - 3 = 4$ degrees of freedom, i.e., is described by 4 sets of independent reactions. These independent reactions are called basis reactions. Any reaction associated with the starting system may be obtained from a linear combination of these basis reactions. A possible basis for each family may be obtained as follows. The last 4 columns of the reaction matrix obtained after application of the Gauss–Jordan algorithm are examined successively. One sets $\sigma = -1$ for the coefficient of compound described in the column under consideration and the coefficients of the three remaining compounds (3 out of the 4 initial compounds) are set equal to zero. The coefficients of the first three compounds are then simply removed from

$$\begin{array}{ccccccc} CH_4 & Cl_2 & HCl & CH_3Cl & CH_2Cl_2 & CHCl_3 & CCl_4 \\ \begin{bmatrix} 1 & 0 & 0 & 1 & 1 & 1 & 1 \\ 0 & 1 & 0 & 1 & 2 & 3 & 4 \\ 0 & 0 & 1 & -1 & -2 & -3 & -4 \end{bmatrix} \end{array}$$

Family number 4

$\sigma_{CCl_4} = -1, \sigma_{CHCl_3} = \sigma_{CH_2Cl_2} = \sigma_{CH_3Cl} = 0$:masters

$\sigma_{CH_4} = 1, \sigma_{Cl_2} = 4, \sigma_{HCl} = -4$:slaves

$$CCl_4 + 4HCl = CH_4 + 4Cl_2$$

Family number 3

$\sigma_{CHCl_3} = -1, \sigma_{CCl_4} = \sigma_{CH_2Cl_2} = \sigma_{CH_3Cl} = 0$:masters

$\sigma_{CH_4} = 1, \sigma_{Cl_2} = 3, \sigma_{HCl} = -3$:slaves

$$CHCl_3 + 3HCl = CH_4 + 3Cl_2$$

Family number 2

$\sigma_{CH_2Cl_2} = -1, \sigma_{CCl_4} = \sigma_{CHCl_3} = \sigma_{CH_3Cl} = 0$:masters

$\sigma_{CH_4} = 1, \sigma_{Cl_2} = 2, \sigma_{HCl} = -2$:slaves

$$CH_2Cl_2 + 2HCl = CH_4 + 2Cl_2$$

Family number 1

$\sigma_{CH_3Cl} = -1, \sigma_{CCl_4} = \sigma_{CHCl_3} = \sigma_{CH_2Cl_2} = 0$:masters

$\sigma_{CH_4} = 1, \sigma_{Cl_2} = 1, \sigma_{HCl} = -1$:slaves

$$CH_3Cl + HCl = CH_4 + HCl$$

Scheme 2

the column associated with the product under consideration. This is illustrated in Scheme 2. Note that the compounds described in the last 4 columns are sometimes called 'masters' because setting $\sigma = -1$ for one of these compounds forces this compound to be the initial compound. The 3 remaining coefficients follow from the value given to the master's coefficient and so are called 'slaves'. For example, in family number 1: master, CH_3Cl: slaves, CH_4, Cl_2, HCl, and so on.

2.6 Significance of the failure of the method

One may ask whether the set of basis solutions obtained in the foregoing section will be the best solution. It is always possible to make linear combinations of basis solutions. To arrive at these new sets, it may seem better to start from the beginning, i.e., to change the presentation of the initial reaction matrix in such a way that a new 'master' is retained. This operation, which is simply a permutation of columns, is easily programmed and may be performed before the application of the Gauss–Jordan algorithm. We note in passing that under certain conditions, this approach does not work because a zero value is obtained for the threshold. This is illustrated with the foregoing example when one attempts to make Cl or HCl the 'master' compound:

$$
\begin{array}{c}
\\
C \\ H \\ Cl
\end{array}
\begin{array}{ccccccc}
CH_3Cl & CH_2Cl_2 & CHCl_3 & CCl_4 & CH_4 & Cl_2 & HCl \\
\left[\begin{array}{ccccccc}
1 & 1 & 1 & 1 & 1 & 0 & 0 \\
3 & 2 & 1 & 0 & 4 & 0 & 1 \\
1 & 2 & 3 & 4 & 0 & 2 & 1
\end{array}\right]
\end{array}
\qquad \text{(IV.14)}
$$

One obtains

$$
\begin{bmatrix}
1 & 0 & -1 & -2 & 2 & 0 & 1 \\
0 & 1 & 2 & 3 & -1 & 0 & -2 \\
0 & 0 & 0 & 0 & 0 & -2 & -2
\end{bmatrix}
\qquad \text{(IV.15)}
$$

The 2 columns corresponding to Cl_2 and HCl are responsible for the failure of the approach because these compounds cannot be used as masters. Every basis reaction is indeed obtained by setting the coefficient of one of the master compounds, equal to -1, thereby causing the coefficients of the remaining masters to vanish. In the present case we would therefore have to set $\sigma_{Cl_2} = -1$ and $\sigma_{HCl} = 0$, which is clearly impossible when one takes into account the other compounds present in the mixture. The only possible solutions are those for which $\sigma_{Cl_2} = 0$ and $\sigma_{HCl} = 0$. This is equivalent to stating that, in this case, $5 - 2 = 3$ independent solutions are possible. They may be obtained as explained previously:

Family number 1

$$\sigma_{CHCl_3} = -1, \qquad \sigma_{CCl_4} = \sigma_{CH_4} = 0: \text{masters}$$

$$\sigma_{CH_3Cl} = -1, \qquad \sigma_{CH_2Cl_2} = 2: \text{slaves}$$

$$CH_3Cl + CHCl_3 = 2CH_2Cl_2$$

Family number 2

$$\sigma_{CCl_4} = -1, \qquad \sigma_{CHCl_3} = \sigma_{CH_4} = 0: \text{masters}$$

$$\sigma_{CH_3Cl} = -2, \qquad \sigma_{CH_2Cl_2} = 3: \text{slaves}$$

$$2CH_2Cl_2 + CH_4 = 2CH_3Cl$$

Family number 3

$$\sigma_{CH_4} = -1, \qquad \sigma_{CHCl_3} = 0, \sigma_{CCl_4} = 0: \text{masters}$$

$$\sigma_{CH_3Cl} = 2, \qquad \sigma_{CH_2Cl_2} = -1: \text{slaves}$$

$$CH_2Cl_2 + CH_4 = 2CH_3Cl$$

2.7 Conclusion

The determination of the number of independent reactions is a necessary preliminary step as soon as the scheme under study reaches a certain level of complexity. This determination involves finding the rank of a rectangular matrix. A program performing this operation has been commercially available for a long time from IBM [10]. Recent reviews [11] emphasize, however, that this problem is of continuing interest and may be extended in several directions. For example, systems including charged or paramagnetic species have been specifically dealt with [12, 13] and thermodynamic criteria for compounds have also been considered [13–17]. In another direction, programs including alphanumeric treatments for searches of independent reactions [12] and those dealing with syntactic analysis of data (elimination of impossible formula) [18] to improve the output quality have been described. A microcomputer program which deals with mixtures has recently been published [19].

IV.3 REVIEW OF KINETIC RULES

3.1 Introduction

When the static stoichiometry of a reaction is well established, the chemist has to deal with the time-dependent variation of concentration for each of the species involved in the reaction mixture. In the next two sections, we will treat the case in which an analytical expression suffices to describe the time-dependent behavior of the species under consideration. This analytical expression may either follow from a rigorous treatment or from a succession of simplifying assumptions. In both cases, the comparison of measured time-dependent concentrations with theoretical values provided by the analytical expression allows one to check whether the initial mechanistic hypothesis fits the experimental data.

3.2   Definition of rate and the laws of kinetics

The rate of a chemical reaction is defined by the change with time of the concentration of a reagent or a product. For the following reaction:

$$\alpha A + \beta B \rightarrow \gamma C + \delta D \tag{IV.16}$$

the rate may be expressed as

$$v = -\frac{1}{\alpha} \cdot \frac{d[A]}{dt} = -\frac{1}{\beta} \cdot \frac{d[B]}{dt} = \frac{1}{\gamma} \cdot \frac{d[C]}{dt} = \frac{1}{\delta} \cdot \frac{d[D]}{dt} \tag{IV.17}$$

The experimental rate law for a reaction involving reagents A and B usually corresponds to the expression:

$$v = [A]^m \cdot [B]^n \tag{IV.18}$$

in which m represents the *partial order* of the reaction with respect to $[A]$, n represents the *partial order* of the reaction with respect to $[B]$, and $m + n$ the *overall reaction order*. The *molecularity* of a reaction describes the number of species involved in the transition state for a one step transformation. The molecularity will be directly connected to m and n only if the reaction under consideration is a one-step reaction. If a large excess of one of the reagents is present, or if one of the species is continuously regenerated, one must note that the apparent order of the reaction will be less then the actual order. In these cases the order is called degenerate. In a kinetic study, a chemist needs to answer the following questions: (1) What is the order of the reaction with respect to each of the reagents? (2) What are the rate constants? (3) Does one mechanistic scheme account for the variation with time, or are there several plausible mechanisms available?

   The determination of the reaction order and the rate constant is described in detail in most kinetic reference books [1, 20–24]. In all cases, one begins by obtaining a series of experimental concentration–time couples for the species under consideration. If possible, it is convenient to know the integrated form of the kinetic law.

3.3   Integrated forms of the kinetic equations

For a given S, (S = species) we note $[S]$ and $[S]_0$ which are respectively, the concentrations at measurement time t and at the initial time t_0. On the other hand, we may set $t_0 = 0$ as if we are able to measure from the initial mixing of the reagents. However, one usually deals with $x = [S]_0 - [S]$, from which follows that $dx/dt = -d[S]/dt$. Table 1 lists the kinetic laws and their integrated forms which are associated with the simplest chemical transformations.

Table 1

Order	Reaction	Kinetic law	Integrated form
0	$A \rightarrow$	$\dfrac{d[A]}{dt} = -k$	$[A] = -kt + [A]_0$
		$\dfrac{dx}{dt} = k$	$x = kt$
1	$A \rightarrow$	$\dfrac{d[A]}{dt} = -k[A]$	$[A] = [A]_0 e^{-kt}$
		$\dfrac{dx}{dt} = k([A]_0 - x)$	$\ln \dfrac{[A]_0}{[A]} = kt$
			$x = [A]_0(1 - e^{-kt})$
			$\ln \dfrac{[A]_0}{[A]_0 - x} = kt$
2	$A + B \rightarrow$ with $[A]_0 = [B]_0$	$\dfrac{d[A]}{dx} = -k[A]^2$	$\dfrac{1}{[A]} - \dfrac{1}{[A]_0} = kt$
		$\dfrac{dx}{dt} = k([A]_0 - x)^2$	$\dfrac{x}{[A]_0([A]_0 - x)} = kt$
	$A + B \rightarrow$ with $[A]_0 \neq [B]_0$	$\dfrac{d[A]}{dt} = -k[A][B]$	$\dfrac{1}{[A]_0 - [B]_0} \ln \dfrac{[B]_0[A]}{[A]_0[B]} = kt$
		$\dfrac{dx}{dt} = k([A]_0 - x)([B]_0 - x)$	
3	$A + B + C \rightarrow$ with $[A]_0 = [B]_0 = [C]_0$	$\dfrac{d[A]}{dt} = -k[A]^3$	$\dfrac{1}{2}\left(\dfrac{1}{[A]^2} - \dfrac{1}{[A]_0^2}\right) = kt$
		$\dfrac{dx}{dt} = k([A]_0 - x)^3$	
n		$\dfrac{d[A]}{dt} = -k[A]^n$	$\dfrac{1}{n-1}\left(\dfrac{1}{[A]^{n-1}} - \dfrac{1}{[A]_0^{n-1}}\right) = kt$
		$\dfrac{dx}{dt} = k([A]_0 - x)^n$	

3.4 Determination of the order for a simple reaction

3.4.1 Differential method

The time lag between concentration measurements is chosen in such a way that the approximation

$$\frac{\Delta[A]}{\Delta t} = \frac{d[A]}{dt} \qquad (IV.19)$$

is valid.

We must then determine k and n values to fit the equation

$$\frac{\Delta[A]}{\Delta t} = -k[A]^n \qquad (IV.20)$$

within the range of experimental precision. This relationship is valid only if the reaction consuming A does not involve other reagents. If this is not the case, then the concentrations of other reagents must be either equal or much greater than $[A]$.

3.4.2 Measurement of the half-reaction time

With the same basic assumption adopted in preceding, section one may construct a set of experiments in which $[A]_0$ changes. If $n = 1$ one finds that $t_{1/2}$ is constant; if $n > 1$ one finds that

$$t_{1/2} = \frac{1}{k(n-1)} \cdot \frac{2^{n-1} - 1}{[A]_0^{n-1}} \tag{IV.21}$$

or equivalently

$$\ln t_{1/2} = (1 - n) \ln [A]_0 + \ln \frac{2^{n-1} - 1}{k(n-1)} \tag{IV.22}$$

From this expression, we see that the reaction order may be obtained by graphical methods or better by linear regression analysis (see Section IV.4).

3.4.3 Direct treatment of time–concentration couples

One tries different integrated rate laws associated with the different reaction orders in order to find the one which best fits the experimental data. A data transformation usually reduces the problem to a simple linear regression analysis. When such a transformation is not possible one must solve a curve-fitting problem.

3.5 Complex reactions that reduce to simple reactions

In practice, chemical reactions are seldom simple reactions, but, in some cases, the assumed mechanism leads to a kinetic expression whose integral is calculable. Two typical cases follow.

A suitable variable change can reduce the reaction studied to a simple one. For example, two species in equilibrium, such as

$$A \underset{k_2}{\overset{k_1}{\rightleftarrows}} B \tag{IV.23}$$

give, with $x = [B]$ and $x_\infty = [B]_\infty$, (in other words, $[B]_\infty$ becomes the concentration of B when equilibrium is reached)

$$\frac{d(x - x_\infty)}{dt} = (k_1 + k_2)(x - x_\infty) \tag{IV.24}$$

Other cases are concurrent reactions, such as

$$\begin{array}{c} & & B \\ & \nearrow & \\ A & & \\ & \searrow & \\ & & C \end{array} \tag{IV.25}$$

and consecutive reactions. The integral calculation is possible without difficulty if there is no molecularity change in any of the elementary steps. Thus, the scheme

$$A \overset{k_1}{\to} B \overset{k_2}{\to} C \tag{IV.26}$$

yields

$$[A] = [A]_0 e^{-k_1 t}$$

$$[B] = \frac{k_1}{k_2 - k_1} [A]_0 (e^{-k_1 t} - e^{-k_2 t}) + [B]_0 e^{-k_2 t} \tag{IV.27}$$

$$[C] = [A]_0 + [B]_0 + [C]_0 - [A] - [B]$$

3.6 Other complex reactions

When an analytical expression cannot be reached by integration from a given kinetic law, several methods may be applied. The first is to modify the experimental conditions in such a way that a simple case is again fulfilled.

For example, one may use a large excess of one of the reagents to modify the apparent reaction order in one step. The resulting simplification may suffice to lead the reaction to a rate law with an integrated form.

Another widely used method is that involved in the *quasi–stationary* state approximation. This approximation assumes that the concentration of an intermediate involved in the reaction mechanism may be considered small and constant throughout the time of the reaction. Actually, one assumes only that the rate of formation of this intermediate is equal to its rate of consumption. In the example (IV.26):

$$\frac{\mathrm{d}[B]}{\mathrm{d}t} = -k_1[A] - k_2[B] \qquad \text{with } k_1[A] = k_2[B] \tag{IV.28}$$

this becomes

$$[B] = \frac{k_1}{k_2} [A] = \frac{k_1}{k_2} [A]_0 e^{-k_1 t} \tag{IV.29}$$

This expression, when compared with the expression obtained without the approximation, (IV.27) shows that the approximation is valid only if $k_2 \gg k_1$.

IV.4   TREATMENT OF KINETIC DATA

4.1   Introduction

The set of experimental data is supposed to fit an expression deduced from a theoretical model. This fit, when valid, provides a set of rate constants.

For this purpose the tools used are all more-or-less connected with the methodology of regression analysis (see Chapter V and refs. [25–28]. Although these methods have been well developed, for some cases it is still the responsibility of the kineticist to decide which model is best suited to his data set.

If we have a function $f(x)$ which can be compared to the experimental data, two main approaches are feasible.

4.2 Mean square estimation problem

4.2.1 Simple linear regression

The simplest case corresponds to a function representing a straight line. For example, consider the reaction between A and B with $[A]_0 = [B]_0$. For a second order reaction, the integrated form of the rate law should be:

$$\frac{1}{[A]} = k \cdot t + \frac{1}{[A]_0} \tag{IV.30}$$

Therefore, if this model is suitable for the experimentally measured set of time versus $[A]$ couples, one should obtain a linear correlation when $1/[A]$ versus t is plotted. The slope of the line and its y-intercept provide, respectively, k and $1/[A]_0$. Because of experimental and model uncertainties, these two pieces of data are only estimates. D. C. Montgomery and A. Peek [27] have published a critical examination of the mean square estimation method as applied to this type of kinetic data treatment.

4.2.2 Linear regression method

In a set of correlated couples, x_i, y_i, where x is the determined variable with negligible error, one may attempt to determine the best $\hat{a}$ and $\hat{b}$ which fit the equation:

$$\hat{y} = \hat{a} \cdot x + \hat{b} \tag{IV.31}$$

in which ^ indicates all values which are associated with an uncertainty.

The classical approach to this problem (see Chapter V), attempts to minimize the sum of the squares of the differences between the experimentally measured data and the values predicted by the theoretical model. The terms whose sums are minimized are called residues:

$$Se = \sum_{i=1}^{n} (y_i - \hat{a}x_i - \hat{b})^2 \tag{IV.32}$$

where Se is the sum of all the residues and in which n experimental determinations have been performed. The optimization follows in the usual way, i.e., one has to calculate:

$$\frac{\mathrm{d}Se}{\mathrm{d}\hat{a}} = 0 \tag{IV.33}$$

and

$$\frac{\mathrm{d}Se}{\mathrm{d}\hat{b}} = 0 \tag{IV.34}$$

This system of two equations with two unknowns yields

$$\hat{a} = \frac{\sum\limits_{i=1}^{n} x_i y_i - \dfrac{1}{n}\left(\sum\limits_{i=1}^{n} x_i\right)\left(\sum\limits_{i=1}^{n} y_i\right)}{\sum\limits_{i=1}^{n} x_i^2 - \dfrac{1}{n}\left(\sum\limits_{i=1}^{n} x_i\right)^2} = \frac{S_{xy}}{S_{xx}} \qquad \text{(IV. 35)}$$

and

$$\hat{b} = \bar{y} - \hat{a}\bar{x} \qquad \text{(IV.36)}$$

where

$$\bar{x} = \frac{1}{n}\sum\limits_{i=1}^{n} x_i \quad \text{and} \quad \bar{y} = \frac{1}{n}\sum\limits_{i=1}^{n} y_i \qquad \text{(IV.37)}$$

are the arithmetic means of x and y.

The algorithms for the calculation of $\hat{a}$ and $\hat{b}$ are so well known that even pocket calculators often have the capability to perform their calculation, although short computer programs are also available [29–33]. Other equivalent expressions may be used to obtain S_{xy} and S_{xx}:

$$S_{xy} = \sum\limits_{i=1}^{n} y_i(x_i - \bar{x}) = \sum\limits_{i=1}^{n} x_i(y_i - \bar{y}) = \sum\limits_{i=1}^{n} (x_i - \bar{x})(y_i - \bar{y}) \quad \text{(IV.38)}$$

$$S_{xx} = \sum\limits_{i=1}^{n} (x_i - \bar{x})^2 = \sum\limits_{i=1}^{n} (x_i - \bar{x}^2) \qquad \text{(IV.39)}$$

A drawback of such complete automation, however, is that one may forget that even a non-linear relationship between x_i and y_i will yield $\hat{a}$ and $\hat{b}$ values when the non-linear data is input to such a program. Fortunately, a critical residual analysis avoids such a mistake and such a treatment is now included in most mean square estimation programs.

4.2.3   Estimation of quality of fit

To estimate the quality of the fit between the experimental data and the theoretical model, one often uses the residual mean square,

$$\hat{\sigma}_y^2 = \frac{1}{n-2}\,Se = \frac{1}{n-2}\sum\limits_{i=1}^{n} (y_i - \hat{a}x_i - \hat{b})^2 \qquad \text{(IV.40)}$$

whose square root defines the standard deviation associated with each y_i.

Another quantity used to evaluate the quality of fit is the coefficient of determination, r^2, defined by

$$r^2 = \hat{a}^2\,\frac{S_{xx}}{S_{xy}} = \frac{S_{xy}\cdot S_{xy}}{S_{xx}\cdot S_{yy}} \qquad \text{(IV.41)}$$

where

$$S_{yy} = \sum\limits_{i=1}^{n} y_i^2 - \frac{1}{n}\left(\sum\limits_{i=1}^{n} y_i\right)^2 \qquad \text{(IV.42)}$$

The value r defines the correlation coefficient between x_i and y_i; its value approaches 1 when x_i and y_i are linearly related. A value close to 1 does not necessarily imply, however, that a valid linear fit has been realized. One must remember that extending the range of the x values may yield a value of r closer to unity even though the model is non-linear (see ref. [27]).

A more trustworthy approach is provided by the statistical tests which allow a sensible estimation of the confidence interval associated with the values of $\hat{a}$ and $\hat{b}$ extracted from the least square approach. One may show that the variance of $\hat{a}$ and $\hat{b}$ are given respectively by:

$$\hat{\sigma}_a^2 = \frac{\hat{\sigma}_y^2}{S_{xx}}$$ (IV.43)

and

$$\hat{\sigma}_b^2 = \hat{\sigma}_y^2 \left(\frac{1}{n} - \frac{\bar{x}^2}{S_{xx}} \right) = \hat{\sigma}_y^2 \left(\frac{\sum\limits_{i=1}^{n} x_i^2}{S_{xx}} \right)$$ (IV.44)

Linear regression programs yield a confidence interval for $\hat{a}$ and $\hat{b}$ based on the assumption that the distribution of scatter $\hat{a} - a$ and $\hat{b} - b$ (where a and b are the true values given by the theoretical model) follow Fisher's 't' distribution with $n - 2$ degrees of freedom. One usually retains a 95% confidence interval for classical kinetic studies, i.e.,

$$\hat{a} - (t_{(1-0.95)/2, n-2})\sqrt{\hat{\sigma}_a^2} \le a \le \hat{a} + (t_{(1-0.95)/2, n-2})\sqrt{\hat{\sigma}_a^2}$$ (IV.45)

$$\hat{b} - (t_{(1-0.95)/2, n-2})\sqrt{\hat{\sigma}_b^2} \le b \le \hat{b} + (t_{(1-0.95)/2, n-2})\sqrt{\hat{\sigma}_b^2}$$ (IV.46)

This 95% value means that we have a 95% probability of the 'true' value being located between the boundaries obtained through these formulae. The 't' value, symbolized as $t_{(1-0.95)/2, n-2} = t_{0.025, n-2}$ in our example, may be found in tables in most statistics books [26, 27]. The usual table entries are the number of degrees of freedom and the probability of being outside the interval. (Thus the 95% probability for confidence interval implies the use of $t_{0.05}$. Because of the symmetry of Fisher's 't' distribution, we use, in practice, twice $t_{0.025}$.

A small confidence interval is an interesting criterion because it provides a precision for the least squares estimators $\hat{a}$ and $\hat{b}$, and therefore an estimation of the goodness of the regression analysis.

The application of the least squares method implies, however, that the errors associated with each y measurement follow a normal distribution whose mean equals 0. Therefore, a constant variance is assumed for these errors and the measurement precision of y_i is assumed independent on the x_i values. The same holds true for $y_i - \hat{a}x_i - \hat{b}$. We must determine whether this normal distribution is well followed for the set of experimental data examined. This may be done by constructing a histogram of the residues or by more elaborate approaches [25]. If the residues are plotted versus x_i or y_i they must be uniformly scattered in a band. If such is not the case, one begins to suspect that the linear model is either not the best suited model or that the variance of measurements is not constant.

4.2.4   Linear transformation

We now return to the example treated in section 4.2.1. The initial model after, rearrangement is

$$[A] = \frac{1}{k \cdot t + \dfrac{1}{[A]_0}} \tag{IV.47}$$

Therefore when $[A]$ is measured, the best fit is related to $1/[A]$ and the statistical properties of the errors discussed in the foregoing treatment are connected to $1/[A]$ and not to $[A]$. This situation biases the statistical treatment of the residues. As a consequence, $\hat{a}$ (k) and $\hat{b}$ $(1/[A]_0)$ will not be the best estimators. A more rigorous approach may be adopted in order to determine the estimators which are associated with the postulated model when the error distribution of the measured parameters is known. This approach is called the method of maximum likelihood estimators [11, 26, 27, 34, 35]. If we assume that errors follow a normal distribution, this method gives the identical expressions for the estimators that the mean squares treatment gave.

The weighted least squares method associates a weight w_i with every measurement. It is to be used when the variance of errors of the variable object is not a constant. One minimizes

$$Se = \sum_{i=1}^{n} w_i(y_i - \hat{a}x_i - \hat{b}) \tag{IV.48}$$

by choosing the variance of y_i for every $1/w_i$. This requires either several measurements of every y to allow an estimation of σ_{y_i}, or to find a relationship between σ_{y_i} and x_i [29]. Dye and Nicely, [36] in their 'General Purpose Curve Fitting Program' give an interesting estimation of the weights for the specific treatment of kinetic data.

4.2.5   Other mean square applications

The least squares method is also used when the expression to be fitted is polynomial. This kind of polynomial regression must always be used with caution. One must realize at the outset that it is always possible to exactly fit a polynomial of $n - 1$ degrees on n distinct points. This fit, however, may not be significant if the precision of measurements does not allow the consideration of the higher terms of the polynomial development.

When a data function of k independent variables may be expressed as:

$$y = a_0 + a_1 x_1 + a_2 x_2 + \cdots + a_k x_k \tag{IV.49}$$

the linear regression must be generalized into a multilinear regression. The estimation of a_i is possible provided that one has access to n measurements for y, each with a different set of independent variables.

These two methods are used in chemical engineering, in data analysis, and for experimental design. The choice of the data set is particularly critical to the significance of the regression [37–39].

Other types of regressions, rather than relying on the least squares criterion, rest on other criteria connected with the distribution of experimental errors. These methods are justified by the maximum likelihood estimation theory [27, 40].

4.3   Non-linear regression

The problem of data analysis grows in difficulty when the kinetic model can no longer be expressed as a linear relationship. The mathematical definition of linearity is as follows: A function $f(a, x)$ is linear with respect to a if $\partial f/\partial a$ is independent of a. Three main methods have been programmed to deal with non-linear kinetic expressions; all of them use iterative procedures and, as such, require more powerful data processing. They also share another feature: one must input initial values for the estimation of the unknown parameters. The process involves the minimization of the residual sum of squares as defined for a linear regression [41]. Bard [42, 43] has reviewed in detail this method, and an annotated bibliography has recently been published by Nash [44].

4.3.1   Linearization of the residue

In this method one develops the residue as a Taylor series including only the first 2 terms. Starting from $f(a, b, x)$, which describes the concentration time dependence of a given species, one retains the following function to be minimized:

$$Se = \sum_{i=1}^{n} [R(a, b, x_i)]^2 \tag{IV.50}$$

where x_i is the measurement time, and

$$R(a, b, x_i) = y_i - f(a, b, x_i),$$

where y_i is the concentration measured at the time x_i.

This minimization is feasible if one assumes that σ_{y_i} is a constant and takes a_0 and b_0 as the initial values for the a and b values. One obtains:

$$R(a, b, x_i) = R(a_0, b_0, x_i) + (a - a_0) \cdot R'_{a_0}(a_0, b_0, x_i)$$
$$+ (b - b_0) \cdot R'_{b_0}(a_0, b_0, x_i) \tag{IV.51}$$

where

$$R'_{a_0}(a_0, b_0, x_i) = \frac{\partial R(a_0, b_0, x_i)}{\partial a_0}$$

and

$$R'_{b_0}(a_0, b_0, x_i) = \frac{\partial R(a_0, b_0, x_i)}{\partial b_0}$$

The residue is now a linear function of a and b in the vicinity of the values a_0 and b_0. Taking the criteria that $\partial Se/\partial a = 0$ and $\partial Se/\partial b = 0$, one obtains

$$2 \sum_{i=1}^{n} R(a, b, x_i) \frac{\partial R(a, b, x_i)}{\partial a} = 0$$

$$2 \sum_{i=1}^{n} R(a, b, x_i) \frac{\partial R(a, b, x_i)}{\partial b} = 0$$

(IV.52)

where $R(a, b, x_i)$ may be replaced by a limited development, which yields:

$$\sum_{i=1}^{n} [R_i + (a - a_0) \cdot A_i + (b - b_0)B_i] \cdot A_i = 0$$

(IV.53)

$$\sum_{i=1}^{n} [R_i + (a - a_0) \cdot A_i + (b - b_0)B_i] \cdot B_i = 0$$

where

$$R_i = R(a_0, b_0, x_i)$$

$$A_i = R'_{a_0}(a_0, b_0, x_i)$$

$$B_i = R'_{b_0}(a_0, b_0, x_i)$$

provided that

$$\frac{\partial R(a_0, b_0, x_i)}{\partial a_0} \simeq \frac{\partial R(a, b, x_i)}{\partial a}$$

$$\frac{\partial R(a_0, b_0, x_i)}{\partial b_0} \simeq \frac{\partial R(a, b, x_i)}{\partial b}$$

(IV.54)

This system of two linear equations provides approximate solutions, which are examined again in the system of equations to arrive at solutions even closer to the 'truth'. These steps are repeated until convergence occurs.

Johnson [30] described this method in detail and wrote the associated resolution program. The program has been used in a general purpose simulation program [36] and rapidly converges if the residues are not too far from linearity. The advantage of the method is that it provides an estimation of the variance for the parameters determinated. Its use, however, demands the knowledge of the expressions for $R'(a, b, x_i)$.

4.3.2 Residue optimization by the gradient method

This method is useful if the expression for the derivatives of the function whose minimum is searched for presents itself in an analytical form. One searches for a minimum for the function:

$$Se = \sum_{i=1}^{n} [y_i - f(a, b, x_i)]^2$$

(IV.55)

or for any other expression associated with the residues provided that $\partial Se/\partial a$ and $\partial Se/\partial b$ may be obtained.

Beginning again from initial input values, a_0 and b_0, one attains better values through:

$$a = a_0 + \varepsilon \frac{\partial Se}{\partial a}$$

$$\text{(IV.56)}$$

$$b = b_0 + \varepsilon \frac{\partial Se}{\partial b}$$

where ε is chosen empirically.

One common drawback of this method is the presence of oscillations near the minimum; recent improvements have minimized this. Beech [29] has written a program which treats data according to this method. (See also the general purpose program published by Ertz in *QCPE* [45].)

4.3.3 Residue optimization by the direct search method

When no analytical expressions are available for the derivatives of the function whose minimum is searched for, more direct methods may be applied. These rely on the search for the directions of the space leading to minimum values of the function investigated [46–48].

Several improvements have been made which minimize the problem of oscillations in the vicinity of the edges. Rosenbrock's method [49, 50] works according to these principles.

Direct search methods may be even used for functions originating from a *numerical integration*. For example, rather than optimizing:

$$Se = \sum_{i=1}^{n} [y_i - f(a, b, x_i)]^2 \qquad \text{(IV.57)}$$

one may use the function

$$Se = \sum_{i=1}^{n} (y_i - f_i)^2 \qquad \text{(IV.58)}$$

where f_i is the result of one integration by a numerical method involving parameters a, b, x_i when the function $f(a, b, x_i)$ does not exist under an analytical form.

This possibility has been exploited in Klaus [35] and Dye and Nicely's programs [36]. One must stress that as in the gradient methods, the direct search method may converge toward a local maximum. To avoid this risk one may begin by using different initial parameters.

4.4 Conclusion

We have seen that the kinetic parameters encountered in this section are described by simple algebraic equations. The adequate computerized treatment is one of regression analysis. Depending on the algebraic equation, methods vary from simple linear regression to function optimization. The chemist can use programs dedicated to kinetic applications described in specialized books [29–33] and reviews such as *QCPE* [51–53]. One can also use general purpose calculation packages available from computer manufacturers [10] or other scientific laboratories [54].

Although these programs are written in FORTRAN, those with a few or no iterative calculations can be adapted to a microcomputer. Unfortunately, non-linear optimization packages need a memory storage and execution speed now available minimally on at least a minicomputer or on a top-level 32-bit desktop computer. Note also that non-linear optimization problems are very critical in chemical engineering, and several articles have been recently published in a special issue of *Computer and Chemical Engineering* [55].

IV.5   KINETIC SIMULATION BY NUMERICAL INTEGRATION

5.1   Introduction

In sections IV.3 and IV.4 we have dealt with the determination of rate constants and reaction order when an analytical expression representing the course of the reaction was available. What can be done when such an expression is not available? The solution rests in numerical integration methods. The classical reaction scheme:

$$A + B \underset{k_{-1}}{\overset{k_1}{\rightleftarrows}} C \overset{k_2}{\rightarrow} D + E \tag{IV.59}$$

demands such an approach.

Knowing the concentration of various species at $t = 0$, we need to calculate the values of these concentrations for discrete values of t, beginning with

$$\frac{d[A]}{dt} = \frac{d[B]}{dt} = -k_1 \cdot [A] \cdot [B] + k_{-1} \cdot [C]$$

$$\frac{d[C]}{dt} = k_1 \cdot [A] \cdot [B] - k_{-1} \cdot [C] - k_2 \cdot [C] \tag{IV.60}$$

$$\frac{d[D]}{dt} = \frac{d[E]}{dt} = k_2 [C]$$

or other differential equations according to the elementary reaction scheme. We must, therefore, solve numerically the foregoing system of differential equations with initial values. This kind of problem is quite general and is found in areas such as heat transmission and flow calculations. This probably explains the reason why the best treatments are found in the area of chemical engineering.

5.2   Result of a numerical integration

To make our explanation simpler, we describe the principle of the method for a single differential equation. Matrix notation allows a straightforward extension to systems of several equations. In our case, y is a function of an independent variable x, where the analytical function, $y(x)$, is unknown but for which we know

— the initial value $y_0 = y(x_0)$ and
— the analytical expression of the derivative of y with respect to x,

$$y'(x, y) = \frac{dy}{dx} = g(x, y) \qquad \text{(IV.61)}$$

(Note that in kinetic treatments, x and y are respectively time and concentration. Owing to the form of the rate laws, we have $dy/dx = g(y)$, which is a particular case.)

For a set of discrete values for:

$$x_0, x_1, x_2, \ldots, x_n$$

we wish to find numerical values of

$$y_1, y_2, \ldots, y_n$$

close to the 'true' value of y

$$y(x_1), y(x_2), \ldots, y(x_n)$$

The difference $y(x_i) - y_i$ is the truncation error or 'true error'. It results from the fact that the numerical method is only approximate. This error is unknown, unless we are dealing with a differential equation whose analytical solution exists, but some methods allow estimation of the error.

5.3 Integration methods

5.3.1 Euler's method

This is a first order, one step method which involves the estimation of $y(x_i)$ using a Taylor series truncated to the first term:

$$y(x_i) \simeq y_i = y_{i-1} + h \cdot y'(x_{i-1}, y_{i-1}) \qquad \text{(IV.62)}$$

where $h = x_{i+1} - x_i$ is the integration step size.

The programming of this method is very simple and it may be found in many tutorial applications [56].

5.3.2 Other single step methods

These methods are said to be k-order methods if one can show that the overall error is at most equal to λh^k, where λ is a constant and h is the integration step size.

If $g_h(x, y)$ is a function such that as h approaches zero, $g_h(x, y)$ approaches $g(x, y) = y'(x, y)$.

One estimation of $y(x_i)$ may be attained by using

$$y(x_i) \simeq y_i = y_{i-1} + h \cdot g_h(x_{i-1}, y_{i-1}) \qquad \text{(IV.63)}$$

The fourth order Runge–Kutta method allows such an approach:

$$g_h(x_j, y_j) = \tfrac{1}{6}(K_1 + 2K_2 + 2K_3 + K_4) \qquad \text{(IV.64)}$$

where

$$K_1 = g(x_j, y_j),$$

$$K_2 = g\left(x_j + \frac{h}{2}, y_j + \frac{h \cdot K_1}{2}\right),$$

$$K_3 = g\left(x_j + \frac{h}{2}, y_j + \frac{h \cdot K_2}{2}\right),$$

and

$$K_4 = g(x_j + h, y_j + h \cdot K_3)$$

Such a method is said to be explicit because the calculation of y_i involves only already calculated data and is equivalent to the calculation of an expression.

Some authors have used implicit Runge–Kutta methods [35, 57, 58]. In these implicit methods, the calculation of y_i involves the resolution of a non-linear equation (or system of equations) such as $y_i = F(y_i)$ using a numerical method.

5.3.3 Multistep methods

These methods use the values of several preceding points to calculate the next point. The formula used, called corrector is:

$$y_i = \sum_{j=1}^{l} a_j \cdot y_{i-j} + h \sum_{j=0}^{m} b_j \cdot g(x_{i-j}, y_{i-j}) \qquad \text{(IV.65)}$$

In this formula, since the term b_0 is always positive, the methods are implicit methods and require an approximative value y_i. This value is obtained by a similar formula, the predictor, in which $b_0 = 0$ (i.e., is explicit). These methods are sometimes called predictor corrector methods. The two best known are

— Adam's k order method derived from formula (IV.65) setting $l = 1$, $m = k - 1$, and

— the backward differentiation formula derived from (IV.65) setting $l = k$, $m = 0$.

5.3.4 Extrapolation methods

Comparing the results obtained for a point calculated using one of the foregoing methods (either single or multistep) with different values for the step size, a better value for this point is attained.

5.4 Accuracy of the numerical integration

The preceding methods are completed by the calculation of the estimated truncation error. If this error is greater than a value fixed by the user, the integration step size is automatically corrected. The simplest programs [10] estimate only the value of the *local error*, whereas the user is generally more interested by the *global error* or 'true error'. This global error is defined as:

$$y(x_i) - y_i \qquad \text{(IV.66)}$$

or if y is a vector

$$\|y(x_i) - y_i\|$$ (IV.67)

whereas the local error for point i is

$$u(x_i) - y_i$$ (IV.68)

or if y and v are vectors

$$\|u(x_i) - y_i\|$$ (IV.69)

in which y_i is the exact solution of another function $u(x)$ such that:

$$u'(x_{i-1}, u_{i-1}) = y'(x_{i-1}, u_{i-1})$$
$$u(x_{i-1}) = y_{i-1}$$ (1IV.70

Contenting oneself with the approximation of the local error is often justified by the fact that studied chemical phenomena are naturally stable. The more rigorous treatment must consider the eigenvalues associated with the Jacobian $J = dy'/dy$ whose real parts must be negative [59].

5.5   Stiff systems

One system of differential equations,

$$y'(x, y) = g(x, y)$$
$$y(x_0) = y_0$$ (IV.71)

in which y and y' are vectors of l elements, has the Jacobian such that the x_i value of the independent variable is

$$J_i = \frac{dy_i'}{dy} = \left(\frac{dy_{i,j}'}{dy_{i,k}}\right)^l_{j,k=1}$$ (IV.72)

which, developed, becomes

$$J_i = \begin{bmatrix} \dfrac{dy_{i,1}'}{dy_{i,1}} & \dfrac{dy_{i,1}'}{dy_{i,2}} & \cdots & \dfrac{dy_{i,1}'}{dy_{i,l}} \\[2mm] \dfrac{dy_{i,2}'}{dy_{i,1}} & \dfrac{dy_{i,2}'}{dy_{i,2}} & \cdots & \dfrac{dy_{i,2}'}{dy_{i,l}} \\[2mm] \vdots & \vdots & & \vdots \\[2mm] \dfrac{dy_{i,l}'}{dy_{i,1}} & \dfrac{dy_{i,l}'}{dy_{i,2}} & \cdots & \dfrac{dy_{i,l}'}{dy_{i,l}} \end{bmatrix}$$ (IV.73)

We assume the condition given in section 5.4, namely that the eigenvalues $\lambda_{i,j}$ of this matrix all display a negative real part. This system of equations is said to be stiff for a given value of x_i if the real parts of the eigenvalues are very different. One quantitatively defines the stiffness as S_i:

$$S_i = \frac{\max \operatorname{Re}(-\lambda_{i,j})}{\min \operatorname{Re}(-\lambda_{i,j})}$$ (IV.74)

Stiffness is encountered for reaction schemes involving steps whose rate constants differ greatly. In such cases, normal integration methods provide poor precision and the step size must be chosen to be very small even during the time when concentrations change very slowly. To avoid an overly time-consuming approach, a specially designed method has to be used. This is especially true when S_i is not a constant over the whole range of time studied (which is usually the case in kinetic studies).

5.6   Choice of the method

The most used programs rely on one of the following methods:

— Runge–Kutta
— Backward differentiation formula
— Adam's
— Extrapolation

The backward differentiation formula is the most efficient for the resolution of stiff systems. The first program using these methods is described in Gear [60, 61]. More recent versions have appeared, such as Fac-Simile [62] and Episode [63].

Comparative studies allow the estimation of the relative efficiencies of these programs [64–66]. Shampine and co-workers [67, 68] recently published a critical review of these methods. One of the important conclusions of this review is that the variable order Runge–Kutta method is usually better than extrapolation methods and, for some cases, as good as Adam's method. Adam's method remains the most efficient but demands greater storage and computation times.

5.7   Implementation of these methods on a digital computer

These numerical methods are computer time-consuming and as such cannot usually be implemented on microcomputers unless they use a compiler as programming language and an arithmetic processor. A trial on a Z80 2-MHz microcomputer using a program in BASIC of a 4th order explicit Runge-Kutta method with a variable step size related to the local error, yields a point every 3 to 5 s for the simulation of the scheme:

$$A \underset{k_{-1}}{\overset{k_2}{\rightleftharpoons}} B \overset{k_3}{\rightarrow} C \tag{IV.75}$$

which involves only 3 differential equations [69].

Furthermore, the user must include in the program a module which computes the derivatives for different values of time and the estimated concentrations, i.e., for $y'(x, y)$. This module, usually a subroutine, has to be modified each time the reaction scheme is modified. This constraint decreases the usefulness of a language such as BASIC, where all the variables are global variables, i.e., accessible from the outside of the subroutine using them (this holds true particularly for BASIC working under CP/M). Therefore, to write the module 'computation of derivatives' one needs to take care to avoid the choice of variables already used by the module 'integration'. In such a case the user must study the integration program or possess detailed documentation describing the names and the roles of the different variables used in the program.

All these drawbacks disappear if one selects a programming language which allows the construction of independent modules. FORTRAN and PASCAL both offer this option. The classical routines of integration in such languages are indeed available in most scientific subroutine libraries. As they are in subroutines, data are transmitted under the form of an argument list representing not only numerical values (initial concentrations, time, integration step size, number of equations etc.) but also the subroutine names (calculation of derivatives and output of results during the calculation). The subroutine 'numerical integration' includes a repetitive part which demands, at every step, several derivative calculations and the storage of the integration result. These two operations, being specific to the problem, must be dealt with by the user, who must write two specific subroutines whose parameters are those needed by the integration routine. We therefore draw attention to the point that the use of commercial integration subroutine libraries implies the knowledge of a programming language, FORTRAN in most cases. The user will also have to write the main program, which reads the data, and to call the integration subroutine. Since a part of these data must also be accessible for the subroutines in charge of computing derivatives and the output of results, this demands the use of COMMON areas (in FORTRAN) or global variables (in PASCAL).

5.8   Use of a complete integration package

Some programs specifically designed to treat systems of differential equations describing kinetic models have been published [36, 70, 71]. To make their utilization easier, some logical analysis of the reaction scheme must be given as input following specific rules. This allows one to include in the package one unique module for the computation of derivatives. The user therefore, no longer has any need to worry about this task; the price is paid usually in longer computation time. This kind of program is interesting if it exactly suits the kinds of kinetic problem studied. Its modification for the development of a specific application demands a complete analysis of the package.

5.9   Numerical integration and other modelizations

All the foregoing sections have focused on treatments involving only the solution of systems of differential equations. If the simulation involves other phenomena such as mass balance, thermodynamic function calculation etc., these phenomena will have to intervene during the repetitive sequence of numerical integration described in section 5.7. Practically, this precludes the use of common integration routines. This situation, usually encountered in chemical engineering, has been well treated in several specifically written programs.

Franck's program [72] is composed of several dozen FORTRAN subroutines communicating through named COMMON. The simulation of chemical process has also given birth to specifically designed languages: FLOWTRAN [73], DYNSYL [74], GRIP [75], and Come's [58] and Edelson's [76] systems are well known.

Another approach simulates, on a digital computer, the complete work of an analog calculator. The best known modeling programs are those developed by IBM and CDC [77, 78]. These programs use a specific language, making them particularly flexible and easy to use. Several examples using the IBM CSMP language are described in *Computer Programs for Chemical Engineering Education* [79]. Recently, Titchener [80] has described a kinetics-oriented program which emulates an analog computer on a microcomputer, using assembly language-coded modules.

5.10 Stochastic methods

Monte Carlo methods may be used in place of numerical integration. These simulate the reaction vessel with an array of integer values, each identifying one of the species involved. The number of locations with identical contents is proportional to the concentration of the corresponding species. They are also randomly distributed in the array. A drawing of lots leads, if favorable, to the transformation of reagents into products. In the same way the Markov chain model has been used.

These methods, although attractive, are at this point used mainly for pedagogical purposes [81–83] in order to simulate a kinetic scheme already solved by a numerical method. The random number generation used repeatedly consumes too much time in the present generation of computers. These methods are, however, fairly efficient in the treatment of statistical thermodynamics.

IV.6   CONCLUSION

Computational applications have always been very important in physical chemistry, and chemical kinetics is no exception. On the other hand, we can see that computerized kinetic studies largely overlap through numerical methods, including the automatic acquisition of data, and the display of results, graphics representation, and so on.

We hope that the tendency for the evolution of integrated software packages and the new generations of computer hardware will handle all steps of kinetic studies, including the manipulation of algebraic expressions, and will foster new insights. Indeed, if some systems exist to day [84, 85] they are implemented on medium size computers and the physical chemist can foresee in the future the disposal of all these tools in a desktop calculator, right next to the measurement installations in his laboratory.

IV.7 REFERENCES

[1] G. G. Hammes, in *Principles of Chemical Kinetics*, Academic Press, New York (1978).

[2] R. K. Boyd, *Chemical Reviews*, **77**(1), 93 (1977).

[3] V. B. Evdokinov, *Russian Journal of Physical Chemistry*, **57**(9), 1297 (1983).

[4] F. Jouget, *J. Ecole Polytechn.*, **21**(1) (1921).

[5] S. R. Brinkley, *J. Chem. Phys.*, **15**, 563 (1947).

[6] G. N. Copley, in *Chemistry*, **41**(9), 22 (1968).

[7] H. P. Meissner, C. L. Kusik and W. H. Dalzell, *Ind. Eng. Chem. Fundom.*, **8**, 659–665 (1969).

[8] J. M. Comes, in *The Use of Computers in Analysis and Simulation of Complex Reactions*, (C. H. Banford and C. F. H. Tipper, eds.), Vol XIV: *Modern Methods in Kinetics*, Chap. 3, Elsevier (1983).

[9] G. R. Blackley, *Chem. Educ.*, **59**(9), 728 (1982).

[10] IBM, System 360 Scientific Subroutine Package.

[11] R. V. Kafarov and V. N. Pisarenko, *Usp. Khim.*, **49**, 193–222 (1980).

[12] J. L. Aime, University Thesis, Marseille (1984).

[13] R. S. Upadhye, *Comp. and Chem. Eng.*, **7**, 87 (1983).

[14] J. Happel and P. H. Sellers, *Adv. Catal.*, **32**, 281 (1983)

[15] Willermaux, in *Genie de la reaction chimique. Conception et fonctionnement des reacteurs*, Technique et Documentation, Paris (1982).

[16] A. G. Turnbull, *Computer Coupling Phase diagrams and thermochemistry*, **7**(2), 137 (1983).

[17] R. Derieux, A. Peneloux, M. Beerli, J. R. Llinas and G. Loridan, *Revue de Chimie Minerale*, **9**, 163–78 (1972).

[18] D. Alron, G. M. Comes, P. Y. Cunin and M. Griffiths, *Comp. and Chem. Eng.*, **3**, 87–89 (1979).

[19] J. L. Ayme, M. Arbelot, M. Chanon, *Comp. and Chemistry*, (to be published).

[20] G. Pannetier and P. Souchay, in *Cinetique Chimique*, Masson, Paris (1961).

[21] A. A. Frost and R. G. Pearson, in *Kinetics and Mechanism*, John Wiley, New York (1961).

[22] N. Emanuel and D. Knorre, in *Cinetique Chimique*, Editions Mir, Moscow (1975).

[23] S. N. Benson, in *Foundation of Chemical Kinetics*, McGraw-Hill, New York (1960).

[24] F. S. Lewis, in *Investigation of Rate and Mechanism of Reactions, Technique of Organic Chemistry*: Vol. IV: J. F. Bunnet, Vol. VII: A. Weissberger, New York (1974).

[25] B. Carnahan, H. A. Luther and J. O. Wilkes, in *Applied Numerical Methods*, John Wiley, New York (1969).

[26] D. M. Himmelblau, in *Process Analysis by Statistical Methods*, John Wiley, New York (1970).

[27]   D. C. Montgomery and F. A. Peak, in *Introduction to Linear Regression Analysis*, John Wiley, New York (1982).

[28]   A. Gourdin and M. Boumahrat, in *Methodes numeriques appliquees*, Technique et Documentation, Paris (1983).

[29]   G. Beech, in *Fortran IV in Chemistry*, John Wiley, London (1975).

[30]   R. J. Johnson, in *Numerical Methods in Chemistry*, Marcel Dekker, New York, (1980).

[31]   D. F. Detar, in *Computer Programs for Chemistry*, Academic Press, New York (1972).

[32]   D. F. Detar, in *Computer Programs for Chemistry*, Academic Press, New York (1972).

[33]   T. I. Isenhour and P. C. Jurs, *Introduction to Computer Programming for Chemists*, Allyn and Bacon, Boston (1972).

[34]   R. H. Fariss and V. H. Law, *Comp. and Chem. Eng.*, **3**, 95 (1979).

[35]   R. Klaus and D. W. T. Rippin, *Comp. and Chem. Eng.*, **3**, 105 (1979).

[36]   J. L. Dye and V. A. Nicely, *J. Chem. Educ.*, **48**(7), 443 (1971).

[37]   V. V. Fedorov, in *Theory of Optimal Experiments*, Academic Press, New York (1972).

[38]   L. Endrenyi, in *Kinetic Data Analysis. Design and Analysis of Enzyme and Pharmaco-kinetic Experiments*, Plenum Press, New York (1981).

[39]   D. J. Currie, *Biometrics*, **38**, 907 (1982).

[40]   P. J. Huber, in *Robust Statistics*, John Wiley, New York (1981).

[41]   D. M. Himmelblau, in *Applied Non-Linear Programming*, McGraw-Hill, New York (1972).

[42]   Y. Bard, in *Non-Linear Parameter Estimation and Programming*, IBM New York Scientific Center (1967).

[43]   Y. Bard, in *Non-Linear Parameter Estimation*, Academic Press, New York (1974).

[44]   J. C. Nash, in *An Annotated Bibliography on Methods for Non-Linear Least Squares Computation*, Mary Nash Information Services, Vanier, Ontario (1976).

[45]   N. F. Ertz, QCPE. **11**, 283, (1975).

[46]   R. Hooke and T. A. Jeeves, *J. Assoc. Comp. Mach.*, **8**, 212 (1962).

[47]   M. J. D. Powell, *Computer J.*, **7**, 155 (1964).

[48]   M. J. D. Powell, *Computer J.*, **7**, 303 (1965).

[49]   H. H. Rosenbrock, *Computer J.*, **3**, 175 (1960).

[50]   H. H. Rosenbrock and C. Sturey, in *Computational Techniques for Chemical Engineers*, Pergamon Press, Oxford (1966).

[51]   R. W. Crossley and E. A. Dorto, *QCPE*, **11**, 179, (1971).

[52]   H. K. Pinnick, *QCPE*, **13**, 394, (1981).

[53]   G. R. Howe, *QCPE*, **11**, 175, (1971).

[54]   J. P. Chandler, *QCPE*, **11**, 307, (1970).

[55]   Large scale optimisation, *Comp. and Chem. Eng.*, **7**(5), (1983).

[56]   A. Perche, Tutoriel d'aide a l'integration numerique, programme 45, Centre documentaire cooperatif informatique—enseignement de la chimie, Université de Nice, 06034—Nice Cedex (1983).

[57]   M. L. Michelsen, *A. I. Ch. E. J.*, **22**, 594–597 (1970).

[58]   G. M. Come, C. Bourlier and C. Guillerm, *J. Chem. Phys.*, **72**, 123 (1975).

[59]   A. Prothero, Estimating the accuracy of numerical solutions to ordinary equations, Conference paper, Manchester (UK), 1978.

[60]   C. W. Gear, in *Numerical Initial Value Problems in Ordinary Differential Equations*, Prentice-Hall, Englewood Cliffs, New Jersey (1971).

[61]   A. C. Hindmarsh and C. W. Gear, Ordinary Differential Equation System Solver, UCID—3001, Rev 2, Lawrence Livermore Laboratory, University of California, Livermore, California (1972).

[62]   E. M. Chance, A. R. Curtiss, I. P. Jones and C. R. Kimby, Fac-Simile, a Computer Program for Flow and Chemistry Simulation, and General Initial Value Problems, AERE Report R. 8775 (1977).

[63]   G. Byrne and A. Hindmarsh, *TOMS*, **1**(1), 79–96 (1975).

[64]   G. Byrne, A. Hindmarsh, K. Jackson and H. Brown, *Comp. and Chem. Eng.*, **1**, 133–147 (1977).

[65]   A. R. Curtis, The Fac-Simile Numerical integrator for Stiff Initial Value Problems, AERE Report R. 9352 (1979).

[66]   W. Enright and T. E. Hull, Comparing numerical methods for the solution of stiff systems of ODE's arising in chemistry, in *Numerical Methods for differential Systems* (L. Lapidus and W. E. Schiesser, eds.), Academic Press, New York (1976).

[67]   L. F. Shampine, What Everyone Solving Differential Equations Numerically shoud Know, Conference paper, Manchester (UK), (1978).

[68]   L. F. Shampine, Variable Order Range-Kutta codes, Sandia Lab. Rept. Sand 78-1652, Albuquerque, New Mexico (1979).

[69]   J. L. Larice, RK2, Programme 48, Centre documentaire cooperatif informatique—enseignement de la chimie, Université de Nice, 06034–Nice Cedex (1983).

[70]   G. D. Byrne, *Comp. and Chem. Eng.*, **5**, 151 (1981).

[71]   Spanget Larsen, *QCPE*, **11**, 433 (1977).

[72]   R. G. E. Franks, in *Modeling and Simulation in Chemical Engineering*, John Wiley, New York (1972).

[73]   F. M. Rosen and A. C. Pauls, *Comp. and Chem. Eng.*, **1**, 11–21 (1978).

[74]   G. K. Patterson and R. B. Rozso, *Comp. and Chem. Eng.*, **4**, 1–20, (1980).

[75]   R. W. Hilst and K. A. Bishop, *Comp. and Chem. Eng.*, **5**, 249–262 (1980).

[76]   D. Edelson, *Comp. and Chem. Eng.*, **1**, 29–33 (1976).

[77]   *CSM Program User Manual*, IBM Corporation, White Plains, (NY) Application Program GH 20 0367-4 (1971).

[78]   *MIMIC, A Digital Simulation Language Reference Manual*, Control Data Corp., Minnesota, Publ. No. 44610400, Rev. D.

[79]   M. Reilly, in *Computer Programs for Chemical Engineering Education*, Vol. II, *Kinetics*, Aztec Publishing Co., Austin, Texas (1972).

[80]   M. R. Titchener, *J. of Comp. Chemistry*, **3**(2), 95 (1982).

[81]   A. F. Para and E. Lazzarini, *J. Chem. Educ.*, **51**(5), 336–342 (1971).

[82]   D. L. Bunker and F. A. Houle, *QCPE*, **11** (1976).

[83]   M. Carlier and A. Perche, Program No. 52, Centre documentaire cooperatif informatique—enseignement de la chimie, Université de Nice, 06034-Nice Cedex (1983).

[84]   M. E. Hangak, *Comp. and Chem. Eng.*, **4**, 223–239 (1980).

[85]   T. G. Koop, E. H. Chimowitz, A. Blouz and I. F. Stutzman, *Comp. and Chem. Eng.*, **5**, 151 (1981).

CHAPTER V

Multivariate Analysis of Chemical Data Sets with Factorial Methods

J. R. Llinas, ESCM and J. M. Ruiz, ESIPSOI

University of Aix-Marseille III, France

V.1 INTRODUCTION

Multidimensional data analysis plays an ever increasing role in many scientific disciplines, including many in the earth sciences, life sciences, sociological sciences and in business management. In chemistry, however, these methods have not exhibited as rapid a development. Although the fundamental basis of the methods was established at the beginning of the century [1–2] and applications of the methods were introduced during the 1930s [3–4], the first applications to chemistry occurred only in the 1960s. Actually, Factor Analysis (FA), Principal Component Analysis (PCA) and Factor Discriminant Analysis (FDA) have become commonly used methods in chemometrics.

The purpose of chemometrics [5, 6] is twofold:

— To provide maximum information through analysis of chemical data
— To design optimal measurement procedures and experiments.

The first aim may be divided into two parts:

(1) The description, classification and interpretation of chemical data
(2) The modelling of chemical experiments, processes and, further, their optimization.

This chapter deals with those multidimensional data analysis methods which may help the chemist in the first part. Their usage has been

200

continuously growing for the following reasons:

— The ever-increasing amount of data which must be treated as a consequence of the progress made in the fields of analytical measurements, including automation and computerization.
— The difficulty in providing for complex chemical experiments and processes good fundamental models which can reproduce experimental results exactly.
— The increasing availability of computers and software tools which can deal with multivariate problems.

Among the various aspects concerning the treatment of chemical data sets, some characteristic applications may be outlined.

— Multicomponent analysis from spectrometric or chromatographic data measured for different mixtures. The goal is the determination of the number of components and, sometimes, identification as well. Factor analysis has been used to solve problems in solution equilibria and complex kinetics.
— The search for non-measurable factors governing physico-chemical properties which would be far too complex to model with comprehensive such as, for example:

(1) chromatographic retention times
(2) NMR chemical shift data
(3) equilibrium and kinetic constants
(4) reaction conversion and selectivity data

The interpretation of these factors may highlight new phenomena or point out physical properties which will help to explain the initial observations.

— The reduction of a chemical data set with a large number of variables (which are often correlated and sometimes redundant) to a smaller one with independent variables. Each observation will be characterized by a smaller number of new variables which can later be useds for modelling studies. This method may be used for natural products with a complex composition and many physico-chemical properties (essential oils, crude petroleum products, etc.) and also for data sets measured during a process.
— The analysis of a multidimensional chemical data set with graphical representations of objects and variables in a vector subspace with a reduced number of dimensions. These representations allow one to have a quick overview of the complete data set in order to classify objects and interpret their position.

Natural products used in food, flavorings and fragrances, and cosmetics are actually extensively analyzed by means of chromatography, and this leads to large data sets. Factor methods may establish relationships between the origin, the species and the quality of a natural product, and its composition.

— The constitution of classes and groups of objects from the measured or calculated values of selected variables. When the discrimination of the different groups is optimum, one may attempt to assign new objects to one of the groups. Many applications may then be considered, such as control of quality, of origin and species, and detection of fraud.

Two kinds of methods are used to describe and classify multidimensional objects: factor analysis [7–11] and pattern recognition [12]. The latter proceeds through classification of objects into groups with evaluation of a neighboring or similarity criterion. The constitution of groups/clusters allows one to outline relationships between variables/objects.

Different techniques may be used in pattern recognition. One of them, the SIMCA method (Soft Independent Modelling of Class Analogy) is, in its purpose, very close to the Factor Discriminant Analysis.

As it will not be treated in this chapter, one may find useful information in the selected literature [13].

In the following discussion, our purpose is to present an introduction to factor analysis methods with all the basic theory and also with realistic applications.

We shall first review the mathematical tools needed to develop the different methods. This section will contain many numerical applications.

Factor Analysis (FA), Principal Component Analysis (PCA) and Factor Discriminant Analysis (FDA) will be then presented with various selected examples which illustrate different kinds of applications. We shall describe in detail how to carefully interpret results and output plots from computer programs and point out some important limitations of these methods.

Since our bibliography contains only the leading references to this topic, the interested reader will also find exhaustive literature in general reviews [14].

V.2 MATRIX CALCULATIONS AND MULTIVARIATE DATA ANALYSIS

The applications of linear algebra to data analysis will be illustrated by the ultraviolet spectroscopic analysis of a complex mixture. At a given frequency, v, the Beer–Lambert law states that the total absorption, A, of a sample containing l absorbing species is given by:

$$A = \sum_{j=1}^{l} \varepsilon_j C_j \tag{1}$$

where ε_j is the molar absorption coefficient for the species j and C_j is the molar concentration of the species j.

If the measurement is performed at n different frequencies, then the single expression above is replaced by a set of n linear equations:

$$A_1 = \sum_{j=1}^{l} \varepsilon_{1j} C_j$$

$$A_i = \sum_{j=1}^{l} \varepsilon_{ij} C_j \tag{2}$$

$$\vdots$$

$$A_n = \sum_{j=1}^{l} \varepsilon_{nj} C_j$$

Matrix notation allows one to write the foregoing set of linear equations in a shorthand notation:

$$\begin{pmatrix} A_1 \\ \vdots \\ A_n \end{pmatrix} = \begin{pmatrix} \varepsilon_{11} & \cdots & \varepsilon_{1l} \\ \vdots & & \vdots \\ \varepsilon_{n1} & \cdots & \varepsilon_{nl} \end{pmatrix} \cdot \begin{pmatrix} C_1 \\ \vdots \\ C_l \end{pmatrix} \tag{3}$$

To simplify the expression further, one writes the absorption matrix, (A), as the matrix product of the extinction coefficient, (ε), and concentration, (C), matrices:

$$(A) = (\varepsilon)(C) \tag{4}$$

The matrix formalism, illustrated here for an example in ultraviolet spectroscopy, can be generalized for any representation of a set of chemical measurements. We leave as an exercise for the reader the application of the method to specific subjects. The great advantage of matrix representations is that a great variety of data sets may be represented in a unique common formulation.

Matrix calculations and its computer applications have allowed the rapid development of multivariate data analysis. In the next section, we recall some mathematical fundamentals to illustrate the example of principal component analysis. Our goal is to provide the reader with an in-depth understanding of the fundamentals of multivariate data analysis.

2.1   Basic elements for matrix calculations

Matrix formalism may be viewed from two standpoints:

(1)   The first is purely operational and is linked to numerical aspects; matrices are mathematical entities on which operations may be performed in a way which somewhat parallels what can be done with numbers.

(2) The second standpoint, more fundamental, connects matrices with vector calculations. Every matrix, indeed, has the significance of a linear application (A) to which is associated for every $\tilde{x}$ vector described in a p-dimensional vector space, a vector $\tilde{y}$ described in a n-dimensional vector space. The coordinates of these vectors are related through:

$$(y) = (A)(x) \tag{5}$$

(y) is the coordinates which define vector $\tilde{y}$, (x) is the coordinates which define vector $\tilde{x}$.

This underlying aspect of matrix calculations provides a better grasp of multidimensional data analysis. This is illustrated in Fig. 1, which shows how a data set becomes a cloud of points through the matrix approach.

LEVEL 1   Data table

	Variable 1	Variable 2
Object 1	2	3
Object 2	2	4
Object 3	3	5
Object 4	4	5
Object 5	4	3

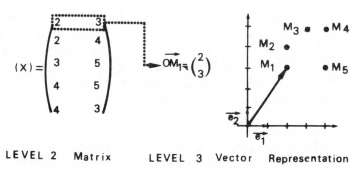

LEVEL 2   Matrix   LEVEL 3   Vector   Representation

Fig. 1   Equivalence between a data table, a matrix and vectors.

This representation allows one a better grasp of some aspects of multivariate data analysis which are difficult to visualize in their multidimensional representation. In order to observe the cloud of points one may elect to remain at the origin, but, if necessary, matrix calculations will allow the chemist to choose any other point as an observational origin and gain a new and hopefully heuristically richer picture of the overall data set.

2.1.1   *Data translations*

Using the previous example, let us translate the initial point of origin $(0, \vec{e}_1, \vec{e}_2)$ to a new one $(0', \vec{e}_1, \vec{e}_2)$. If 0 is translated to M_1, the initial data will then be described by the new coordinates with respect to the new origin:

$$\overrightarrow{OM}_i = \overrightarrow{OM}_1 + \overrightarrow{M_1M}_i \tag{6}$$

To express this vector addition in matrix terms, one uses matrix addition (Fig. 2). The new data matrix constitutes a new representation of the same cloud of points.

Therefore, using the matrix formalism, we are placed in a situation of being an observer who can choose the position from which observation takes place. Among the infinity of possible choices, however, we will see that a specific picture is obtained by placing the origin at the center of gravity of the cloud of points G the coordinates of which are $(\overline{X})$, i.e., the mean values of the two variables (Fig. 3).

2.1.2   *Data rotations*

To observe the cloud of points, the observer may gain further insights either by moving around the stationary cloud or by turning the cloud itself while keeping the point of observation stationary. Observing a globe-map of the earth while rotating it about its axis intersecting the poles clearly illustrates the advantages of such an operation. The analogy would be more apt with a more complex globe which contains a mechanism allowing the choice of several different axes of rotation. In factor data analysis this kind of operation is very often performed. One of the key steps in the operation is the

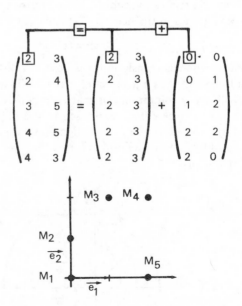

Fig. 2   Representation of object after translation by matrix addition.

$$(x) \quad = \quad (\bar{x}) \quad + \quad (x - \bar{x})$$

$$\begin{pmatrix} 2 & 3 \\ 2 & 4 \\ 3 & 5 \\ 4 & 3 \\ 4 & 5 \end{pmatrix} = \begin{pmatrix} 3 & 4 \\ 3 & 4 \\ 3 & 4 \\ 3 & 4 \\ 3 & 4 \end{pmatrix} + \begin{pmatrix} -1 & -1 \\ -1 & 0 \\ 0 & 1 \\ 1 & 1 \\ 1 & -1 \end{pmatrix}$$

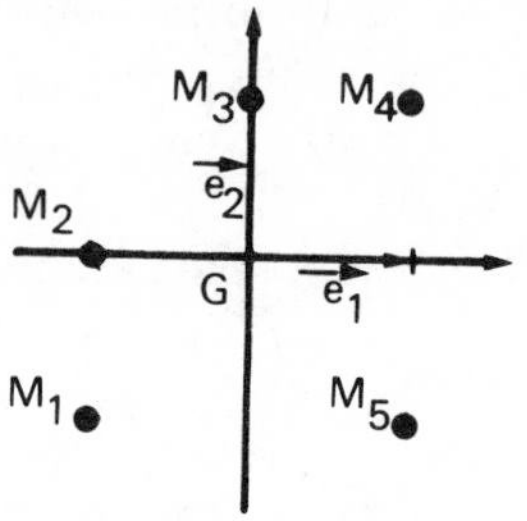

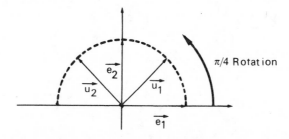

Fig. 3   Determination of a centered data matrix.

calculation of the rotation axis associated with the richest heuristic description of the data set. One of the reasons matrix calculations are so versatile and powerful in data analysis lies precisely in the amazing efficiency of programs which deal with the translation and rotation of data sets.

This operation is performed with a matrix called the transformation matrix (U). This matrix relates the coordinates of vector $\tilde{v}$ in the old basis (x) to the corresponding ones in the new basis (y) according to the relationship:

$$(x) = (U)(y) \tag{7}$$

Therefore, whereas translation was described as an addition of matrices, rotation is described as a product of matrices. To determine matrix (U), we simply write every column of this matrix as consisting of the coordinates of each vector of the new basis expressed in the old basis. We illustrate this proposition by expressing a $\pi/4$ rotation about an axis perpendicular to the plane defined by $\tilde{e}_1$ and $\tilde{e}_2$ (Fig. 4). During this transformation $\tilde{e}_1$ becomes $\tilde{u}_1$ and $\tilde{e}_2$ becomes $\tilde{u}_2$:

Fig. 4   Transformation of the initial basis by a $\pi/4$ rotation.

To find the new coordinates as a function of the old ones we introduce a new matrix operation: inversion.

$$(y) = (U)^{-1}(x) \tag{8}$$

For our example, (U) and $(U)^{-1}$ are, respectively,

$$\begin{pmatrix} \sqrt{2}/2 & -\sqrt{2}/2 \\ \sqrt{2}/2 & \sqrt{2}/2 \end{pmatrix} \quad \text{and} \quad \begin{pmatrix} \sqrt{2}/2 & \sqrt{2}/2 \\ -\sqrt{2}/2 & \sqrt{2}/2 \end{pmatrix} \tag{9}$$

The columns of (U) are the coordinates of $\check{u}_1$ $(\sqrt{2}/2, \sqrt{2}/2)$ and $\check{u}_2$ $(-\sqrt{2}/2, \sqrt{2}/2)$. Note that the product $(U)(U)^{-1}$ gives the identity matrix:

$$\begin{pmatrix} \sqrt{2}/2 & -\sqrt{2}/2 \\ \sqrt{2}/2 & \sqrt{2}/2 \end{pmatrix} \cdot \begin{pmatrix} \sqrt{2}/2 & \sqrt{2}/2 \\ -\sqrt{2}/2 & \sqrt{2}/2 \end{pmatrix} = \begin{pmatrix} 1 & 0 \\ 0 & 1 \end{pmatrix} = (I) \tag{10}$$

For example, the operation of $(U)^{-1}$ on the original coordinates of the point M_1 $(-1, -1)$ yields the new coordinates after a $\pi/4$ rotation:

$$(y) = \begin{pmatrix} \sqrt{2}/2 & \sqrt{2}/2 \\ -\sqrt{2}/2 & \sqrt{2}/2 \end{pmatrix} \cdot \begin{pmatrix} -1 \\ -1 \end{pmatrix} = \begin{pmatrix} -\sqrt{2} \\ 0 \end{pmatrix} \tag{11}$$

To generalize this rotation calculation to a data set, and more precisely to the matrix $(X - \bar{X})$, we must first of all introduce a new transformation: the transposition.

$$(X - \bar{X}) \xrightarrow{\text{transposition}} (X - \bar{X})'$$

$$(X - \bar{X})' = \begin{pmatrix} -1 & -1 & 0 & 1 & 1 \\ -1 & 0 & 1 & 1 & -1 \end{pmatrix} \tag{12}$$

$(X - \bar{X})'$ is called transposed matrix of $(X - \bar{X})$ such as $(X - \bar{X})_{ij} = (X - \bar{X})_{ji}$.

Moreover, we shall use a property of the (U) matrix specific to the transformation matrix in which the unit vectors are all orthogonal and normalized to one:

$$(U)^{-1} = (U)' \tag{13}$$

The new data matrix, (Y), is obtained using one of these relations:

$$(Y)' = (U)^{-1}(X - \bar{X})'$$
$$(Y)' = (U)'(X - \bar{X})' \tag{14}$$

$$\boxed{(Y) = (X - \bar{X})(U)}$$

and is written thus:

$$(Y) = \begin{pmatrix} -\sqrt{2} & 0 \\ -\sqrt{2}/2 & \sqrt{2}/2 \\ \sqrt{2}/2 & \sqrt{2}/2 \\ \sqrt{2} & 0 \\ 0 & -\sqrt{2} \end{pmatrix} \qquad (15)$$

2.2 Information reduction

In the foregoing section we have seen that matrix calculations may transform an initial data matrix into a new one which expresses the same data on a different basis (Fig. 5).

For a two-dimensional space, such a transformation is not very interesting from a physical point of view. The (X) matrix is the direct representation of the data table and these values have a physical significance. For the chemist, the significance could be concentrations of products in various solutions, NMR chemical shifts, temperature, etc. After translation and rotation, we have completely lost the physical significance of the initial data table. The new variables are linear combinations of the old ones. Of course, we are now in the middle of the cloud of points, but what for, if we have no physical tool with which to observe or if we do not understand the significance of the new axes and the significance of the new coordinates of points?

In a n-dimensional space, however, such transformations may help in extracting, from the initial data matrix, less accurate but nevertheless more understandable information. Before examining more specifically this point in

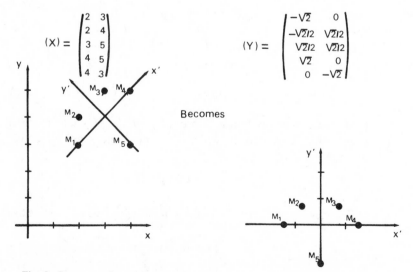

Fig. 5 Representation of the transformed data matrix after translation and axis rotation.

the next section, we must be precise in what is meant by information reduction. Consider the matrix ($X1$) obtained from (X) by addition of one more column.

$$(X) = \begin{pmatrix} 2 & 3 \\ 2 & 4 \\ 3 & 5 \\ 4 & 5 \\ 4 & 3 \end{pmatrix} \qquad (X1) = \begin{pmatrix} 2 & 3 & 7 \\ 2 & 4 & 8 \\ 3 & 5 & 11 \\ 4 & 5 & 13 \\ 4 & 3 & 11 \end{pmatrix} \qquad (16)$$

We shall imagine that the variables in the three columns of ($X1$) result from independent observations performed on each of five objects. In this case, the third measurement hopefully brings additional data to study and better discriminate between the five objects. In fact, however, one can easily determine that such is not the actual result because:

$$\text{variable (3) = 2} \times \text{variable (1) + variable (2)}$$

If we had, before making more measurements, been aware of this relationship, we would not have wasted our time adding this redundant third variable. In mathematical terms one expresses this situation by stating that even though a 3-dimensional space apparently is needed to describe our set of objects, a 2-dimensional subspace may contain as much information. The ($X1$) matrix is said to be a rank 2 matrix with only two independent variables. A chemical example is given in Section 3.3.3 with a dimension reduction from 4 to 2. Now consider matrix ($X2$), obtained from ($X1$) under the assumption that the three variables are experimental results which include a random error.

$$(X1) = \begin{pmatrix} 2 & 3 & 7 \\ 2 & 4 & 8 \\ 3 & 5 & 11 \\ 4 & 5 & 13 \\ 4 & 3 & 11 \end{pmatrix} \qquad (X2) = \begin{pmatrix} 2.1 & 3.2 & 7.2 \\ 1.8 & 4.1 & 7.9 \\ 3.2 & 5.3 & 10.9 \\ 4.1 & 4.7 & 12.9 \\ 3.9 & 3.2 & 11.4 \end{pmatrix} \qquad (17)$$

In this new situation, the correlation between the third variable and the first two variables is no longer perfect, although it still exists. Going from a 3-dimensional space to a 2-dimensional subspace in describing the five objects would lead to some loss in information (but not much) (cf. Section 3.3.1). In Section 2.3 we shall see that the major feature of data analysis within the matrix framework is precisely to extract as much information as possible from a given data set by taking into account possible redundancies and by using optimally the transformations and searching for the best representation subspace.

2.3   Determination of a vector subspace which fits the data set

In Sections 2.1 and 2.2, we have shown how to change the point of reference of a data set using transformation matrices and to search for the vector subspace which expresses the relevant information present in the data most efficiently. To attain our next objective, the search for the best subspace in a general situation, we need only to define one more important property of the spaces in which we are going to work. These spaces are Euclidean spaces because their basis vectors are orthonormal. This property means precisely that the vectors are orthogonal to each other and their norm equals unity.

We are now in the proper position to locate the best representation subspace for a given data set. To do so, we will use a least squares criterion. The sum of least squares that we must minimize is illustrated in Fig. 6 for a set of n points.

This figure illustrates our search for an axis containing a vector $\check{u}$ included in the subspace and crossing a point A (an axis can be defined by a point and a vector). In terms of mechanics, s is simply the moment of inertia, J_Δ of the elementary particles associated with each of the n points (each assumed to have the same uniform weight equal to unity)

$$J_\Delta = \sum_{i=1}^{n} m_i \cdot d_i^2 = \sum_{i=1}^{n} 1 \cdot d_i^2 \qquad (18)$$

Huygens' or Steiner's theorem state that the moment of inertia with respect to an axis can be expressed as follows:

$$J_\Delta = J_{\Delta_G} + ma^2 \qquad (19)$$

Δ_G is an axis parallel to Δ but crossing the center of gravity and a is the distance between Δ and Δ_G.

The axis searched for must, therefore, intersect the center of gravity, G, which thus identifies points A and G.

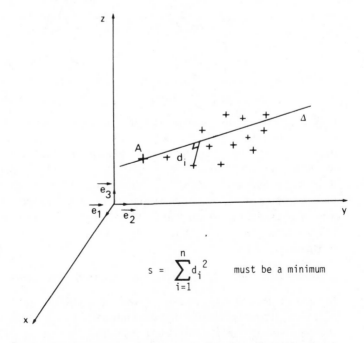

Fig. 6   Search of the first principal factor of the first principal inertia axis.

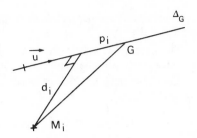

Fig. 7   Representation of the link between the criteria of minimization and maximization.

We now place a data point with respect to one axis crossing G, as in Fig. 7. As a consequence of the Pythagorean theorem as generalized to a complete data set, one obtains the expression:

$$\sum_{i=1}^{n} \|\overrightarrow{GM_i}\|^2 = \sum_{i=1}^{n} d_i^2 + \sum_{i=1}^{n} p_i^2 = \text{Constant} \tag{20}$$

This expression states that minimizing the sum of the squares of the distances leads to maximizing of the sum of the squares of the projections. We must, therefore, search for the maximum of the second term of the foregoing expression.

In vector and matrix notation, this second term may be written as:

$$S = \sum_{i=1} (\overrightarrow{GM_i} \cdot \tilde{u})^2 = ((X - \bar{X})(u))'(X - \bar{X})(u)$$

$$\boxed{\Rightarrow \ S = (u)'(X - \bar{X})'(X - \bar{X})(u)} \tag{21}$$

To find (u), we must solve a problem of maximization under constraint.

$$\begin{cases} \max & S = (u)'(X - \bar{X})'(X - \bar{X})(u) \\ \text{constraint } C = 1 - (u)'(u) = 0 & \text{i.e. } (u)'(u) = 1 \end{cases} \tag{22}$$

The constraint states that the vector $\tilde{u}$ has a norm equal to unity. The application of the Lagrange multiplier is appropriate to find an optimum in the presence of a constraint. We must first set:

$$L = S + \lambda C = (u)'(X - \bar{X})'(X - \bar{X})(u) + \lambda(1 - (u)'(u)) \tag{23}$$

Then, we must apply this relationship with respect to each of the components of $\tilde{u}$ and finally to cancel each derivative. This yields:

$$2(X - \bar{X})'(X - \bar{X})(u) - 2\lambda(u) = 0 \tag{24}$$

or, equivalently,

$$\boxed{(X - \bar{X})'(X - \bar{X})(u) = \lambda(u)} \tag{25}$$

This relation is fundamental in data analysis. It leads to the solution of the maximization problem. Only some particular vectors are solutions of this equation.

The set of these vector solutions is called the set of eigenvectors of the matrix $(X - \bar{X})'(X - \bar{X})$. The values of λ are called eigenvalues. Since at this point we may calculate, as in a classical derivation, the whole set of extrema, the next step is to locate among these solutions the one which provides a maximum for the expression S. This expression takes a particular value for every eigenvector $\hat{u}$:

$$S = (u)'((X - \bar{X})'(X - X)(u)) = (u)'\lambda(u) = \lambda \qquad (26)$$

To maximize S we must retain the largest of all the eigenvalues given by the equation.

At this point we may summarize: The one-dimensional subspace which fits the data according to the criterion of information maximization is defined by the eigenvector, $\hat{u}$, associated with the largest eigenvalue, of the matrix $(X - \bar{X})'(X - \bar{X})$. This result may be generalized: the q-dimensional subset which fits the initial data according to the criterion of information maximization is defined by the q eigenvectors associated with the q largest eigenvalues. Every supplementary eigenvector brings complementary information until the dimension of the subspace attains the dimension of the data matrix (X). When this dimension has been reached, the overall information contained in the data matrix has been extracted.

2.3.1 Constraints on axis

The method described above, which is at the basis of Principal Component Analysis as performed on centered and non-reduced data, leads to an axis, plane, or hyperplane containing the center of gravity of the cloud of points. We could also have modified the approach such that the axis, plane, or hyperplane also contains the origin of the point of reference. In such a case the principal factors would have been defined by the relation

$$(X)(X)'(v) = \delta(v) \qquad (27)$$

where X is the matrix of non-centered data rather than by

$$(X - \bar{X})'(X - \bar{X})(u) = \lambda(u) \qquad (28)$$

in which $(X - \bar{X})'$ is the matrix of centered data. The vectors $\hat{v}_1$, $\hat{v}_2$ are obviously different from vectors $\hat{u}_1$, $\hat{u}_2$ (Fig. 8).

The data matrix on which one performs the calculation of principal factors must therefore be examined with respect to the chemical system under study.

The study of the subspace solution for the fitting problem is easier when the properties of the matrix $(X - \bar{X})'(X - \bar{X})$ are well understood. This matrix exhibits interesting properties which may be classified as:

— properties linked to role of this matrix in multidimensional statistics
— properties linked to the matrices having the characteristics of square, symmetric and positive matrices.

2.3.2 Properties of the $(X - \bar{X})'(X - \bar{X})$ matrix in multidimensional statistics

We have not yet discussed statistics in a chapter whose purpose is clearly the statistical treatment of data. This will be understood by showing that the

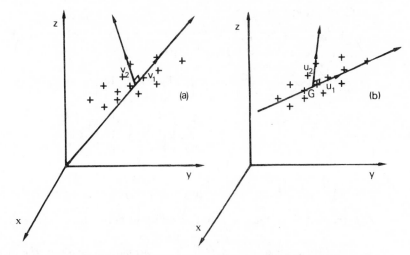

Fig. 8 Principal factors for pure (a) and centered (b) variables.

properties of the matrices are directly connected with most of ⁣.ıe fundamental information dealt with in statistics.

The product of a matrix by its transpose yields directly the most used statistical parameters: variance, covariance, and coefficients of correlation. This proposition holds true provided that one uses:

— centered variables: $(X_{ij} - \bar{X}_j)$
— centered and reduced variables: $(X_{ij} - \bar{X}_j)/\sigma_j$

where $\bar{X}_j$ and σ_j are the mean and standard deviation of the variable j. To illustrate this point let us extract various pieces of statistical data from the data set discussed in Section 2.1.

$$`(X)'(X) = \begin{pmatrix} 2 & 2 & 3 & 4 & 4 \\ 3 & 4 & 5 & 5 & 3 \end{pmatrix} \cdot \begin{vmatrix} 2 & 3 \\ 2 & 4 \\ 3 & 5 \\ 4 & 5 \\ 4 & 3 \end{vmatrix} = \begin{pmatrix} 49 & 61 \\ 61 & 84 \end{pmatrix} \qquad (29)$$

Matrix of variance–covariance:

$$\frac{1}{n}(X - \bar{X})'(X - \bar{X}) = \frac{1}{5}\begin{pmatrix} -1 & -1 & 0 & 1 & 1 \\ -1 & 0 & 1 & 1 & -1 \end{pmatrix} \begin{vmatrix} -1 & -1 \\ -1 & 0 \\ 0 & 1 \\ 1 & 1 \\ 1 & -1 \end{vmatrix}$$

$$= \begin{pmatrix} 0.8 & 0.2 \\ 0.2 & 0.8 \end{pmatrix} \qquad (30)$$

Therefore, 0.8 is the variance of the first and the second variables and 0.2 is the covariance between them.

Correlation matrix:

$$\frac{1}{n}\left(\frac{X-\bar{X}}{\sigma}\right)'\left(\frac{X-\bar{X}}{\sigma}\right) = \frac{1}{5}\begin{pmatrix} -1.12 & -1.12 & 0 & 1.12 & 1.12 \\ -1.12 & 0 & 1.12 & 1.12 & -1.12 \end{pmatrix}$$

$$\times \begin{vmatrix} -1.12 & -1.12 \\ -1.12 & 0 \\ 0 & 1.12 \\ 1.12 & 1.12 \\ 1.12 & -1.12 \end{vmatrix} \tag{31}$$

$$\frac{1}{n}\left(\frac{X-\bar{X}}{\sigma}\right)'\left(\frac{X-\bar{X}}{\sigma}\right) = \begin{pmatrix} 1 & 0.25 \\ 0.25 & 1 \end{pmatrix}$$

Therefore, 0.25 is the correlation coefficient between the two variables.

2.3.3 Properties of the $(X-\bar{X})'(X-\bar{X})$ matrix linked to the fact that the matrix is square, symmetric and positive.

Square, symmetric and positive matrices display some properties particularly useful in the analysis of data:

— the eigenvalues are real; they are also positive or equal to 0.
— the number of non-zero eigenvalues equals the rank of the matrix.
— two eigenvectors associated with two distinct eigenvalues are orthogonal.

To illustrate these properties and show their importance in data analysis, let us take the variance–covariance matrix obtained in Section 2.3.1 and determine the eigenvalue solutions of the least squares fit:

$$\begin{pmatrix} 0.8 & 0.2 \\ 0.2 & 0.8 \end{pmatrix} \tag{32}$$

The two eigenvalues are obtained by solving the equation:

$$\begin{pmatrix} 0.8-\lambda & 0.2 \\ 0.2 & 0.8-\lambda \end{pmatrix} = 0 \Leftrightarrow (0.8-\lambda)^2 - (0.2)^2 = 0 \tag{33}$$

which yields $\lambda_1 = 1$ and $\lambda_2 = 0.6$

λ_1 and $_2$ are both real and positive. The rank of the matrix must be 2 since two non-zero eigenvalues are associated with the system. The components of the eigenvectors associated with each of the eigenvalues are obtained from the definition of the eigenvector, as follows:

For the first eigenvalue:

$$\begin{pmatrix} 0.8 & 0.2 \\ 0.2 & 0.8 \end{pmatrix} \cdot \begin{pmatrix} a_1 \\ b_1 \end{pmatrix} = 1 \begin{pmatrix} a_1 \\ b_1 \end{pmatrix} \Rightarrow a_1 = b_1 \tag{34}$$

For the second eigenvalue:

$$\begin{pmatrix} 0.8 & 0.2 \\ 0.2 & 0.8 \end{pmatrix} \cdot \begin{pmatrix} a_2 \\ b_2 \end{pmatrix} = 0.6 \begin{pmatrix} a_2 \\ b_2 \end{pmatrix} \Rightarrow a_2 = -b_2 \tag{35}$$

We note that the two vectors associated with each of the two eigenvalues are indeed orthogonal (i.e., their scalar product $= 0$). In these two sets of vectors, we may choose two normed vectors which therefore constitute an orthonormal basis

$$\vec{u}_1 = \begin{pmatrix} \sqrt{2}/2 \\ \sqrt{2}/2 \end{pmatrix} \qquad \vec{u}_2 = \begin{pmatrix} -\sqrt{2}/2 \\ \sqrt{2}/2 \end{pmatrix} \tag{36}$$

vectors $\vec{u}_1$ and $\vec{u}_2$ are in fact the vectors defined in Section 2.1 and the coordinates of the data matrix with respect to this point of reference have been calculated:

$$(Y) = (X - \bar{X})(U) \tag{37}$$

2.4   Quantification of information

Having developed one of the main objectives of data analysis, information reduction, we must now answer the question: how much information have we extracted during the previous operation? We can use the example from Section 2.2 to answer this question.

Stating that a 2-dimensional subspace is sufficient to represent a 3-dimensional data set is equivalent to affirming that the two axes of this subspace allow one to include 100% of the information. Then we must evaluate the quantity of information covered by each axis of the subspace fitted to the data. In the previous section, we showed that for every eigenvector solution of the S maximization problem, S took a specific value equal to the eigenvalue associated with the vector being considered in 3-dimensional space.

$$\begin{array}{ll} \text{along } \vec{u}_1 & S_1 = \lambda_1 \\ \text{along } \vec{u}_2 & S_2 = \lambda_2 \\ \text{along } \vec{u}_3 & S_3 = \lambda_3 \end{array} \tag{38}$$

with $\lambda_1 \geq \lambda_2 \geq \lambda_3 \geq 0$

Let us examine a data point with respect to these three eigenvectors, $\vec{u}_1, \vec{u}_2, \vec{u}_3$ (Fig. 9).

$$\|\overrightarrow{GM}_i\|^2 = p_1^2 + p_2^2 + p_3^2$$

Therefore:

$$\sum_{i=1}^{n} \|\overrightarrow{GM}_i\|^2 = \sum_{i=1}^{n} p_1^2 + \sum_{i=1}^{n} p_2^2 + \sum_{i=1}^{n} p_3^2$$

$$\sum_{i=1}^{n} \|\overrightarrow{GM}_i\|^2 = S_1 + S_2 + S_3 = \varphi \tag{39}$$

$$\sum_{i=1}^{n} \|\overrightarrow{GM}_i\|^2 = \lambda_1 + \lambda_2 + \lambda_3$$

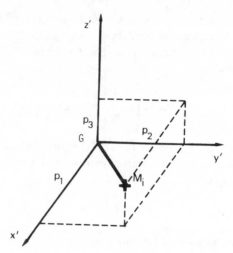

Fig. 9   Decomposition of the total variance.

We have seen in Section 2.3 that

$$J_G = \sum_{i=1}^{n} \|\overrightarrow{GM}_i\|^2 \tag{40}$$

but, on the other hand, the total variance for a population of n objects is:

$$\frac{1}{n} \sum_{i=1}^{n} \|\overrightarrow{GM}_i\|^2 \tag{41}$$

Thus, from the foregoing we can note the following points:

— in a mechanical approach, the moment of inertia with respect to the center of gravity equals:

(1)   on one hand, the sum of the moments of inertia with respect to the principal inertia axis;

(2)   on the other hand, the sum of the eigenvalues of the matrix

$$(X - \bar{X})'(X - \bar{X}).$$

— within a statistical approach, the total variance equals:

(1)   on one hand, the sum of the partial variances along each of the principal axes

(2)   on the other hand, the sum of the eigenvalues of the matrix

$$(X - \bar{X})'(X - \bar{X})$$

divided by n, or to the sum of the eigenvalues of the variance–covariance matrix. These points allow us to quantify the information in terms of percentage of variance associated with each principal axis, with respect to the total variance (Fig. 10):

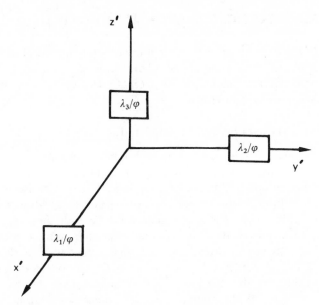

Fig. 10   Part of the total information associated with each factor.

In a more general view:

$$\varphi = \sum_{i=1}^{r} \lambda_i \qquad (42)$$

where r, the rank of matrix $(X - \bar{X})'(X - \bar{X})$, is the number of non-zero eigenvalues.

If we work in a 3-dimensional space, defined by $\hat{u}_1, \hat{u}_2, \hat{u}_3$, the quantity of information extracted by the fitting will be:

$$(\lambda_1 + \lambda_2 + \lambda_3)/\varphi \qquad (43)$$

To summarize, we may now take an overall look at the initial data matrix, (X), and its transform, (Y):

$$(X) = \begin{pmatrix} 2 & 3 \\ 2 & 4 \\ 3 & 5 \\ 4 & 5 \\ 4 & 3 \end{pmatrix} \qquad (Y) = \begin{pmatrix} -\sqrt{2} & 0 \\ -\sqrt{2}/2 & \sqrt{2}/2 \\ \sqrt{2}/2 & \sqrt{2}/2 \\ \sqrt{2} & 0 \\ 0 & -\sqrt{2} \end{pmatrix} \qquad (44)$$

Variance	0.8	0.8	1 $(=\lambda_1)$	0.6 $(=\lambda_2)$
Total variance		1.6		1.6
% Information	50%	50%	62.5%	37.5%

The matrix fundamentals of factors analysis have now been briefly covered; we now move on to practical applications of such an approach.

V.3 PRACTICAL FACTOR ANALYSIS (FA)

In this section, we shall present the general method of factor analysis and its application to some chemical problems. Our approach will essentially be practical, and further developments may be found in the general book of Malinowsky and Howery [9].

3.1 The data matrix. Variable transformation

As seen in Section 2.1, a data table constitutes a data matrix with n rows and p columns. Usually, the rows are associated with the concept of objects or observations, and the columns with variables or characters as evaluated for each object or observation. Every object or observation, i, is represented by a point in a p-dimensional space with the coordinates $(X_{i1}, \ldots, X_{ip})$.

In chemistry, this table may either contain homogeneous data (i.e., molar fractions of constituents) or heterogeneous data (i.e., density, temperature, yield or selectivity). In the latter case, it may be necessary to normalize the variables so that the comparison of objects involves a pseudo-homogeneous data set and the comparison is, in some way, fairer. To do so, various transformations are possible:

Centering initial variables:

$$X_{ij} \to x_{ij} = (X_{ij} - \bar{X}_j)/\sqrt{n} \tag{45}$$

where $\bar{X}_j$ = average of the jth variable correction factor allowing the identification of the inertia matrix $(x'x)$ with the variance–covariance matrix.

(1) Consequence: All the variables, x_j, have an average of zero. This transformation does not affect the range of variation.

(2) Interest: When the average values of the p variables have very different magnitudes, this process makes their comparison easier.

To work on non-centered variable forces the first eigenvector to cross over the origin of the initial point of reference. If this point is an implicit data point in the matrix data set no problem occurs, but if not, the fitting vector subspace will not be optimal (see Fig. 8). This problem is analogous to a linear regression in which the linear model $Y = aX$ is used, but where the experimental data is better fitted using $Y = aX + b$.

Reducing initial variables:

$$X_{ij} \to x_{ij} = X_{ij}/\sigma_j\sqrt{n} \tag{46}$$

where σ_j: standard deviation of the jth variable.

(1) Consequence: All the variables x_j have a σ_j equal to 1, and thus the same range of variation.

(2) Interest: This type of transformation, although less frequently used, allows an homogenization of a data set which presents a widely dispersed range. The minimization is performed with respect to an axis crossing over the origin, but the initial shape of the cloud of points is modified.

Centering and reducing the initial variables:

$$X_{ij} \rightarrow x_{ij} = (X_{ij} - \bar{X}_j)/\sigma_j\sqrt{n} \tag{47}$$

(1)   Consequence: All the variables x_j have the same average value ($\bar{x}_j = 0$) and the same standard deviation ($\sigma_j = 1$). Under these conditions, the inertia matrix $(x'x)$ is also the correlation matrix.

(2)   Interest: This transformation, which is frequently used, allows a homogenization of data and the obtainment of an inertia axis crossing over the center of gravity with optimal minimization criteria (cf. demonstration Section 2.3).

While centering and reducing the variables gives identical values for average and inertia, it also modifies the shape of the cloud of points. For example, in a data matrix composed of a complex mixture (reaction medium, essential oil, etc.), this transformation associates a weight factor to the concentration of minor constituents which is identical to that of the major products. It should be clear that there is some risk in such a transformation, since the measurement precision is indeed not as good for the components present in small concentrations as for those present in larger quantities.

One important limitation in the use of factor methods may be avoided by a transformation of variables: the eigenvectors obtained are linear combinations of initial variables, whereas physico-chemical models are seldom linear and simply additive. To obtain linear models, one may apply operators on some variables, such as $\log(X)$, $1/X$, e^X, etc.

Example: In the study of a reaction, the yield (R) is rarely linear with the function of duration t, or of temperature T. The application of a kinetic model may reveal a linear relationship between $\log R$, $\log t$ and $1/T$.

Another possibility is to generate from the initial variables new variables which can impart new information to the problem.

Example: In the analysis of a data table containing the composition of a complex mixture [15], one may regroup some known constituents, calculate the ratios between constituents, evaluate the composition in terms of percent.

The transformations applied to the data matrix will be a very important step in determining the treatment which follows. The choice is guided by the origin, the nature of the data and the existence of explicative physico-chemical models. Whereas for a table of homogeneous data, the normalization of variables is not compulsory [16], it becomes highly desirable for variable sets of very different nature and order of magnitude. For these cases, the centering and reduction of variables considerably simplifies the interpretation of results.

Throughout the following, (x) will represent the transformed matrix (X).

3.2   Diagonalization of the matrix $(x'x)$

Classical algorithms perform the searches for the eigenvectors and eigenvalues of the inertia matrix $(x'x)$ of order (p, p):

$$(x'x)(U) = (\lambda)(U) \tag{48}$$

where (λ) is a diagonal matrix of order (p, p), where each diagonal element is an eigenvalue.

(U) is a square matrix of order (p, p), where each column contains the coefficients of an eigenvector.

$$
(\lambda) =
\begin{matrix}
\text{eigenvalues} \\
\begin{pmatrix}
\lambda_1 & 0 & \cdots & 0 & \cdots & 0 \\
0 & \lambda_2 & \cdots & 0 & \cdots & 0 \\
. & . & \cdots & . & \cdots & . \\
0 & 0 & \cdots & 0 & \cdots & \lambda_p
\end{pmatrix}
\end{matrix}
\qquad
(U) =
\begin{matrix}
\text{eigenvectors} \\
\begin{pmatrix}
u_{11} & u_{12} & \cdots & u_{1j} & \cdots & u_{1p} \\
u_{21} & u_{22} & \cdots & u_{2j} & \cdots & u_{2p} \\
. & . & \cdots & . & \cdots & . \\
u_{p1} & u_{p2} & \cdots & u_{pj} & \cdots & u_{pp}
\end{pmatrix}
\end{matrix}
\tag{49}
$$

The eigenvalues, λ_j, are ranked in decreasing order and represent the variances of the associated eigenvector $\tilde{u}_j$, that is, the part of the information inertia described by this eigenvector.

$$
\lambda_1 \geq \lambda_2 \geq \cdots \geq \lambda_j \geq \cdots \geq 0 \tag{50}
$$

Since the diagonalization of $(x'x)$ keeps the trace invariant, the total inertia is conserved:

$$
\varphi = \sum_{j=1}^{p} (x'x)_{jj} = \sum_{j=1}^{p} \lambda_j \tag{51}
$$

With centered variables:

$$
(x'x)_{jj} = \sigma_j^2 \Rightarrow \varphi = \sum_{j=1}^{p} \sigma_j^2
$$

With reduced and centered variables:

$$
(x'x)_{jj} = 1 \Rightarrow \varphi = p
$$

The fraction of the variance or inertia of the cloud of initial data represented by the jth axis is defined by the following ratio (see Section 2.4).

$$
\frac{\lambda_j}{\sum_{j=1}^{p} \lambda_j} \tag{52}
$$

The principal factors (or axes or components) are associated to the larger values of λ_j: there is a concentration of information in one or several factors.

Zero values for λ_j exhibit linear relationships which exist between the variables constituting the initial data table because the rank of the $(x'x)$ matrix will be lower than p.

The eigenvector $\tilde{u}_j$ is defined by the p components u_{kj}, $k = 1, 2, \ldots, p$ which determine the contribution of each initial variable k in the construction of the jth factor.

The sign of the eigenvector is not significant and may vary depending upon the diagonalization process used: the sense of the factor axes may be inverted from one procedure to another one.

The p eigenvectors are orthogonal, therefore:

$$
(U)^{-1} = (U)' \tag{53}
$$

These eigenvectors will allow one to compute the new coordinates (y) of the objects in the new vector space according to:

$$(y) = (x)(U) \tag{54}$$

The coordinates of object i on the axis u will be:

$$y_{ij} = \sum_{k=1}^{p} x_{ik} u_{kj} \tag{55}$$

3.3 Reproduction of the initial data matrix

When one uses UV spectroscopy to quantitatively determine the composition of a mixture, one records the spectra of p distinct mixtures composed of l unknown constituents, the concentrations of which are also unknown. Each spectrum is defined by n absorptions measured at n different wavelengths.

A_{ij} = Absorption at the frequency v_i for the mixture j

One of the goals of factor analysis of this data set is to decompose each absorption into the sum of the contributions from each compound.

$$A_{ij} = \varepsilon_{i1} C_{1j} + \cdots + \varepsilon_{il} C_{lj}$$

$$A_{ij} = \sum_{k=1}^{l} \varepsilon_{ik} C_{kj} \tag{56}$$

$$(A) = (\varepsilon)(C)$$

One will therefore try to analyze the data matrix $A(n, p)$ into the product of two matrices $\varepsilon(n, l)$ and $C(l, p)$, and first, the number of products present in the mixtures must be determined. This is possible only in cases where $p \geqslant l$.

Factor analysis of the data matrix (x) allows us to determine p new variables u_j which are linear combinations of p initial variables and which present a maximum variance. One may reproduce the initial matrix (x) with these new variables:

$$(y) = (x)(U)$$
$$(y)(U)' = (x)(U)(U)'$$
$$(y)(U)' = (x)(U)(U)^{-1} \tag{57}$$

$$\boxed{(y)(U)' = (x)}$$

Therefore, every element of the (x) matrix may be recalculated from the eigenvectors (U) and the coordinates in the new basis (y)

$$x_{ij} = \sum_{k=1}^{p} y_{ik} u'_{kj}$$

$$\boxed{x_{ij} = \sum_{k=1}^{p} y_{ik} u_{jk}} \tag{58}$$

The matrix (x) may be decomposed in an infinity of ways into a product of two matrices of dimension (n, l) and (l, p). One ignores *a priori* those which possess a physical significance: matrix of concentrations of mixtures, molar absorption of l pure products. The decomposition obtained from eigenvectors of the inertia matrix $(x'x)$ allows the representation of the total variance with a restricted number of factors, but has only a small probability of having a physical meaning.

The eigenvectors (U) obtained are therefore purely mathematical tools (one may call them abstract factors) to which the interpretation of results attempts to associate a physical meaning.

The goal of the factor analysis of the table (x) is:

(1)   the determination of the number of principal factors by the reproduction of the initial matrix (x): l such that $1 \leqslant l \leqslant p$

(2)   the interpretation and the search for physical significance to be associated with the principal abstract factors from the analysis of the eigenvectors coefficients

(3)   the search for factors derived from principal factors using rotation operators and which possess a meaningful physical significance.

In the example cited, it is sometimes possible to proceed up to the matrices (ε) and (C) starting from matrices (y) and (U).

3.3.1   *Determination of principal factors*

The factors are ranked in order of decreasing variances. The selection of principal factors is an important step of multidimensional data analysis because the main goal remains the reduction of the dimension of the vector space of the p initial variables.

Several quantitative criteria have been developed to rigorously define the number of principal factors, primarily using:

— λ_j values by elimination of the least significant values
— reconstruction of the initial data within the limits of experimental error.

3.3.1.1   *Selection of λ_j*   The l initial eigenvalues λ_j are added in percent of explained inertia according to:

$$\sum_{j=1}^{l} \frac{100 \, \lambda_j}{\sum_{k=1}^{p} \lambda_k} \tag{59}$$

One selects the value of l in such a way that this value becomes greater than 80, 90 or 95% of the total inertia. When one operates on centered and reduced variables, one saves only the eigenvalues greater than or which approaches 1 (i.e., the variance of every reduced variable).

3.3.1.2   Reconstruction of the initial data   Let us suppose that in using an appropriate criterion, one saves only the l initial eigenvectors with maximum variance, and discards the $(p - l)$ following factors. The data matrix (x) can then no longer be reproduced exactly:

$$(x) = (y)(U)'$$

$$x_{ij} = \sum_{k=1}^{p} y_{ik} u_{jk}$$

(60)

One calls x_{ij}^* the value of x_{ij} reproduced using only the first l factors

$$x_{ij}^* = \sum_{k=1}^{l} y_{ik} u_{jk}$$

such that

$$x_{ij} = x_{ij}^* + e_{ij}$$

(61)

in which e_{ij} is the error associated with the information contained in the $(p - l)$ unused factors

$$e_{ij} = \sum_{k=l+1}^{p} y_{ik} u_{jk}$$

(62)

The Root Mean Square error (RMS) for the data matrix reproduced with l factors is given by:

$$\text{RMS} = \sqrt{\sum_{i=1}^{n} \sum_{j=1}^{p} e_{ij}^2 / np} = \sqrt{\sum_{i=1}^{n} \sum_{j=1}^{p} (x_{ij} - x_{ij}^*)^2 / np}$$

$$\text{RMS} = \sqrt{\sum_{i=1}^{n} \sum_{j=1}^{p} \sum_{k=l+1}^{p} y_{ik} u_{jk}}$$

(63)

To obtain the optimal values of l, one performs successive reproductions of the matrix (x) with 1, 2, 3, $\cdots$, l factors until the residual error e_{ij} becomes, at every point, smaller than the estimated experimental error.

Malinowsky and Howery [9] have established different criteria which allow the determination of the exact number of principal factors beginning with the experimental errors of the data studied; furthermore their treatment also applies to cases where the errors are unknown (in this case, the variance of secondary factors provides the needed information).

Other applications of the reconstruction of the data matrix are:

— determination of the experimental error of a data set, using the assumption that principal factors allow the reconstruction of 'pure data' [17]
— the comparison of various analytical techniques [18]
— the factor analysis of the residual matrix which may allow the recognition of an internal structure in the partition of errors within the initial data matrix.

Deviation from additive rules may be revealed for some observations, thus, in quantitative spectroscopic analysis, matrix effects may be showed.

Therefore, one defines a set of l principal factors and $(p - l)$ secondary factors which represent only the residual errors e_{ij}.

3.3.2 Significance of the number of principal factors

The number of principal factors is the actual dimension of the factor subspace and the rank of the initial data matrix. It is also the number of linear relationships which exist between the initial variables, and help to explain the major part of the information enclosed in the actual data table, residual information $(x - x^*)$ being associated with a noise or an experimental error.

Indeed, the existence of a linear combination between variables leads to the presence of a negligible or zero eigenvalue; therefore one factor is eliminated.

Additionally the number of principal factors may have a physical significance:

3.3.2.1 In quantitative analysis

In the analysis of spectroscopic or chromographic data of mixtures, the number of principal factors corresponds to the number of constituents present in all the mixtures studied, provided that there is additivity of the pure compounds spectra [19].

This technique (ref. 9, p. 141) has been applied successfully to data originating from:

— absorption spectroscopy,
— mass spectroscopy,
—chromatography,
— optical rotatory dispersion

In the case of several products not completely resolved by gas chromatography, but studied by GC–MS, this approach may be used to recognize how much components are present under one of the poorly resolved GC peaks, provided that mixtures with different compositions may be sampled. The factor analysis of the mass spectra of these samples, usually allows the determination of the number of non-solved products [20]. The same method has been used in liquid chromatography coupled to UV spectroscopy [21] and in the analysis of solid state NMR spectra [22].

3.3.2.2 In qualitative interpretation of physico-chemical phenomena

The search for principal factors allows the study of complex physical phenomena to be modelled which are represented by data set obtained from a great number of experiments.

In the analysis of GC or LC retention times for various solutes in different solvents the solute–solvent interactions will be determined by a restricted number of factors.

Factor analysis (ref. 9, p. 187) has led to the determination and identification of principal factors associated with well-known physical phenomena involved in solute–solvent types of interactions:

— LONDON dispersion forces (molecular weight, number of carbon atoms, etc.)
— Polar interactions (dipole moments, dielectric constants, etc...)

A comparable approach has been applied to the NMR chemical shifts for ^1H, ^{19}F and ^{29}Si nuclei (ref. 9, p. 171) and particularly for ^{13}C [23].

In other instances, the physical chemist does not search for the physical significance of principal factors, but simply uses the reduction of the initial data table as an 'analytical tool'. The reduction of the dimension of the vector space describing the n starting objects, allows the characterization of every individual point by a number of components far smaller than the number of initial variables. Even when the components have no physical meaning, they may be conveniently used in applications where the characterization of multidimensional objects is necessary:

— graphical representation for studying the topology of the complete set of objects (principal component analysis, see Section 4)
— modelling studies in which it is necessary to identify different multidimensional objects but where a restricted set of variables is desirable.

Discriminant factor analysis provides a specific example of such an approach. When the number of starting variables is too high, one may apply discriminant analysis on principal components (see Section 5).

3.3.3   Example of an application to quantitative analysis

The study by Ritter and co-workers [24] illustrates the foregoing kind of application to mass spectrometry applied to the quantitative analysis of mixtures [16, 25]. Table 1 contains the intensity values of twenty selected m/e peaks collected for four different mixtures of cyclohexane and cyclohexene.

Table 1   Digitized intensity values cyclo-hexane/cyclohexene mixtures

m/e	% Cyclohexane			
	80%	60%	40%	20%
27	2.3	3.2	3.4	2.1
28	1.2	1.3	1.3	0.7
29	1.1	1.1	1.1	0.5
39	3.9	5.8	6.8	4.9
40	0.7	1.0	1.0	0.6
41	8.6	10.5	10.5	6.0
42	3.5	3.7	3.1	1.1
43	1.6	1.7	1.4	0.4
51	0.5	1.1	1.5	1.0
53	0.9	1.7	1.7	1.7
54	3.6	8.0	11.3	10.1
55	5.1	5.4	4.5	1.9
56	14.2	14.4	11.4	3.6
67	4.4	11.1	16.0	15.1
68	0.5	0.8	1.1	0.8
69	4.1	4.3	3.3	1.0
79	0.4	0.8	1.2	1.0
81	0.5	1.0	1.6	1.4
82	1.5	4.1	6.1	5.4
84	10.5	11.8	8.5	2.4

Table 2 Eigenvalues and eigenvectors for cyclohexane/cyclohexene mixtures

n	Eigenvalue	% Information	Cumulative information	Eigenvector			
1	1035.8	82.2	82.2	0.41	0.56	0.59	0.39
2	222.7	17.7	99.9	−0.63	−0.27	0.24	0.68
3	0.7	0.1	100	0.57	−0.76	0.23	0.14
4	0.2	0.0	100	−0.30	−0.12	0.72	−0.59

The data have been centered and the matrix $(x'x)$ obtained is therefore a variance-covariance matrix.

$$(x'x) = \begin{pmatrix} 264.7 & 279.4 & 218.7 & 73.2 \\ 279.4 & 348.7 & 333.1 & 190.8 \\ 218.7 & 333.1 & 379.6 & 281.5 \\ 73.2 & 190.8 & 281.5 & 266.5 \end{pmatrix} \tag{64}$$

One notes that very strong correlations (covariances) between the four columns of the table exist; these are as important as the variances associated with the four variables (diagonal elements). The diagonalization of this

Table 3 Difference between experimental and reconstructed data (values smaller than 0.1 are set equal to 0)

m/e	With one factor				With two factors			
	20%	40%	60%	80%	20%	40%	60%	80%
27	0.1	0	0.1	0	0	0	0	0
28	−0.2	0	0.1	0.1	−0.1	0.1	0	0
29	−0.2	0	0	0.1	−0.1	0.1	0	0
35	0.7	0.4	−0.3	−0.7	0	0.1	−0.1	0
40	0.1	0	0	−0.1	0	0	0	0
41	−1.0	−0.2	0.3	1.0	−0.1	0.2	−0.1	0
42	−1.0	−0.4	0.4	1.1	0	0	0	0
43	−0.4	−0.2	0.1	0.5	0	0	0	0
51	0.5	0.1	−0.2	−0.3	0.1	0	−0.1	0.1
53	0.4	0	0.1	−0.7	−0.1	−0.2	0.3	−0.2
54	3.4	1.5	−1.4	−3.6	0	0	0	0
55	−1.5	−0.5	0.6	1.4	−0.1	0.1	0.1	−0.1
56	−5.0	−1.8	1.8	5.0	−0.2	0.3	−0.1	−0.1
67	5.4	2.3	−2.0	−5.9	0	−0.1	0.1	−0.1
68	0.3	0.2	−0.1	−0.3	0	0	0	0
69	−1.3	−0.6	0.5	1.4	0	0	0	0
79	0.4	0.2	−0.2	−0.5	0	0	0	0
81	0.5	0.3	−0.2	−0.6	0	0.1	0	0
82	2.9	0.1	−0.9	−2.1	0.1	0	0.1	0.1
84	3.4	−2.2	1.6	4.2	0.4	−0.1	0.1	0.1

Table 4   RMS error for cyclo-
hexane/cyclohexene MS data

Number of factors	RMS error
1	1.423
2	0.098
3	0.045
4	0.021

inertia matrix allows one to obtain four factors, ranked according to decreasing values of their associated eigenvalues (Table 2).

It is apparent that two eigenvalues are not significantly different from zero. The number of principal factors therefore equals 2, which tells us the number of pure products present in the mixture studied. The reproduction of the matrix of initial data is realized with 1 and 2 principal factors. The errors calculated on each point are displayed in Table 3.

One notes that only 5% of the residual errors are greater than 0.25 when one uses two factors; this contrasts with the reproduction using only one factor, which usually yields errors greater than the experimental errors. Table 4 shows the impressive decrease in the residual mean square, which drops from 1.42 (only 1 factor used) to 0.1 (two factors).

This example (Table 4) shows that the RMS is not a sufficient criterion for the selection of the number of principal factors. A good analysis must, therefore, rely on several criteria: RMS, eigenvalues, individual errors e_{ij}, etc.

This kind of approach is well illustrated by the analysis of seven binary mixtures of cyclohexane-hexane based on 18 m/e peaks in each mass spectrum [24]. The factor analysis of Table 5 shows that three factors are present. The study of the experimental conditions revealed that the source of the instrument may have been contaminated by nitrogen.

A further factor analysis performed by suppressing, in the initial data set, the intensity associated with m/e = 28 gives the expected two factors.

Table 5   RMS error for cyclohexane/hexane MS data

Number of factors	Complete data set		With suppression of m/e = 28	
	λ	RMS error	λ	RMS error
1	1059.1	1.467	1058.4	1.432
2	333.9	0.287	333.4	0.111
3	17.6	0.094	0.7	0.079
4	0.5	0.069	0.3	0.057
5	0.3	0.048	0.2	0.034
6	0.2	0.021	0.1	0.013

3.4 Physical significance of factors

The matrices (y) and $(U)'$ obtained from the decomposition of the matrix (x) have, in principle, no physical meaning and may be considered as simply mathematical tools.

One may nevertheless attempt to interpret the abstract factors according to two main lines of reasoning:

(a) By analyzing the contribution of the initial variables to the construction of every factor axis. For this, one relies on the coefficients of eigenvectors corrected by the associated eigenvalue.

$$u_{ij} \rightarrow u_{ij}\sqrt{\lambda_j} \text{ (factor loading)} \qquad (65)$$

The combination of variables obtained for every factor may sometimes be associated with physical factors. In a recent study on the classification of organic solvents, Chastrette *et al.* [26] performed a factor analysis of a data set which described 83 solvents with 8 physico-chemical data, either calculated or measured. The results of this analysis are enclosed in Table 6. Four principal factors have been extracted from this data set; they represent 91.7 % of the total variance.

The analysis of the coefficients of the first factor shows the important role played by molecular refraction, index of refraction, and HOMO energy. The authors, therefore, interpret this factor as a polarizability factor. The second factor is mainly composed of the Kirkwood function, the molecular dipole moment and the boiling point with strong positive coefficients; the authors, therefore, interpret it as a factor describing the polarity of the solvents. The third factor exhibit a great coefficient for the LUMO energy of the solvents and it follows that it may be identified as the electric acceptor variable (or electron affinity) of the solvents. The δ parameter of Hildebrandt contributes

Table 6   Principal factors

Factor	1	2	3	4
% Explained variance	39.7%	31.0%	11.6%	9.3%
Kirkwood function	−0.53	0.74	0.07	−0.21
Molar refraction	0.91	0.02	0.24	−0.10
Dipolar moment	−0.27	0.85	0.24	−0.40
Hildebrandt parameter	−0.55	0.56	−0.12	0.58
Refractive index	0.82	0.27	−0.19	0.34
Boiling point	0.46	0.78	0.28	0.18
HOMO energy (EHT)	0.81	0.23	0.25	−0.19
LUMO energy (EHT)	−0.37	−0.40	0.82	0.03

to factors 1, 2 and 4. The variance represented by the two first factors (39.7 and 31 %) suggests that polarizability and polarity play an equally important role in the classification of solvents.

(b) By transforming the abstract initial factors $(U)'$ into actual factors (U^*) which are postulated by the physico-chemist. To do so, one performs the appropriate rotations in the vector space, which are represented by the matrix products:

$$(U)' = (T)(U^*)' \tag{66}$$

where $(U^*)'$ matrix of factors having a physical meaning and (T) transformation matrix of dimension (l, l).

In this case, the matrix (y) must also be transformed to find the corresponding decomposition of matrix (x):

$$(x) = (y)(U)'$$
$$(x) = (y)(T)(U^*)' \tag{67}$$

One defines then the matrix (y^*) such that

$$(y^*) = (y)(T)$$
$$(y) = (y^*)(T)^{-1} \tag{68}$$

then

$$(x) = (y^*)(T)^{-1}(T)(U^*)'$$
$$= (y^*)(U^*)'$$

Two criteria are usually retained to determine these transformations:

— the intuitive knowledge of one or several measurable parameters (vectors) which could be associated with abstract factors. One then determines the experimental values for the parameters and then one has to calculate the rotation matrix which transforms the abstract factors into real factors. This method has been developed by Malinowsky and Howery [9] and is called Target Transformation Factor Analysis (TTFA).
— one may search for factor axes on which some initial variables present maximum values of coefficients. The constraints which follow from this choice permit the determination of the transformation matrix. The new factor axis may then be associated with one or several actual variables. One should remember that according to the performed transformations, the generated factors may or may not remain orthogonal.

V.4 PRINCIPAL COMPONENT ANALYSIS

Principal Component Analysis (PCA) is a factor method which allows one to describe a data set in a space of lower dimension by taking into account only the first principal components or factors [1, 3, 11].

4.1 Objectives and methods

One can establish simple geographical representations of objects and variables with the goal of:

— studying the proximity of the various objects in order to make classifications and to find atypical objects,
— analyzing the position of objects in various graphical representations,
— displaying new objects in representations characterizing the whole population,
— searching for the meaning of the principal component factors.

The initial data set is composed of the values taken by p variables evaluated for n objects. Generally, data are transformed into centered variables and more often into centered and reduced variables:

$$X_{ij} \rightarrow x_{ij} = (X_{ij} - \bar{X}_j)/\sigma_j \sqrt{n} \qquad (69)$$

In this latter case, Principal Component Analysis is said to be standardized. Consequently, the matrix $(x'x)$ is the correlation matrix of the initial variables. The significance of the coordinates of these variables on the principal factors will be the correlation coefficient.

The diagonalization of the matrix $(x'x)$ will give:

— p eigenvalues classified in decreasing range:

$$\lambda_1, \lambda_2, \ldots, \lambda_p \qquad (70)$$

— p associated eigenvectors:

$$\vec{u}_1, \vec{u}_2, \ldots, \vec{u}_p \qquad (71)$$

Usually, variables and objects are plotted in two graphical representations, the axis of which are the two first principal components.

The percent of variance explained will be:

$$100 \frac{(\lambda_1 + \lambda_2)}{\sum_{j=1}^{p} \lambda_j} \qquad (72)$$

If this value is too far from 100%, it will be necessary to examine the projections of objects and variables on the other factors $\vec{u}_3, \vec{u}_4, \ldots,$.

For standardized PCA, it will be suitable to study all components whose associated eigenvalues are large or approach 1. In the following sections, we investigate only representations on the two initial axes, in order to simplifying the presentation.

4.2 Graphical representation of variables

The contribution of each variable k to the construction of a principal factor j is evaluated from the coefficient u_{jk} of the eigenvector $\vec{u}_j$. But, as the weight of each eigenvector $\vec{u}_j$ used to represent the cloud of points is not the same, one considers as in factor analysis, the associated eigenvalue λ_j and more precisely the square root of the eigenvalue, i.e., the standard deviation of the cloud of points on the axis considered (cf. Section 2.4).

Thus, in the basis $(\tilde{u}_1, \tilde{u}_2)$, the coordinates of the variable k will be:

$$\frac{u_{1k}\sqrt{\lambda_1}}{\|\tilde{u}_1\|}, \quad \frac{u_{2k}\sqrt{\lambda_2}}{\|\tilde{u}_2\|}$$

where   (73)

$$\|\tilde{u}_j\| = \sqrt{\sum_{k=1}^{p} u_{jk}^2}$$

For standardized PCA λ, these coordinates are equal to the correlation coefficients between the variable k and the principal factors 1 and 2, r_{1k} such as:

$$-1 \leq r_{1k}, r_{2k} \leq 1 \Rightarrow r_{1k}^2 + r_{2k}^2 \leq 1 \tag{74}$$

The point associated to variables are thus displayed in an ellipse for PCA and in a circle (the correlation circle) for standardized PCA

— Points near the circle are perfectly represented in this basis and the associated variables participate strongly in the construction of the two first principal factors and (of course) very weakly for the others.
— Points near the center of the circle are poorly represented and it will be necessary to represent them with other factors.

For standardized PCA, distances between points can be interpreted in terms of correlations. The distance between two variables is defined as follows:

$$d_{kk'}^2 = \sum_{i=1}^{n} \left[\frac{x_{ik} - \bar{X}_k}{\sigma_k \cdot \sqrt{n}} - \frac{x_{ik'} - \bar{X}_{k'}}{\sigma_{k'} \cdot \sqrt{n}} \right]^2$$

$$d_{kk'}^2 = \frac{\sum_{i=1}^{n} (x_{ik} - \bar{X}_k)^2}{n\sigma_k^2} + \frac{\sum_{i=1}^{n} (x_{ik'} - \bar{X}_{k'})^2}{n\sigma_{k'}^2} - 2\frac{\sum_{i=1}^{n} (x_{ik} - \bar{X}_k)(x_{ik'} - \bar{X}_{k'})}{n\sigma_k\sigma_{k'}}$$

$$d_{kk'}^2 = 2 - 2\rho_{kk'} = 2(1 - \rho_{kk'}) \tag{75}$$

In fact, a maximum correlation appears whenever:

— variables are very close together,

$$\rho_{kk'} = 1 \Rightarrow d_{kk'} = 0 \tag{76}$$

— variables are diametrically opposed

$$\rho_{kk'} = -1 \Rightarrow d_{kk'} = 2 \tag{77}$$

For instance, in Fig. 11, variables X_1, X_2, X_3, X_4, and X_6 are well represented by the first two factors, and X_1, X_2 and X_3 are strongly correlated with each other, but not at all with X_4. Variable X_5 is poorly

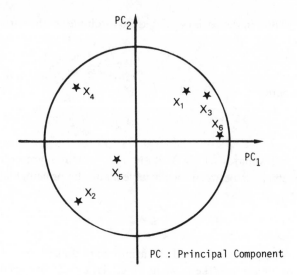

Fig. 11   Graphical variable representation.

represented in this plane and cannot be studied there. Finally, variable X_6 is strongly correlated with the first axis as shown by its large value on this axis.

In the search for physical significance, one will notice that the role of X_2, X_3, X_4 and X_6 in the construction of these axes and their interpretation will depend on the nature of these variables.

4.3   Graphical representation of objects

The graphic representation of objects will allow one to study the structure of the population and to determine the existence of groups, atypic objects, etc.

One calculates the coordinates y_{ij} of each object i in relation to the new basis, constituted by the principal components, from the initial data x_{ik} and the components of the eigenvector (cf. 3.2).

$$y_{i1} = \sum_{k=1}^{p} x_{ik} u_{k1} / \|\tilde{u}_1\|$$

$$y_{i2} = \sum_{k=1}^{p} x_{ik} u_{k2} / \|\tilde{u}_2\| \qquad (78)$$

$$\vdots \qquad \vdots \qquad \vdots$$

Since the first component has the greatest variance, points will have a maximum dispersion on this axis. If one considers only the first two coordinates, y_{i1} and y_{i2}, part of the information is not taken into account: that which is associated with the coordinates y_{i3}, y_{i4}, etc. Doing so, one obtains the projection of the cloud of points on to the two first components. The real distance between two objects i and i' may therefore be greater:

$$(y_{i1} - y_{i'1})^2 + (y_{i2} - y_{i'2})^2 < \sum_{j=1}^{p} (y_{ij} - y_{i'j})^2 \qquad (79)$$

One can evaluate the quality of a graphical representation of an object and its position with respect to the others in two ways:

— qualitatively by evaluating the graphical representation in other bases: $(\tilde{u}_1, \tilde{u}_3), (\tilde{u}_2, \tilde{u}_3)$, etc.
— quantitatively by studying the angle between the object M_i and its projection on to the plane of the first two components:

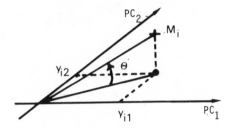

In this case:

$$\cos^2 \theta = \frac{y_{i1}^2 + y_{i2}^2}{\sum_{j=1}^{p} y_{ij}^2} \qquad (80)$$

$$\cos^2 \theta = \cos^2 \theta_1 + \cos^2 \theta_2$$

where θ_1 and θ_2 are the angles of $\overrightarrow{OM}_i$ with respect to the two axes

When $\cos \theta$ approaches 1, the representation with the axis 1 and 2 is perfect; on the other hand, when $\cos \theta$ approaches 0, the representation is very poor.

The representation will allow one to establish a mapping of objects. One will then easily observe groups of objects and/or peculiar objects which would have been indistinguishable or only apparent with great difficulty from the data table alone.

New objects may be positioned in this graph from their initial measured variables $X_1^{(S)}, X_2^{(S)}, \ldots, X_p^{(S)}$

Their coordinates may be computed from:

$$y_j^{(S)} = \sum_{k=1}^{p} \frac{X_k^S - \bar{X}_k}{\sigma_k \sqrt{n}} u_{jk} / \|\tilde{u}_j\| \qquad (81)$$

where $y_j^{(S)}$ are the coordinates of the new object on the jth axis.

Therefore, it will be possible to analyze their position with respect to the initial population, and to see their proximity to other objects, to groups, to classes of objects, etc.

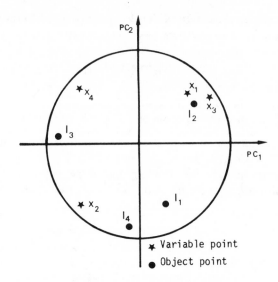

Fig. 12   Graphic representation of objects and variables.

The positions of objects may be interpreted as a function of the initial variable by superposition of the two graphical representations of objects and variables on the same graph (Fig. 12).

Objects I_1, I_2, I_3, I_4 are well distinguished; I_3 is differentiated from the others due to the first axis while the second axis differentiates I_2 from I_1 and I_4.

If the representation in this basis is considered acceptable, one will deduce that:

— object I_3 will likely be characterized by a large value of X_2 and X_4 and a low value of X_3 because of its very negative coordinate on the first axis

— object I_2 will likely be characterized by high values of X_1 or X_3 and a low value of X_2.

4.4 Application example

To illustrate this method, we can examine treated data from a study performed in our laboratory [27] where the goal was the optimization of the experimental conditions of a halogenation reaction. The halogen was added during a known time (variable 1: TIME) to an organic compound A having a defined initial concentration (variable 2: CONC) in a solvent. The mixture was maintained at a given temperature (variable 3: TEMP) and the reaction was carried out in the presence of a catalyst at a certain concentration (variable 4: CATA).

For each trial, we measured the amount of substrate consumed, in percent (variable 5: REACT) and, as the reaction gave two halogenated products B and C, the selectivity for B (variable 6: SELECT).

There were thus six variables, four being active variables and two being responses. Twenty trials were performed initially; sixteen of them selected from an experimental design and four supplementary ones inside the experimental area (Table 7). Later additional trials were performed and were treated as supplementary objects.

TIME: 2 h–6 h

TEMP: 0°C–96°C

CATA: 0.015–0.08

CONC: 0.05%–0.5%

SELECT: 0%–91%

REACT: 40%–96%

PCA has been performed on centered and reduced data.

From the correlation matrix (Table 8) some remarks can be outlined.

The percent of substrate consumed (REACT) is positively correlated to the concentration of catalyst (CATA) and to the selectivity (SELECT). The very low correlation between the four active variables is inherent in the structure of experimental design: these variables are nearly orthogonal.

Table 7   Experimental results of a halogenation reaction matrix

	TIME	TEMP	CATA	CONC	SELECT	REACT
E01	6.0	96.0	0.05	0.5	91.4	94.0
E02	6.0	0.0	0.05	0.5	36.1	83.0
E03	6.0	0.0	0.02	0.1	0.0	45.0
E04	6.0	96.0	0.02	0.1	0.9	46.0
E05	2.0	0.0	0.02	0.5	0.0	45.0
E06	2.0	0.0	0.05	0.1	0.0	65.0
E07	2.0	96.0	0.02	0.5	1.0	48.0
E08	2.0	96.0	0.05	0.1	5.5	69.0
E09	2.0	0.0	0.02	0.1	0.0	43.0
E10	2.0	96.0	0.05	0.5	30.2	84.0
E11	2.0	96.0	0.02	0.1	0.4	50.0
E12	6.0	0.0	0.02	0.5	0.0	53.0
E13	6.0	96.0	0.02	0.5	13.1	59.0
E14	6.0	96.0	0.08	0.5	0.0	96.0
E15	3.0	96.0	0.05	0.05	0.0	45.0
E16	4.0	20.0	0.02	0.17	7.2	57.0
E17	4.0	96.0	0.05	0.5	51.4	85.0
E18	4.0	20.0	0.015	0.1	0.0	40.0
E19	3.0	20.0	0.08	0.25	45.1	74.0
E20	4.0	96.0	0.035	0.10	5.2	78.0
EA	3.0	50.0	0.08	0.5	44.0	97.0
EB	4.0	70.0	0.06	0.4	30.0	81.0
EC	5.0	60.0	0.04	0.8	39.0	82.0

Table 8   Correlation matrix

	TEMP	CONC	CATA	REACT	SELECT
TIME	−0.002	0.283	0.006	0.246	0.237
TEMP		0.113	0.208	0.343	0.192
CONC			0.199	0.489	0.455
CATA				0.767	0.455
REACT					0.675

From the eigenvalues, it can be seen that 90% of the total variance is concentrated in the four first principal components (Table 9) and 46% by the first one only. The relative independence of the four active variables does not allow one to further reduce the number of principal components.

Loading factors for each component are, in this case, the correlation coefficients between the initial variables and the principal factors. They were used to interpret the principal components.

One will notice that the two variables to be explained (REACT and SELECT) appear significantly only in the first factor, with factor loading 0.94 and 0.81 and with two other variables: CATA (0.74) and CONC (0.64).

This factor expressed the relationship between the two responses and these two active variables, i.e., the quantity of substrate consumed and the selectivity of this reaction increases with the concentration of the catalyst and at a lower level with the concentration of compound A.

It may also be mentioned that the selectivity (SELECT) appears also, alone in the fifth component (0.56): this is a part of the information which is not related to the other variables.

The second axis is mainly described by the duration of the addition of halogen (0.75). Thus, this sample is an external variable to the other variables and has almost no influence on the selectivity or the substrate consumption. Axes 3 and 4 show weak links between active variables.

The analysis of the graphical representation on Fig. 13 in the basis of the first two components (PC1 and PC2), which concentrates 65% of the total

Table 9   Eigenvalues and principal components (factor loadings)

No	λ	% Variance	TIME	TEMP	CONC	CATA	REACT	SELECT
1	2.75	45.9	0.34	0.40	0.64	0.74	0.94	0.81
2	1.14	19.1	0.75	−0.45	0.42	−0.41	−0.12	0.09
3	0.85	14.1	0.25	0.80	0.06	−0.35	−0.09	−0.12
4	0.67	11.2	−0.49	−0.00	0.58	−0.26	−0.09	0.10
5	0.45	7.5	−0.06	0.01	−0.26	−0.24	−0.09	0.56
6	0.13	2.2	−0.04	−0.04	−0.06	−0.19	0.29	−0.06

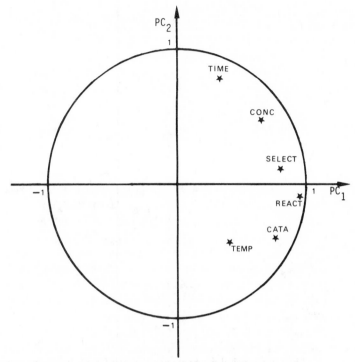

Fig. 13   Graphical representation of variables for a halogenation study.

information, corroborates the following observations:

— the proximity of variables REACT and SELECT
— the absence of correlation between these two variables and the variables TIME and TEMP
— a strong correlation of REACT and SELECT with the first principal component PC1
— a poor representation of the temperature as shown by its distance from the correlation circle.

In the graphical representation of objects (Fig. 14) we shall only analyze the position of those which have high coordinates on the first principal component (PC1). The positive correlation of this compound with selectivity and the substrate consumed allows us to associate these objects with high values of SELECT and/or REACT.

When substrate consumption is high, these points present high coordinates on axis PC1.

E01: This experiment is characterized by a simultaneous maximum of the selectivity and of the substrate consumed.

E02, E10, E14, E17, E19: These experiments are characterized by a high consumption of the substrate but only a moderate selectivity (or even a low one (E14)). One can distinguish three groups characterized by:

— a high duration of the halogen addition: E02.
— a low duration of the halogen addition: E10, E19.
— a moderate duration for the halogen addition: E14, E17.

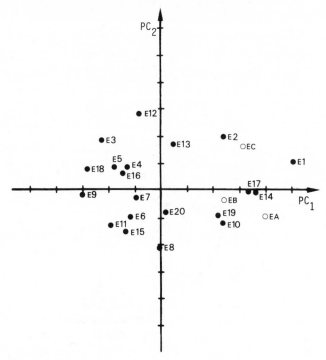

Fig. 14   Graphical representation of objects for a halogenation reaction study.

The analysis of the quality of the representation, on Table 10, of these points nevertheless shows the relative aspects of this classification.

In fact, trials E14 and E19 are poorly represented by this graphical display and are thus relatively far away from trials E17 and E10.

From these various observations, one may make the following conclusions:

— the two studied responses REACT and SELECT seem to be linked only to the two variables CATA and CONC
— the active variables TEMP and TIME do not seem to have any effect on the reaction in the area explored

Table 10   Quality of the representation of objects

	$\cos^2 \theta_1$	$\cos^2 \theta_2$	$\cos^2 \theta$
E01	0.80	0.03	0.83
E02	0.45	0.34	0.79
E10	0.51	0.21	0.72
E14	0.56	0.00	0.56
E17	0.90	0.00	0.90
E19	0.32	0.11	0.42

— the selectivity cannot be completely explained by the present variables: there must be, therefore, 'hidden' variables which have not been taken into account.

An attempt to model REACT and SELECT with the active variables from the initial data matrix confirmed these observations. The best models obtained from a stepwise multilinear regression were.

$$REACT = 30.3 + 629.6 \times CATA + 32.4 \times CONC$$

$$R^2 = 0.71, \text{ residual standard deviation: } 10.6$$

$$SELECT = -15.8 + 453.6 \times CATA + 46.5 \times CONC$$

$$R^2 = 0.35, \text{ residual standard deviation: } 20.9$$

These adjustments were quite insufficient to set up a reliable model.

Additional experiments E_A, E_B, E_C are displayed on the graphical representation of objects and may therefore be compared with other experiments.

4.5 Scope and limitations

PCA is a descriptive method which is used to describe, clarify, and display multidimensional objects. Since it has the possibility of plotting graphical representations of objects and variables, this method is the most widely used. A physical significance may be assigned to each principal component and further, the position of each object and variable in the plotted diagrams may be interpreted in terms of variable values and significance of axis.

This method is more and more used to describe data set from various topics:

— compositions and physico-chemical properties of essential oils [28] and heavy oils derived from distillation and cracking petroleum [29].
— composition and sensory or quality tests of foods, beverages and aromas [30].
— environment chemistry, biomedical sciences [14].

However, some limitations must be overcome in order to avoid mistakes in the interpretation of results.

— the validity of the analysis of the position of objects or variables on the plotted diagrams is related to the quality of their representation. Quality criteria may be controlled:
>   for variables, the position towards the circle of correlation
>   for objects, the cos θ value
— the use of linear models for establishing principal components is restrictive because physico-chemical phenomena are seldom linear [31]. In these cases, mathematical transformation may be performed on initial variables.
— the sensitivity of the coefficients of the principal components must be studied according to the sets of objects selected for the initial data matrix. For this, some objects may be added or suppressed from the initial population in order to check the stability of the coefficients of the principal components.

V.5 FACTOR DISCRIMINANT ANALYSIS (FDA)

The methods described in the previous sections do not presuppose any particular inhomogeneity of the objects studied. One assumes that the objects belong to a rather homogeneous group and searches for the principal factors which give the best representations for distinguishing the objects from each other. Two types of observations are often made during a principal component analysis:

— some atypical objects may occur with one or several variables which exhibit respectable differences from the remainder of the group. Their presence may affect the calculation of the variance–covariance matrix. Therefore, it may be suitable to eliminate these objects in a subsequent analysis in order to study a more homogeneous population.
— the observed population is obviously inhomogeneous and several groups of objects are present, for example:

   (a)   Type of substitution in chemical structures
   (b)   Nature of the activity of a chemical compound
   (c)   Origin, species of natural products
   (d)   Quality criteria of natural products
   (e)   etc.

At this time, a new problem arises. We have to find the factors which contribute to distinguishing the various groups of objects as well as possible. These discriminant factors are different from the principal factors. In point of fact, they are established with different constraints.

The main objective of principal component analysis was to find linear combinations of variables which allow one to distinguish objects as well as possible. The objective of discriminant analysis is to find other linear combinations of variables whose objective is to distinguish groups of objects or the centers of gravity of the various groups encountered.

Two kinds of objectives may be attained:

(a)   For known objects already classified into general groups, one may:
      — Discriminate these groups in an optimal way by using new variables called discriminant factors.
      — Plot the objects and the center of gravity of their groups using the most discriminant axis.
      — Analyze the contribution of every initial variable to the discrimination of groups and select the more efficient ones.
(b)   For new objects, one must make assignments to previously chosen groups using initial variables.

5.1 From canonical analysis to discriminant analysis

Canonical analysis is, among the various methods of multidimensional data analysis, certainly one of the most attractive from a theoretical point of view, but is also a rather unusual method because the interpretation of results is quite difficult. It nevertheless presents two particular cases which are frequently used:

— Multilinear regression,
— Factor Discriminant Analysis (FDA)

Canonical analysis allows one to simultaneously study two different sets of variables evaluated for the same objects. One must treat two data tables: (X_1) of dimensions (n, p_1) and (X_2) of dimensions (n, p_2). This situation is frequently found in chemistry where one may have to analyze quantitative data such as:

— Compounds for which one has independently evaluated structural parameters and spectroscopic parameters or biological activities, etc.
— Reactions or processes characterized by several trials with different values for the active variables (temperature, pressure, concentration, solvent, and measured results (conversion rates, selectivity,...) as was seen in the data described in Section 4.4.

In canonical analysis, one calculates couples of linear combinations of variables from data set X_1 and from data set X_2 which present correlation coefficient maxima between themselves. The new variables are called 'canonical factors'. The relations between the two data sets are studied with these factors previously set in decreasing order of correlation coefficient.

The special case where one of the two tables reduces to only one variable is a frequent situation in which one must model this variable called 'response' as a function of the variables from the other data set. If the variable to be modelled is quantitative, the derived method of canonical analysis will be the multilinear regression. A linear combination of variables must be found to model the quantitative response.

If the variable to be modelled is a qualitative one which can only take predefined values (acidic/basic, position of a substituent, type of chemical or biological activity, quality criteria, etc.), then the derivative method of canonical analysis is called Factor Discriminant Analysis (FDA). The objects studied are first classified into various groups according to a known criterion.

The data set is composed of p variables evaluated for n objects. A supplementary column contains new data associated with each object: an integer value defining the group to which the object is initially assigned. In this way, m groups are constructed. These data make up the reference data set from which the discriminant factors will be computed. Generally another data set is used with new objects not previously assigned to a group and which can be used to check the discriminant power of the calculated factors.

5.2 Theoretical basis for discriminant analysis [4, 11]

It was stated in Section 2.4 that the total variance can be decomposed into the sum of the variances of the initial variables, and was equal to the sum of variances of the principal components. The basic principle of discriminant analysis results from a similar operation: the total variance (V) of a set of objects belonging to several groups will be decomposed into two terms:

(1) Variances associated with each group of objects. The sum of these m variances is called the 'intra-group variance' (VI).
(2) Variances associated with m centers of gravity of the different groups. This term characterizes their dispersion and their weight (population of the group) and is called the 'inter-group variance' (VG).

$$(V) = (VI) + (VG)$$

The variables are always centered and sometimes reduced. Thus, the variances may be computed according to: $(V) = (x)'(x)$.

$$(V)_{jj'} = \frac{1}{n} \sum_{i=1}^{p} x_{ij} x_{ij'} \tag{82}$$

To calculate (VI) and (VG), the matrix (G) of the centers of gravity is needed.

$$(G)_{kj} = g_{kj} = \sum_{i=1}^{n_k} x_{ij}^{(k)}/n_k \tag{83}$$

where g_{kj} = coordinates of the center of gravity kth group on the jth variable. This is the mean value of the jth variable for the objects assigned to kth group.
n_k = population of the kth group
$x_{ij}^{(k)}$ = elements of the kth group

Therefore

$$(VG) = (G)'(G)$$
$$(VG)_{jj'} = \sum_{k=1}^{m} \frac{n_k}{n} \cdot g_{kj} \cdot g_{kj'} \tag{84}$$

$(n_k/n$ is a weighting factor)

$$(VI) = \frac{1}{n}(x - G)'(x - G)$$

$$(VI)_{jj'} = \frac{1}{n} \sum_{k=1}^{m} \sum_{i=1}^{n_k} (x_{ij}^{(k)} - g_{jk})(x_{ij'}^{(k)} - g_{j'k})$$

From the invariance property of the total variance, the aim of Discriminant Analysis is clear: to find the best method of differentiation between the groups, the inter-group variance (VG) must be maximized for a constant total variance (V). In a parallel way, the intra-group variance (VI) must be minimized to reduce distances between the objects of a group and the center of gravity. The two procedures are equivalent but the first is generally used for the calculation of discriminant factors.

5.3 Method of calculation

For principal component analysis in Section 2.3, the maximization problem under constraint leads to the search for eigenvectors of the total variance–covariance matrix $(V) = (x)'(x)$.

$$(V)(U) = (\lambda)(U) \tag{85}$$

In discriminant analysis, one must maximize the inter-group variance (VG) with respect to the total variance (V), so the ratio $(VG)/(V)$. In matrix notation, the following expression must be maximized:

$$\lambda = \frac{(u)'(VG)(u)}{(u)'(V)(u)} \quad \text{with } 0 \le \lambda \le 1 \tag{86}$$

From the properties of matrix derivation, the following relations may be obtained:

$$2((u)'(V)(u))(VG)(u) - 2((u)'(VG)(u))(V)(u) = 0 \tag{87}$$

where $(u)'(V)(u)$ and $(u)'(VG)(u)$ are quadratic forms, i.e., scalar numbers. Thus

$$(VG)(u) = \frac{(u)'(VG)(u)}{(u)'(V)(u)} \cdot (V) \cdot (u)$$

$$\tag{88}$$

$$(VG)(u) = \lambda(V)(u)$$

and multiplication by $(V)^{-1}$ gives

$$\boxed{(V)^{-1}(VG)(U) = (\lambda)(U)} \tag{89}$$

This expression shows that factor discriminant analysis is a special case of principal component analysis, being performed on specific objects associated with the center of gravity of the different groups as weighted by each group member and with a particular rule called D^2 Mahalanobis's distance $(V)^{-1}$ [4].

As for principal component analysis the eigenvectors (U) and eigenvalues (λ) must be obtained from diagonalization of $(V)^{-1}(VG)$. In fact, the matrix $(V)^{-1}(VG)$ is not symmetrical and its order is p. For simplification and speed in the diagonalization procedure, some transformations are convenient:

$$(VG) = (G)'(G)$$
$$(G)'(G)(U) = (\lambda)(V)(U)$$

(W) is defined according to

$$(U) = (V)^{-1}(G)'(W) \tag{90}$$

Substitution of (U) gives:

$$(G)'(G)(V)^{-1}(G)'(W) = (\lambda)(V)(V)^{-1}(G)'(W)$$
$$(G)' \cdot (G)(V)^{-1}(G)'(W) = (\lambda)(G)'(W)$$
$$(G)(V)^{-1}(G)'(W) = (\lambda)(W)$$

Thus, the eigenvalues (λ) and the eigenvectors (W) may be computed from the diagonalization of the matrix $(G)(V)^{-1}(G)$, which is symmetrical and the order of which is m the number of groups, and generally, $m > p$. The eigenvectors (U) of the $(V)^{-1}(VG)$ matrix may be derived from (W).

The initial variables are always centered, thus, the weighted sum of the coordinates of all the centers of gravity is zero. Therefore, one eigenvalue will always be zero and we obtain $(m - 1)$ eigenvectors associated with $(m - 1)$ eigenvalues ordered in decreasing sequence:

$$1 \geq \lambda_1 \geq \lambda_2 \geq \cdots \geq \lambda_{m-1} \geq 0 \tag{91}$$

λ_q is called the discriminant power of the qth factor. A value approaching 1 shows an optimum discrimination, while a value approaching 0 shows a poor discriminant effect.

5.4 Discriminant factors

The choice of the factors used for graphical representation of objects is obvious in two cases:

$m = 2$: When only two groups are found there is only one discriminant factor corresponding to the line joining the two centers of gravity.

$m = 3$: When three groups are present only two discriminant factors are constructed in the plane defined by the three centers of gravity.

When $m > 3$, generally the two discriminant factors associated with the highest values of λ_q are selected for graphical representation, but other axes with values of λ_q close to 1 must be studied because they may realize a subsidiary discrimination between two groups not sufficiently separated by the first two factors.

5.5 Graphical representations

The representation of variables is the same as that realized in principal component analysis, the variables may be plotted with the correlation circle to study the quality of their representation by the discriminant factors. Therefore, the coordinates on these axes are the inter-group coefficients of correlation.

So, on the qth axis, the coordinates of the jth variable will be:

$$\frac{\sqrt{\lambda_q}}{\sqrt{(VG)_{jj}}} \cdot \sum_{j=1}^{p} (V)_{jj'} u_{qj'} \tag{92}$$

The objects can be plotted in the coordinate system defined by the first two discriminant factors, the coordinates are computed according to the following expression:

$$y_{iq} = \sum_{j=1}^{p} x_{ij} u_{jq} \tag{93}$$

In the same way, the coordinates of the centers of gravity and of the new unassigned objects may be calculated. For the qth discriminant factor, the coordinate of the center of gravity of the kth group is:

$$y_q^k = \sum_{j=1}^{p} g_{kj} u_{jq} \tag{94}$$

5.6   Assignment of an object to a group

The affiliation of an object to a group can be made in either of two ways:

— a distance criterion.
— a probabilistic assignment with risk evaluation.

The first method is generally used. For each object, the m distances to the m centers of gravity are calculated according to:

$$d^2(x_i, G^k) = \sum_{q=1}^{m-1} (y_{iq} - y_q^k)^2 \qquad k = 1, \ldots, m \tag{95}$$

the ith object will be assigned to the lth group if

$$d^2(x_i, G^l) = \text{Min}\{d^2(x_i, G^k), \qquad k = 1, 2, \cdots, m\} \tag{96}$$

For all objects of the reference set, an affiliation matrix (A) is calculated; the general element A_{k1} is the number of objects initially assigned to the kth group and that the discriminant analysis assigned to lth group. Thus, the well-assigned objects appear along the matrix diagonal (A_{jj}) and the number of wrongly assigned objects is the sum of the off-diagonal elements. The percentage of well-assigned objects is:

$$100 \cdot \sum_{j=1}^{m} A_{jj}/n \tag{97}$$

5.7   Application example

This example has been described by Wold and Sjostroem (ref. 5, p. 243) in application of their classification method SIMCA. The Carbon-13 NMR chemical shifts of the carbons of a norbornane skeleton, mono-substituted on C-2, are used to determine the endo or exo substitution.

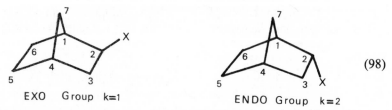

$$(98)$$

EXO   Group   k=1   ENDO   Group   k=2

The data matrix is composed of chemical shift differences in ppm between substituted compounds and the norbornane itself, measured for the seven carbons. Nineteen compounds have been studied and thus, seven characters evaluated for each one (Table 11).

PCA and FDA have been performed on a centered and reduced data matrix. The results from PCA are contained in Table 12. 72% of the total variance is represented by the first two principal axes.

The first component is dominated by the variables, $\Delta\delta C_4$, $\Delta\delta C_5$, $\Delta\delta C_7$, while the second is strongly correlated to $\Delta\delta C_6$ with exclusion of all other variables. The representation of objects (Fig. 15) shows, however, that a separation occurs between the two groups endo and exo, from a mixed effect of the two first components.

Table 11   Reference data set

Compounds		$\Delta\delta C_1$	$\Delta\delta C_2$	$\Delta\delta C_3$	$\Delta\delta C_4$	$\Delta\delta C_5$	$\Delta\delta C_6$	$\Delta\delta C_7$	Group
Exo	Me	6.7	6.7	10.1	0.5	0.2	−1.1	−3.7	1
Exo	NH$_2$	8.9	25.3	12.4	−0.4	−1.2	−3.1	−4.4	1
Exo	OH	7.7	11.3	12.3	−1.0	−1.3	−5.2	−4.1	1
Exo	CN	5.5	1.0	6.3	−0.3	−1.5	−1.6	−1.3	1
Exo	COOH	4.6	16.7	4.4	−0.2	−0.3	−1.0	−1.8	1
Exo	COOR	5.1	16.1	4.2	−0.4	−1.1	−1.4	−2.1	1
Exo	CH$_2$OH	1.8	15.1	4.4	−0.2	0.2	−0.7	−3.3	1
Exo	F	5.6	65.8	10.0	−1.9	−1.8	−7.5	−3.4	1
Exo	Et	4.6	14.8	8.4	0.3	−0.8	0.5	−3.3	1
Exo	OAc	5.2	47.3	10.0	−0.8	−1.3	−5.2	−2.9	1
Endo	Me	5.4	4.5	10.6	1.1	0.5	−7.7	0.2	2
Endo	NH$_2$	6.8	23.3	10.5	1.2	0.6	−9.5	0.3	2
Endo	OH	6.3	42.4	9.5	0.9	0.2	−9.7	−0.9	2
Endo	CN	3.4	0.1	5.5	0.2	−0.7	−4.9	0.0	2
Endo	COOH	4.2	16.2	2.1	0.9	−0.6	−4.8	1.9	2
Endo	COOR	4.0	15.9	2.2	0.7	−0.7	−5.0	1.7	2
Endo	CH$_2$OH	1.7	12.8	4.0	0.4	0.2	−7.2	1.4	2
Endo	Et	3.2	12.5	7.4	0.8	0.5	−7.3	1.6	2
Endo	OAc	3.9	45.5	7.2	0.2	−0.4	−8.8	−1.1	2

Table 12 Principal components from PCA

	λ	%	$\Delta\delta C_1$	$\Delta\delta C_2$	$\Delta\delta C_3$	$\Delta\delta C_4$	$\Delta\delta C_5$	$\Delta\delta C_6$	$\Delta\delta C_7$
PC1	3.25	46.4	−0.66	−0.66	−0.64	0.82	0.72	−0.18	0.86
PC2	1.78	25.4	−0.40	−0.40	−0.61	−0.36	−0.42	0.86	−0.20
PC3	1.19	17.0	0.46	−0.56	0.37	0.36	0.31	0.45	−0.30
PC4	0.50	7.1	0.38	−0.17	−0.11	0.10	−0.43	−0.07	0.33
...	...	...	...	...	...	...	...	...	...

In the first step of factor discriminant analysis, general statistics are computed for each variable in each group: mean, variance and intra-group correlation matrix. Thus, some remarks may be made about obvious differences between the two groups:

— the mean values of $\Delta\delta C_2$, $\Delta\delta C_6$, $\Delta\delta C_7$, are quite different in the two groups.
— A strong correlation is observed in the exo group between the variables $\Delta\delta C_2$, $\Delta\delta C_4$, $\Delta\delta C_6$ while on the other hand, in the endo group, $\Delta\delta C_6$ is correlated to $\Delta\delta C_3$ and $\Delta\delta C_5$.

These observations outline the specific part of the Carbon-13 chemical shift of C-6 in the discrimination of the two groups. Further, to distinguish the two groups, only one factor is calculated (Table 13).

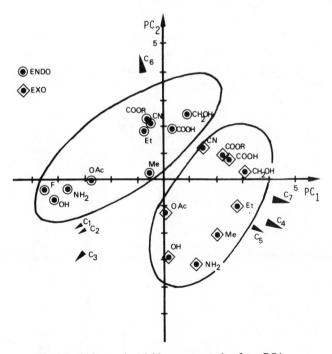

Fig. 15 Object and variables representation from PCA.

Table 13   Discriminant factor

Discriminant power	$\lambda = 0.965$
	Eigenvector coefficient
$\Delta\delta C_1$	-0.17
$\Delta\delta C_2$	-0.23
$\Delta\delta C_3$	-0.27
$\Delta\delta C_4$	0.64
$\Delta\delta C_5$	-0.30
$\Delta\delta C_6$	-0.92
$\Delta\delta C_7$	-0.19

The discriminant power is quite good (0.965) and the greatest contributions to the factor come from $\Delta\delta C_6$ and $\Delta\delta C_4$. This can be readily analyzed in terms of γ effects following substitution on C-2. All the different compounds in the reference data set are well classified, as the two groups are constructed with a low intra-group variance. This is verified by the calculated coordinates of objects (Table 14) which are very close to the center of their group. Furthermore, there is no overlapping of the two clouds of points.

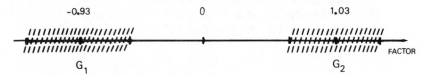

To check the discriminant power of this factor, new compounds have been introduced and are, in fact, well classified (Table 15) even though they exhibit some structural differences from the initial norbornane skeleton (exocyclic double bonds).

Table 14   Coordinates of compounds on the discriminant axis

Group	1(Exo)	2(Endo)
Center of gravity	-0.93	1.03
CH_3	-1.02	1.27
C_2H_5	-1.04	0.92
CH_2OH	-0.89	1.14
COOH	-1.03	0.98
COOR	-0.75	0.96
CN	-0.54	0.67
OH	-1.17	1.17
OAc	-0.72	0.99
NH_2	-1.19	1.21
F	-0.96	—
Min	-1.19	0.67
Max	-0.54	1.27

Table 15 Coordinates and assignment of new compounds

	$\Delta\delta C_1{}^a$	$\Delta\delta C_2$	$\Delta\delta c_3$	$\Delta\delta C_4$	$\Delta\delta C_5$	$\Delta\delta C_6$	$\Delta\delta C_7$	Coordinate on factor 1	Group affected
	5.1	3.1	2.2	0.4	-2.1	-3.2	0.5	1.02	2 (Endo)
	6.6	7.0	10.1	0.2	-1.2	0.5	-3.7	-1.16	1 (Exo)
	endo 4.1	4.2	7.0	0.7	0.5	-7.4	0	1.07	2 (Endo)
	endo 5.6	4.9	7.0	0.2	-1.1	0.2	-3.9	-1.36	1 (Exo)

^{a} Deviations observed with the parent unsubstituted compound.

It must be added that some compounds cannot be well assigned, i.e.

— Bromo C-2 substituted norbornane skeletons.
— 1-Methyl and 7-Methyl substituted norbornanes.

Some effects were not taken into account in the reference data set and therefore classification cannot be good, particularly in the latter case where δ steric interactions between the substituent and methyl groups are present.

5.8   Scope and limitations

Factor Discriminant Analysis (FDA) is a complementary method for PCA and FA, in those cases where inhomogeneity of aggregative type is observed in the population of objects studied. The first step is mainly descriptive:

— Search for factors and variables which allow an optimal discrimination between the different groups.
— Graphical representation of objects with new coordinates on discriminant axis.

Then the objective of the following step is the diagnosis and decision in affiliation of initial and new objects to the different groups on the basis of the initial variable values and the computed coordinates on discriminant axis.

FDA has been used in differentiation of classes of foods and beverages from chromatographic data [15, 32]. Origin and quality criteria of essential oils used in flavors and fragrances have been readily modelled [28, 33]. Further, this method has also been used to interpret pyrolysis mass spectra [34].

A good interpretation of FDA results requires:

— the establishment of steady discriminant factors in a well-defined population of objects in the reference data set.
— the use of statistical tests to accept or deny the hypothesis of the assignment of an object to a group, because the criterion of minimum distance to the different centers of gravity may sometimes be mistaken. Indeed, with this criterion good assignment of initial objects may be realized while the computed discriminant factors are not satisfactory, with poor discriminant power and high residual intra-group variance.

Some constraints on the number of objects (n) and variables (p) must be taken into account:

— n must be very much higher than p.
— the populations of the different groups must have the same magnitude.
— the number of variables must not be too large.

In the latter case, a reduction of the number of variables must be performed by either of the following methods.

— the FDA may be performed with substitution of the initial variables by principal components issued from a previous PCA performed on the same

data set [35]. Thus the number of principal components used in FDA must be selected according to the percent of variance represented. Then, the physical interpretation of discriminant axis becomes more difficult.
— the initial variables may be selected according to their ability to discriminate groups of objects. Different methods have been used in which variables are introduced consecutively one by one. The selection of the new variable is performed according to the increase of the discriminant power of the discriminant functions. This method, called stepwise discriminant analysis, is widely used [36].

Finally, some improvements in FDA methods may be pointed out. An initial classification of objects must be performed by the analyst. The association of FDA with other classification methods (segmentation, clustering etc.) will give an automatic procedure of group constitution before the discriminant analysis. This procedure may be iterative with optimization of the classification.

V.6   GENERAL CONCLUSION

Factor methods are now widely used for chemical data analysis. These methods are mainly descriptive and allow considerable reduction in the size of data sets with a minimal loss of information, and synthetic graphical data representations.

Although these methods do not have the modelling power of regression analysis, they may be used to identify:

— the number of components in multicomponent mixtures as analyzed by UV-visible, fluorescence, IR, mass spectrometry, chromatography (FA).
— the real/physical factors governing experimental data (TTFA).
— the group to which a new object may be assigned from among all the initial groups into which the initial set of objects have been classified (FDA).

Actually, the chemist who wants to apply these metods to his data sets has a large choice among the computer softwares available. Many programs have been written for large, mini and, recently, microcomputers. The reader will find a non-exhaustive list of these in the Appendix.

However, it must not be overlooked that a good interpretation of results cannot be made without the knowledge of the physico-chemical models which govern the experiments performed and the data set obtained. Thus, the human touch will still be indispensable in extracting useful information from lists of results and plotted diagrams.

The chemist's intervention occurs at different steps:

— selection of an initial data set which correctly represents the whole population of objects studied.
— choice of suitable data transformations.

— search for a physical meaning of the calculated factors.
— interpretation of the relative positions of objects.
— group constitution.

For the near future, some general trends may be discerned, at two levels: applications and methods (or tools).

Some of the general fields of application to chemistry, which continue to grow:

— the analytical methods which give complex spectra such as IR, UV-visible, fluorescence, solid-state NMR and where only a small part of all the information contained is actually extracted and used. Factor analysis may be applied to the complete frequency domain for multicomponent analysis, spectra comparison and library searching, chromatographic detection, etc.
— the analysis of complex industrial processes where large amounts of data are available and for which no pure fundamental model can be established. Factor analysis of these data sets will be the first step in modelling these processes.
— the studies of the relationship between structure and physico-chemical properties, such as reactivity, biological activity of organic, inorganic and bioorganic compounds.
— the approach of environment chemical problems in relation with geographical and climatic parameters.

On the other hand, the methods will become easier to use by the non-specialist.

— the different software available will be more and more integrated with many supplementary tools for data introduction, data set pretreatment, factor analysis, modelling, optimization and pattern recognition. They will become available on personal computers for the convenience of the chemist. They will often become part of automated systems of data acquisition and treatment in physico-chemical analysis.
— the search for new starting variables calculated from experimental starting data with random or selected mathematical relationships. These may be automatically explored to check the factor linearity assumption.
— the determination of errors associated with factor loading [37] will help the analyst to define the limits encountered in the use of the factors for object representation. Moreover, statistical tests must be developed to be used in the decision of the assignment of a new object to one group.
— the simultaneous treatment of multiple data sets will allow one to analyze combinations of different analytical methods in multicomponent analysis [38], and also, the time dependent data sets for investigation of the evolution of chemical processes.

Finally, the introduction of artificial intelligence concepts will help the analyst in the interpretation of results, of an object's position, and further, in the automatic generation of groups and clusters of objects.

V.7 REFERENCES

[1]   K. Pearson, *Phil. Magn.*, **2** (11), 559 (1901).

[2]   C. Spearman, *J. Amer. J. Psychol.*, **15**, 72 (1904); ibid., **15**, 201 (1904).

[3]   H. Hoteling, *J. Educ. Psy.*, **24**, 417 (1933); ibid., **24**, 498 (1933).

[4]   R. A. Fisher, *Ann. Eugen. Lond.*, **7**, 178 (1936); ibid., **10**, 422 (1940);
      P. C. Mahalanobis, *Proc. Nat. Inst. Sciences India*, **12**, 49 (1936).

[5]   B. R. Kowalski, *Chemometrics: Theory and Applications*, ACS Sympo-
      sium series, American Chemical Society, Washington, D.C. (1977).

[6]   D. L. Massart, A. Dijkstra and L. Kaufmann, *Evaluation and Optimisa-
      tion of Laboratory Methods and Analytical Procedures*, Elsevier, Am-
      sterdam (1978).

[7]   L. L. Thurnstone, *Multiple Factor Analysis*, Chicago University Press
      (1947).

[8]   H. H. Harman, *Modern Factor Analysis*, Chicago University Press
      (1967).

[9]   E. R. Malinowsky and D. G. Howery, *Factor Analysis in Chemistry*,
      John Wiley (1980).

[10] R. Gnanadesikan, *Methods for Statistical Data Analysis of Multivari-
      ate Observations*, Wiley, New York (1977); K. V. Mardia, J. T. Kent
      and J. M. Bibby, *Multivariate Analysis*, Academic Press, New York
      (1979).

[11] J. M. Romeder, *Méthodes et Programmes d'analyse discriminante*,
      Dunod (1973); Berthier, J. M. Bouroche, *Analyse de données multidi-
      mensionnelles*, PUF (1981); L. Lebart, A. Morineau and J. P. Fenelon,
      Traitement des données statistiques, Dunod (1982); T. Foucart, *Analyse
      Factorielle*, Masson (1982).

[12] K. Varmuza, *Pattern Recognition in Chemistry*, Springer-Verlag, New
      York (1980); C. Albano, W. J. Dunn, U. Edlund, E. Johansson, B.
      Norden, M. Sjostrom and S. Wold, *Anal. Chim. Acta*, **103**, 429 (1978);
      J. MacDonald, *Amer. Lab.*, **2**, 31 (1977); D. B. Pratt, C. B. Moore,
      M. C. Parsons and D. L. Anderson, *Res. Dev.*, **2**, 53 (1978); G. A.
      Erickson, R. W. Gerlach, C. J. Jochum, and B. R. Kowalski, *Cereal
      Food World*, 383 (1981); A. J. Stuper, W. C. Brugger and P. C. Jurs, ref.
      5, p. 165.

[13] S. Wold and M. Sjostrom, ref. 5, p. 243; S. Wold, *Pattern Recognition*,
      8, 127 (1976); S. Wold, *Technometrics*, **20**, 397 (1978); M. Sjostrom and
      B. R. Kowalski, *Anal. Chim. Acta*, **112**, 11 (1979); M. P. Derde, D.
      Coomans and D. L. Massart, *Anal. Chim. Acta*, **141**, 187 (1982); S.
      Wold, C. Albano, W. J. Dunn, K. Esbenben, S. Hellberg, E. Johansson
      and M. Sjostrom, *Proc. IUFOST*, Applied Science Publication, Lon-
      don (1983); J. C. W. G. Bink and H. A. Van't Klooster, *Anal. Chim.
      Acta*, **150**, 53 (1983); S. Wold, C. Albano, W. J. Dunn, K. Esbenben, S.
      Hellberg, E. Johansson, W. Lindberg and M. Sjostrom, *Analysis*, **12**
      (10), 477 (1984).

[14] L. A. Currie, J. J. Filliben and J. R. De Voe, *Anal. Chem.*, **44**, 497R
      (1972); P. S. Schoenfeld and J. R. De Voe, *Anal. Chem.*, **48**, 403R

(1976); B. R. Kowalski, *Anal. Chem.*, **52**, 122R (1980); I. E. Frank and B. R. Kowalski, *Anal. Chem.*, **54** 232R (1982); M. F. Delaney, *Anal. Chem.*, **56**, 261R (1984).

[15] D. Wencker, M. Louis, A. Patris, P. Laugel and M. Hasselmann, *Analysis*, **9**(10), 498 (1981).

[16] R. W. Rozett and E. M. Petersen, *Anal. Chem.*, **47**, 1301 (1975).

[17] E. R. Malinowski, *Anal. Chem.*, **49**(4), 606 (1977); E. R. Malinowsky, ref. 5, p. 53.

[18] R. N. Carey, S. Wold and J. O. Westgard, *Anal. Chem.*, **47**(11), 1824 (1975).

[19] W. F. Maddams, *Applied Spectroscopy*, **34**(3), 245 (1980).

[20] M. A. Sharaf and B. R. Kowalski, *Anal. Chem.*, **53**, 518 (1981); H. B. Woodruff, P. C. Tway and L. J. Cline Love, *Anal. Chem.*, **53**, 81 (1981).

[21] M. MacCue and E. R. Malinowsky, *J. Chromatogr. Sci.*, **21**, 229 (1983).

[22] D. W. Kormos and J. S. Waugh, *Anal. Chem.*, **55**, 633 (1983).

[23] K. B. Wiberg, W. E. Pratt and W. F. Bailey, *Tetrahedron Letters*, **49**, 4861 (1978); ibid., *J. Org. Chem.*, **45**, 4936 (1980); B. Eliasson, D. Johnels, S. Wold and U. Edlund, *Acta Chim. Scand. ser. B*, **b36**, 155 (1982); M. Azzaro, S. Geribaldi, B. Videau and M. Chastrette, *Org. Magn. Reson.*, **22**(1), 11 (1984).

[24] G. L. Ritter, S. R. Lowry and T. L. Isenhour, *Anal. Chem.*, **48**(3), 591 (1976).

[25] E. R. Malinowsky, M. MacCue, *Anal. Chem.*, **49**(2), 284 (1977); F. J. Knorr and J. H. Futrell, *Anal. Chem.*, **51**(8), 1236 (1979); A. Lorber, *Anal. Chem.*, **56**, 1004 (1984).

[26] M. Chastrette, M. Rajzmann, M. Chanon and K. F. Purcell, *J. Amer. Chem. Soc.*, **107**(1), 1 (1985).

[27] G. Buono and B. Baron, Laboratory of Ecole Supérieure de Chimie de Marseille, private communication.

[28] J. Touche, M. Derbesy and J. R. Llinas, *Rivista Italiana*, **63**(6), 31 (1981); R. Cantarel, *Parfums, Cosmétiques, Aromes*, **61**, 73 (1985).

[29] M. Matsubara, *Japan Petroleum Institute*, **23**, 539 (1980).

[30] D. H. Palmer, *J. Sci. Fd. Agric.*, **25**, 153 (1974); H. Martens, Y. Solberg, L. Roer and E. Vold, *Potato Res.*, **18**, 515 (1975); M. Moll, R. Flayeux, J. M. Lehuede and D. Bazard, *Bios*, **8**(9), 32 (1977); M. Moll, T. Vinh and R. Flayeux, *Flavor of Foods and Beverages*, 329 (1978); T. Aishima, *Agric. Biol. Chem.*, **43**(8), 1711 (1979); ibid., **43**(9), 1935 (1979); J. F. Clapperton and J. R. Piggot, *J. Inst. Brew.*, **85**, 271 (1979); W. O. Kwan and B. R. Kowalski, *J. Food Sci.*, **45**, 213 (1980); A. C. Noble, R. A. Flath and R. R. Forrey, *J. Agric. Food Chem.*, **28**(2), 346 (1980).

[31] C. J. Jochum and B. R. Kowalski, *Anal. Chim. Acta*, **133**, 583 (1981).

[32] J. J. Powers amd E. S. Keith, *J. of Food Sci.*, **33**, 207 (1968); L. L. Young, R. E. Bargmann and J. J. Powers, *J. of Food Sci.*, 35, 219 (1970); J. P. Noel, *Bios*, **3**, 148 (1972); A. Dravniek and C. A. Watson, *J. of Food Sci.*, **38**, 1024 (1973); A. Dravniek, H. G. Reilich, J. Whitfield and C. A. Watson, *J. of Food Sci.*, **34**, 38 (1973); R. C. Lindsay, *The*

Brewer Digest, **52**, 44 (1977); T. Aishima, *Agric. Biol. Chem.*, **43**(9), 1905 (1979); P. Schreier and L. Reiner, *J. Sci. Food Agric.*, **30**, 319 (1979); M. S. Cabezudo, M. Herraiz and C. Llaguno, *The Quality of Foods and Beverages*, 225 (1981); E. M. Gaydou, E. Ralambofetra, L. Rakotovao and J. R. Llinas, *Analysis* (1985).

[33] E. M. Gaydou, J. P. Bianchini, R. Randriamiharisoa and J. R. Llinas, *J. Agric. Food Chem.* (1986) (in preparation).

[34] W. Windig, J. Haverkamp and P. G. Kistemaker, *Anal. Chem.*, **55**, 81 (1983).

[35] R. Hoogerbrugge, S. J. Willigg and P. G. Kistemaker, *Anal. Chem.*, **55**, 1710 (1983).

[36] J. M. Romeder, C. Guille, Y. Hecht, C. Garcon and M. H. Giard, *Path. Biol.*, **19**(19), 871 (1971); C. R. Rao, *Linear Statistical Inference and its Applications*, John Wiley and Sons, p. 574 (1973).

[37] P. M. Owens, R. B. Lam and T. L. Isenhour, *Anal. Chem.*, **54**, 2344 (1982); B. A. Roscoe and P. K. Hopke, *J. Radioanal. Chem.*, **70**, 483 (1982).

[38] C. J. Appellof and E. R. Davidson, *Anal. Chim. Acta*, **146**, 9 (1983).

V.8 APPENDIX: SOME MULTIDIMENSIONAL SOFTWARES AVAILABLE

FACTANAL (FA, TTFA)	E. R. Malinowski *et al.* Program 320, Cl.CP.E. Indiana University, Bloomington, Ind. (1976).
BMD (Package)	University of California Press, Berkeley. W. J. Dixon, ed (1968).
ARTHUR (Package)	A. M. Harper *et al.*, ref. 5, p. 14.
SIMCA (Pattern Recognition)	S. Wold and M. Sjostrom, ref. 5, p. 243.
FACANAL (FA, TTA)	P. C. Gillette, J. P. Lando and J. L. Koenig, *Anal. Chem.*, **55**, 630 (1983).
SPSS (Package)	N. H. Nie *et al.*, *Statistical Package for Social Sciences*, McGraw-Hill, New York (1975).

BASIC Software (PCA, FDA, ...)   T. Foucart, *Analyse factorielle*,
                                        Masson (1982).

FORTRAN Software (PCA)   L. Lebart, A. Morineau and J. P.
                                        Fenelon, *Traitement des données
                                        statistiques*, Dunod (1982).

ANDON (PCA, FDA)   J. R. Llinas, Computer Center,
                                        Ecole Supérieure de Chimie de
                                        Marseille (1983).

CHAPTER VI

Computer Aids to Crystallography

Marcel Pierrot

University of Aix-Marseilles, France

VI.1 INTRODUCTION

During the last few years, crystal structure analysis has become a powerful tool for the investigation of molecular conformations. Largely because of the advances in computer facilities, the study of a molecular structure by X-ray diffraction is now a fairly automatic procedure. Experimental methods for measuring the X-ray diffraction pattern of a crystal as well as methods for solving and refining the structure are very efficient, but not fully automatic. In most cases, X-ray data can be rapidly and correctly interpreted in terms of molecular conformation. However, the procedure may fail even in the absence of systematic experimental errors, for instance, in the case of symmetry ambiguity or with twinned or disordered crystals. Nevertheless, X-ray analysis is the main source of structural information of more or less complex molecules and therfore it is valuable for chemists to acquire a basic knowledge of the methods of X-ray crystallography [1, 2].

The aim of this chapter is to provide to non-specialists a minimum of crystallographic information and to draw attention to the computational aspects of the different steps of a crystal structure analysis.

VI.2 PRINCIPLES AND METHODS IN CRYSTALLOGRAPHY

2.1 Geometric considerations

In contrast with amorphous solids or liquids, the crystalline state is characterized by a regular packing of molecules. Atoms or molecules in a crystal are

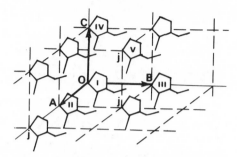

Fig. 1   The crystal lattice and the unit cell parameters.

arranged in such a way that they can be described with the help of a three-dimensional lattice, which is obtained as follows:

First, an origin O must be chosen which can be attached to any atom of a molecule, for instance to the jth atom of molecule I (see Fig. 1). Let A, B and C be the centers of the same j^{th} atom of molecules II, III and IV which are the nearest neighbors of molecule I. Then $\vec{a}$, $\vec{b}$ and $\vec{c}$ are the vectors defined by $\overrightarrow{OA}$, $\overrightarrow{OB}$ and $\overrightarrow{OC}$ respectively.

Now, any molecule of the three-dimensional crystal can be obtained from molecule I merely by a translation to the extremity of the vector:

$$p\vec{a} + q\vec{b} + r\vec{c}$$

where p, q and r are integers and the only condition at the moment is that $\vec{a}$, $\vec{b}$ and $\vec{c}$ are not coplanar. Thus a, b and c are the base vectors of a three-dimensional lattice called the crystal lattice. Obviously, the geometry of the lattice is independent of the choice of the origin and depends only on the relative arrangement of the neighboring molecules, e.g., the molecular packing.

The unit cell is the solid that can be built from the cell parameters a, b and c and which includes the molecular motive (composed of one or several equivalent molecules). A very realistic picture of a crystal could be given by the spatial packing of identical unit cells: in each cell the same motive has rigorously the same orientation.

A lattice plane, Fig. 2, contains at least three lattice points for which p, q and r are integers. In fact, a lattice plane is a member of a plane family composed of an infinity of parallel and equidistant planes, Fig. 3. In a three-

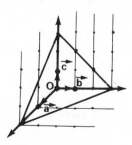

Fig. 2   A lattice plane.

Fig. 3   A family of (*hkl*) planes.

dimensional crystal lattice an infinity of plane families can be defined. In order to denote the orientation of a plane with respect to the unit cell axes, crystallographers use the Miller indices (*hkl*). Miller indices are three integers which are proportional to the reciprocals of the intersections of the plane with the axes *a*, *b* and *c*.

The distance between two consecutive planes of an (*hkl*) family is the equidistance d_{hkl} which can be easily calculated from unit cell parameters. The Miller indices (*hkl*) and the equidistances d_{hkl} are characteristic of a plane family and will be referred to later.

It is well known that according to the external geometry of the unit cell and to its related symmetry, a crystal lattice belongs to one of the seven crystal systems ranging from the triclinic system, to the cubic system. Moreover, taking into account the internal symmetry of the unit cell which links the equivalent parts of the molecular motive, crystal systems are subdivided into 230 space groups which are described in the *International Tables of Crystallography* [3].

The asymmetric unit cell is that fraction of the cell from which the unit cell can be completed by the play of the symmetry operators of the group. Therefore, the study of a crystal structure is limited to the knowledge of that part of the motive contained in the asymmetric unit.

The determination of the unit cell parameters and of the space group as well as the number, Z, of molecules contained in the cell is a prerequisite for starting the structure determination [4]. For that purpose it is sometimes advisable to undertake a preliminary study by taking diffraction pictures on a camera. However, in most cases cell parameters and space groups can be directly determined with the diffractometer. The Z number can be easily checked by measuring the crystal density.

2.2   The interaction between X-rays and crystals

2.2.1   The Bragg Law

Let us see now what happens when a crystal is set in an X-ray beam. First of all the incident beam of intensity I_0 will be partly absorbed according to the Beer–Lambert law:

$$I = I_0 \exp[-\mu\rho x] \tag{1}$$

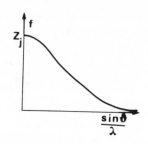

Fig. 4 The atomic scattering factor f_j.

where I is the intensity after crossing a thickness X of a crystal where the
density is ρ and the mass absorption coefficient is μ. For a given radiation, μ
only depends on the nature of the atoms and can reach very high values for
heavy atoms (HA) having high atomic numbers. In such cases, absorption
can be the source of systematic errors in diffraction measurements and care
must be taken. From a practical point of view it is necessary to minimize the
absorption effects. Crystals of small dimensions, a few tenths of millimeters or
less, with an isotropic external shape must be used. In some cases it will be
necessary to apply an absorption correction (see 4.2 Data Reduction).

X-rays are a form of electromagnetic radiation having wavelengths in the
range of the interatomic distances in crystals; the energy $h\nu$ of the X-ray
photon is close to the atomic bond energy. Consequently, although the
theoretical principles of X-ray and ordinary optics are the same, the
applications of X-rays optics differ from those of radiations such as visible or
UV [5, 6].

This is the case for the interaction between X-rays and the periodic lattice
of a crystal. Since Thomson's study, it is known that the interaction between
X-rays and molecular material is of electronic origin. The amplitude of the
waves diffused by an atom j is called the atomic scattering factor f_j. The
variation of f_j versus the diffusion angle θ is shown in Fig. 4. At $\theta = 0: f_j = Z_j$
the atomic number of atom j.

What is interesting in the phenomenon of the X-ray crystal interaction is
that, due to the agreement between the incident wavelength λ and the period
of the crystal lattice, the waves diffused by individual atoms can interfere and
give rise to an interference lattice. The interference lattice, also called the
diffraction lattice, is composed of a multitude of diffraction beams. As shown
by the Bragg relation, a diffracted beam can also be seen as a beam reflected
by a lattice plane (hkl), Fig. 5, and the Bragg relation gives the angles of the
reflection:

$$2d_{hkl} \sin \theta = n\lambda \qquad (2)$$

where n is an integer.

The Bragg angle 2θ is the angle between the incident and the reflected
(diffracted) beam. The beam reflected by the plane (hkl) is also the beam
diffracted in the direction hkl.

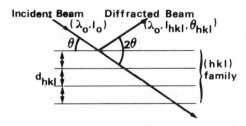

Fig. 5 The Bragg law.

The Bragg relation is fundamental as it shows that a crystal plane can reflect the X-rays, but the reflection only occurs for some incident angles, those angles for which the Bragg relation is met: X-ray reflection is a selective reflection. This relation also shows that distances in the crystal lattice can be obtained from the measurement of the Bragg angles, if the wavelength λ is known.

As there are an infinity of plane families, one can understand that the number of beams which can be reflected by a crystal is very large, the only limiting condition is:

$$\sin \theta \leq 1 \text{ e.g. } d_{hkl} \geq n\lambda/2$$

2.2.2 The structure factors and the electron density function

The direction of the diffracted beams depends only on the distances d_{hkl}; the intensity of the beams will depend on the nature of the atoms and on the position of the atoms in the unit cell. It can be shown that the amplitude of the wave diffracted in the direction hkl is the structure factor F_{hkl} given by the relation:

$$F_{hkl} = \sum_{j=1}^{n} f_j \exp[-2\pi i(hx_j + ky_j + lz_j)] \tag{3}$$

where x_j, y_j and z_j are the fractional coordinates of the atoms in the unit cell.

In the general case, the structure factor is a complex number defined by a module F_{hkl} and a phase φ_{hkl} (Fig. 6).

The intensity is the square of the amplitude and is given by:

$$I_{hkl} = |F_{hkl}|^2 \tag{4}$$

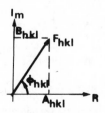

Fig. 6 Representation of F_{hkl} in the complex plane.

When the crystal structure possesses an inversion center one can easily show that the structure factor is no longer a complex number but becomes real, the phase φ being equal to 0 or π and:

$$F_{hkl} = \pm |F_{hkl}| \qquad (5)$$

Looking back to the expression (3) for the structure factor one can say that the wave diffracted in the direction hkl is the resultant of the waves diffused by the individual atoms of the cell. Let us consider now the unit cell as formed by a set of infinitesimal units of volume dv. Each volume dv contains some electron distribution which can be described by a function $\rho(xyz)$. This function $\rho(xyz)$ is the electron density function giving the number of electrons in dv. For instance, $\rho(xyz) = Z_j$ if x, y and z are the coordinates of atom j; $\rho(xyz)$ is equal to zero where the probability to find an electron is nul.

One can now define the structure factor as the result of waves diffused by units dv and the analogy with the first expression for F_{hkl} leads to the next expression:

$$F_{hkl} = \int_v \rho(xyz) \exp[-2\pi i(hx + ky + lz)]\, dv \qquad (6)$$

which gives to F_{hkl} the form of a Fourier transform, a well known formulation in chemistry:

$$F_{hkl} = FT\{\rho(xyz)\}$$

and by using the inversion property of the Fourier transform:

$$\rho(xyz) = FT(F_{hkl}) = k \int_{v*} F_{hkl} \exp[2\pi i(hx + ky + lz)]\, dv* \qquad (7)$$

where K is a constant and $v*$ a reciprocal volume $1/v$.

Since F_{hkl} is nul when h, k and l are not integers, one can substitute a discrete summation for the integral. In taking into account that the structure factor modules in two opposite directions given by indices hkl and $-h - k - l$ are equal (Friedel law) one obtains for $\rho(xyz)$ the form of a real number:

$$\rho(xyz) = k \sum_h \sum_k \sum_l F_{hkl} \cos 2\pi(hx + ky + lz) \qquad (8)$$

Maps of electron density calculated with this formula give a picture of the three-dimensional structure of a molecule. Unfortunately, they can not be calculated directly from the experimental data. Actually, what is measured by X-ray diffraction is the intensity I_{hkl} and for each (hkl) reflection the module of the structure factors is only known:

$$F_o = \sqrt{I_{hkl}} \qquad \text{(O is written for observed)}$$

since the phases φ are not experimentally accessible. Crystallographers have developed methods in order to solve the phase problem.

2.3 Methods for solving the phase problem

If the phases φ_{hkl} cannot be measured it appears obvious that they have to be estimated or calculated in some way.

2.3.1 *From the phasing model to the molecule*

Suppose that among the n atoms of the molecular structure to be solved, the atomic positions of m atoms ($m < n$) are known. The means for finding these atomic positions will be seen later. But now the contribution of the m atoms to the structure factors can be calculated. For each reflection (hkl) we have:

$$F_c^m = \sum_{j=1}^{m<n} f_j \exp[-2\pi i(hx_j + ky_j + lz_j)] = |F_c^m| \exp[-i\varphi_c^m] \qquad (9)$$

If the phasing model (the structure of the m atoms) is not too far away from the real structure, then for most of the reflections the angles φ_c^m are a good approximation of the real phases φ_{hkl} and can be used for further calculations in the expression (8) of the electron density. Two electron density maps can be calculated [7], both with φ_c^m as phases:

— The observed electron density map uses the F_o as amplitudes for teach term of the Fourier series:

$$\rho_o(xyz) = \sum_h \sum_k \sum_l F_o \cos 2\pi(hx + ky + lz - \varphi_c^m) \qquad (10)$$

and shows the m atoms plus the ($n - m$) unknown atoms, or part of them;
— Difference-Fourier has the differences $\Delta F = F_0 - F_c$ as Fourier coefficients:

$$\Delta\rho(xyz) = \sum_h \sum_k \sum_l \Delta F \cos 2\pi(hx + ky + lz - \varphi_c^m) \qquad (11)$$

and only reveals the ($n - m$) lacking atoms, or part of them.

The number of new atoms which can appear on the ρ_o or $\Delta\rho$ maps depends on the quality of the φ_c^m phases and also on the number of terms used in the Fourier calculation. If the number m' of atoms located after this step is still less than n, another Fourier analysis must be undertaken using $F_c^{m'}$ and $\varphi_c^{m'}$ until m' reaches n. Provided the phases $\varphi_c^{m'}$ can be improved and the number of Fourier terms can be gradually increased to the total number of observed reflections, the coordinates of the n atoms of the molecule can be rapidly obtained.

However, if the number m of atoms cannot be increased, then the starting phasing model has to be revised.

A classical tool to check the gradual improvement of the resolution is to calculate the residual R-factor defined as:

$$R = \frac{\sum_r |F_0 - |F_c||}{\sum_r F_0} \qquad (12)$$

where Σ_r means that the summation is done with all reflections (hkl).

Values of 0.35 to 0.40 are commonly observed for the R-factor at the beginning of the work. Theoretically each improvement in the resolution should correspond to a gain in the R-factor, which can reach very low values, sometimes less than 0.05, when the structure is finished.

Once the molecule is completed by Fourier analysis one has to find the best set of coordinates for the n atoms. According to the Legendre principle (1806) this can be done by minimization of the square-difference:

$$M = \sum_r (F_o - |F_c|)^2.$$

optimum parameters of the F_c function are obtained in the least-squares method [1] by setting the F_c partial derivatives to zero. The resulting equations are solved in an iterative matrix calculation which will be discussed later.

At this point in the discussion one must emphasize the fact that if the starting phasing model is correct, Fourier analysis and the least-squares method will lead more or less quickly to the final refined structure. But the problem which remains is:

2.3.2   How to obtain a phasing model

For this purpose two different approaches are available depending on the nature of the molecule under study.

First: the Heavy Atom (HA) method. Suppose that the molecule contains one or several HA's: atoms with an atomic number Z equal to or greater than, say, 20-25 electrons. Then the position of these atoms in the unit cell can be obtained from the interpretation of a modified electron density function in which the structure factors (modules and phases) are replaced by intensities, the phases vanishing. Therefore this function, called the Patterson function [8], can be directly calculated from experimental data. As long as the number of HA's is small and their atomic number rather high, the Patterson function can be solved easily.

This method is very useful, for instance, in organo-metallic chemistry [9, 10] and is still the only method used to solve the structure of biological macromolecules [11 to 14].

However, the interpretation of the Patterson function becomes rapidly tedious when the number of HAs is too large and even becomes impossible when the atomic number is not high enough.

Second: Direct methods. Direct methods consist of the determination of crystal and molecular structures directly from the intensities of the diffracted beams. They imply the use of procedures involving the direct evaluation of the phases of diffracted amplitudes without the need of any special structural information. The mathematical complexity of these methods compensates for the considerable success toward the goal of making crystal structure determination a fairly routine operation. The discovery of special mathematical relationships and of their useful practical aspects took place over several years and has led to the development of modern crystal structure analysis. The main concepts of these methods may be summarized as follows.

The basic property leading to the practical solution of the phase problem is the non-negativity of the electron density distribution in the crystal unit cell. This was utilized by Karle-Hauptman [15] to obtain the main formula for phase determination, which was preceded by the derivation of a complete set of inequalities by Harker and Kasper [16]. Harker and Kasper inequalities are fairly simple relations which express the phases of structure factors in terms of measured magnitudes.

The next logical step was to consider the problem of developing the probabilistic aspects of the phase relationships by introduction of the joint probability distribution as a mathematical tool [17, 18]. The result of this work was the derivation of the formula labeled Σ_1, Σ_2. These formulas are written in a form that permits probabilistic conclusions to be drawn by the introduction of the normalized structure factors E. The determination of the E's is based on Wilson's statistics [19] which presumed a random distribution of the atoms in the unit cell. Besides having useful statistical properties, the normalized structure factors E represent scattering from point atoms and thereby enhance the probabilistic implications of phase formulas.

However, a period of more than ten years elapsed between the elaboration of the basic relationships and the publication of the structure of L-arginine dihydrate, the first non-centrosymmetric crystal structure solved by using direct methods [20], thus indicating the difficulties in applying the theory in practice.

Before giving the most useful relationships for practical solution of crystal structure, let us consider the Miller indices $h_1k_1l_1$, $h_2k_2l_2$ and $h_3k_3l_3$ of three reflections which are related in such a way that $h_3 = h_1 - h_2$ $k_3 = k_1 - k_2$ and $l_3 = l_1 - l_2$. For instance, $h_3k_3l_3 = 322$; $h_1k_1l_1$ and $h_2k_2l_2$ are 423 and 101 or 353 and 021 respectively.

H will be used for $h_1k_1l_1$, K for $h_2k_2l_2$ and $H - K$ for $h_3k_3l_3$; H, K and $H - K$ will be the indices of three reflections involved in phase relationships and will be called 'triples'.

For centrosymmetric crystals where the phases are either 0 or π, or equivalently the signs of the structure factors are either $+$ or $-$, the Σ_2 formulas are [17]:

$$sE_H \sim s(E_K E_{H-K}) \tag{13a}$$

or for several contributors:

$$sE_H \sim s\left\{\sum_{K_r}(E_K E_{H-K})\right\} \tag{13b}$$

where the symbols s and $\sim$ mean 'the sign of' and 'almost equal to'; K_r refers to a summation on all K and $H - K$ reflections restricted to the largest magnitudes. Expressions related to (13a) and (13b) have appeared in a variety of forms as Sayre's relation [21] or as formulas derived by algebraic investigations [22 to 24].

The probability associated with the Σ_2 relationships was calculated by Woolfson [25] and shows that the larger the magnitude of E_H, E_K and E_{H-K} the higher will be the probability of a correct sign assignment.

The application of the method can be understood with the image of the snowball effect as illustrated in Fig. 7. So is a nucleus of a few (three to five) known phases which will be referred to below as the initial set of phases. The triples which can be built up from these reflections are used in the Σ_2

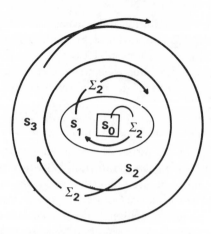

Fig. 7   Snowball effect in phase determination.

relationships, hence the signs of new reflections can be determined. These new reflections and those of the nucleus form the S_1 shell to which Σ_2 is applied and leads to the determination of new signs. This procedure is repeated until the sign of a large number of reflections can be determined. However, the process is dependent upon the sequential determination of phases: each shell in the ball, each step in the chain of events must be correct, otherwise this should collapse. In order to insure the greatest probability that a sign or a phase is correct, a phase determination:

— should be initiated with reflections corresponding to the largest E magnitudes;
— is limited to the set of the highest E, about 10 to 15% of the total number of reflections.

For non-centrosymmetric crystals where the phases φ of structure factors can have any value between $-\pi$ and $+\pi$, two phase relationships are particularly important:

— The Σ_2 relationships [15, 20, 26]:

$$\varphi_H \sim \langle \varphi_K + \varphi_{H-K} \rangle_{K_r} \tag{14}$$

— The tangent formula [18]:

$$\tan \varphi_H = \frac{\sum_K E_K E_{H-K} \sin(\varphi_H + \varphi_{H-K})}{\sum_K E_K E_{H-K} \cos(\varphi_H + \varphi_{H-K})} \tag{15}$$

The variance associated with the determination of φ can be derived from a probability formula of Cochran [27] and is given by Karle and Karle [26]. Larger magnitudes of E lead to smaller variances of the determined phase, a situation comparable to the higher reliability of sign determination associated with higher magnitudes of E in centrosymmetric crystals.

Details on computational procedures will be given later. Only two points remain to be discussed before leaving these general considerations.

(i) The initial set of phases. Direct phasing processes must be started with a limited number of known phases. It can be shown that these phases result:

— from the conditions that fix the origin of the unit cell and the enantiomorph for non-centrosymmetric crystals [28, 29];
— from auxiliary formulae referred to as Σ_1, Σ_3 which are space group dependent [30 to 32].

The choice of the phases included in the initial set is perhaps the most important element influencing the success or failure of a direct solution of the phase problem. What criteria does a good starting set fulfil and how does one find the better combination of starting reflections? This requires very careful study [33] and is usually achieved by adopting selection criteria based on probability and frequency of interactions:

— the reflections should have high E values;
— the reflections should be involved in a large number of Σ_2 interactions, particularly with other reflections with high E values, i.e. in reliable Σ_2 relationships.

(ii) Output of the results. The phases resulting from a direct phase determination process are used in a Fourier calculation where the Fourier coefficients are the normalized structure factors E. Depending on the quality of the phase determination the so-called E-maps [34] may reveal a more or less important part of the molecular motive which will serve as the phasing model for completing the structure.

VI. 3   DIFFRACTOMETERS AND THE COLLECTION OF DATA

Once a crystal has been obtained, the study of its three-dimensional structure can begin. A diffractometer [35] is generally employed for the measurement and the collection of the diffracted intensities.

3.1   General Description

Diffractometers were one of the first analytical instruments to be controlled by computer. Prior to this, manual machines had shown that intensity measurements made with detectors were more accurate than film methods. The tedious film recording–measuring work was replaced by a repetitive sequence of setting and measuring operations. Data collection with a single crystal diffractometer requires that the intensity of thousands of Bragg reflections must be measured. For each reflection, crystal and detector must be precisely oriented. Following the development of computers as instrument controllers, it was natural to automate this extremely repetitive procedure. Automatic diffractometers are now sophisticated machines, in most cases manufactured by private companies.

This paragraph is mainly concerned with:

— the goniometer which holds the crystal and the detector;
— the computer which pilots the goniometer and collects data.

The X-ray source is generally a sealed tube with Cu or Mo as metal-anticathode. The high voltage generator must provide maximum reliability and safety and ensure optimal stabilization of the high voltage and tube current, e.g., less than 0.01 V variation for a line fluctuation of 10%. The incident beam is monochromatized by an absorption filter or by a crystal monochromator. It is worthwhile to note, even by average users not directly concerned with these problems, that the technical manual edited by the manufacturers must not only contain a clear description of the different parts of the diffractometer (goniometer, high-tension generator, electronic detector...) but also all the details relative to their utilization in optimum conditions. It is important to realize that this will determine the quality of the measurement.

The manual must also give a procedure for the alignment of the goniometer, particularly in the case where a monochromator is employed. This procedure is generally performed with the help of a test-crystal which is also used as a reference for periodical checks of the installation.

3.1.1 The Goniometer

Whatever the make, all diffractometers share some common building principles, particularly the fact that the incident and diffracted beam are in the horizontal plane: the equatorial plane. The bulky X-ray source is obviously a fixed direction and the detector can only move in the equatorial plane around the vertical θ-axis, Fig. 8.

Therefore any (hkl) plane will be a diffracted plane if:

(1)   it can be set vertically;
(2)   the incident angle is equal to the Bragg angle for this particular (hkl) plane.

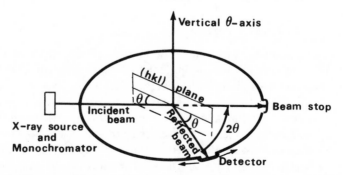

Fig. 8   The equatorial plane of the goniometer containing the incident and reflected beams.

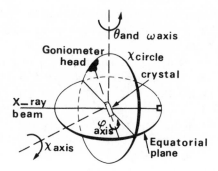

Fig. 9   The Eulerian geometry of the goniometer. The goniometer head is moved along the χ-circle by the φ-axis, the φ-axis is the rotation axis of the goniometer head.

When these conditions are fulfilled, the detector has to be moved to the 2θ position and then the intensity of the diffracted beam can be measured.

In order to achieve these setting conditions the rotation of the crystal around three axes is necessary and can be described in the Eulerian geometry, Fig. 9. The crystal is held by a glass capillary to a goniometer head, a device which also allows the user to center the crystal at the intersecting point of the axis. The goniometer head is linked to a vertical circle and can move around the χ-axis, which is always perpendicular to the χ-circle, Fig. 10. The crystal can be rotated around the φ-axis which is the axis of the goniometer head.

Finally, the φ-circle holding the goniometer head can move around a vertical ω-axis.

All diffractometers but one have an Eulerian goniometer with a physical χ-circle. The Kappa geometry of the CAD4 diffractometer [36] is rather difficult to describe. It is easier to give the crystal setting of the Kappa geometry in the Eulerian geometry and then use the geometrical relationships between Kappa and Eulerian geometries.

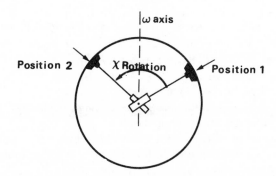

Fig. 10   The χ-circle, moving the goniometer head from Position 1 to Position 2 is achieved by a rotation around the χ-axis which is normal to the χ-circle.

As shown on Fig. 8, the reflecting (*hkl*) plane bisects the 2θ angle. Setting the crystal at the bisecting position is the usual method for observing the diffracted beam.

There are several ways to record the intensity diffracted by a plane. In the usual methods, the goniometer angles are first set at values corresponding to the bisecting position for the (*hkl*) plane. Then the crystal is slightly tilted from the Bragg position while the diffracted peak is recorded by the detector.

In the ω scan, the detector is stationary and the crystal moves through the ω-axis. This technique is often used for crystals with large unit cells, like protein crystals, for instance, where there is a probability of overlapping reflections.

In the $\theta:2\theta$ scan, the detector and the crystal move simultaneously, the crystal around the θ-axis at half the speed of the detector moving around the ω-axis. Usually the entire peak is scanned from the background on the low $\theta - \delta\theta$ side to the background on the high $\theta + \delta\theta$ side. By making a scan range somewhat larger than necessary this method will produce an accurately integrated intensity in spite of slight instrumental errors in any of the axes or errors in centering the crystal, which is assumed to be always bathed in an uniform X-ray beam.

3.1.2 The computer

Early automatic diffractometers were built at IBM Research [37] and at Oak Ridge National Laboratory [38]. The programs developed by Busing and Levy [39] were written on a DEC-PDP-5, which is compatible with the PDP-8 series of minicomputer. As a result, most of the diffractometers of the first generation were supplied with software which was an adaptation of the original Busing and Levy programs written in the computer's assembly language. For instance, the Syntex P 1 programs fit in 4096 16-bit words of core memory and the P 2 use 8192 words of core [40]. Although it is always possible to modify any of the instructions of a machine language program, few crystallographers are familiar with the assembly language of a minicomputer. In order to achieve greater flexibility in computer controlled diffractometers, some manufacturers have decided to deliver FORTRAN diffractometer systems. In this second generation of automatic diffractometers, it is necessary to use a disk for program overlays and storage. The Syntex P 2 FORTRAN version now requires 24 K of core memory and a 1.2 K word disk. The disk is also used to store the data, which can be transferred to a magnetic tape for processing. In order not to sacrifice efficiency, parts of the programs dealing directly with instrument operations are still in assembly language.

More recently, the power of minicomputers has increased to the point where all calculations performed by crystallographers can be made on a computer whose size is comparable with that used by modern diffractometers. On some of the latest commercially available diffractometers, a microcomputer is used as an interface to drive the axis positioning motors and to open and close the X-ray shutter ... while the minicomputer does more sophisticated calculations to determine, for instance, where the axes should be positioned. Data collection being slow compared with the speed of minicomputers, a system with both a micro and a minicomputer has a large excess of mini-CPU time. In this situation, which will be examined later, the minicomputer can be time-shared with some other tasks.

3.2   Control routines and crystallographic programs

In controlling a diffractometer, there are relatively few specific operations requiring machine code handlers: aperture of the disk, opening and closing the beam shutter and beam attenuator, control of the scaler and timer and control of the goniometer axes. For driving the axes, stepping motors or regular motors and encoders, which are absolute position devices, can be used. The best arrangement is probably the use of stepping motors, allowing variable scans, and encoders as a check.

Once a crystal has been mounted on a goniometer head and optically centered on the goniometer, all the operations, from the unit cell calculations to the data collection process can be done without, or almost without, manual intervention. Programs give instructions to the operator via a terminal or accept instructions from the operator in a convenient interactive manner. Two steps are now considered.

3.2.1   Determination of the crystal orientation and of the unit cell parameters

Busing and Levy [41] have shown that the θ–φ–ω–χ angles can be obtained from the (hkl) indices of the reflection and from a 3×3 orientation matrix A. Vice versa, the nine elements of the matrix A and the cell parameters a, b, c, α, β, and γ can be obtained from the accurate measurement of the setting angles of several reflections. The Busing–Levy relationships are used in the routine procedure for orienting a crystal on a goniometer and determining the components of the matrix A.

(1)   A few reflections (15 to 25) can be found by a systematic search in the diffraction space. By scanning over sufficiently large ranges of θ, φ and χ, some moderate or strong reflections with an intensity greater than a preset minimum are detected.

(2)   The centers of these reflections are determined in several ways depending on the type of diffractometer. The actual alignment of a reflection involves adjusting the instrumental angles in a systematic way so that the diffracted beam passes through the center of the detector aperture. The procedure is very well described by Busing [42].

(3)   The refined angular settings are then stored in a list. The problem is then to assign Miller indices to these reflections (e.g. to attach the reflections to a crystal lattice) and thus find the unit cell parameters.

Several indexing algorithms have been given by Philips [43], Sparks [44] and Mighell and Rodgers [45] so as to overcome the possibility of an incorrect indexing. However, these methods require some skill in the interpretation of the results, for it is possible to miss the true solution as a submultiple of the one found. It is therefore wise to check the results by:

(a) using another program which will try to reduce the primitive cell or to transform it in the appropriate Bravais lattice [1];
(b) verification of the symmetry;
(c) eventually taking photographs on a diffraction camera.

Difficulties can be encountered with crystals of poor quality or when the sample is not a monocrystal (a twin or polycrystals) or when it cannot be properly fixed at the center of the goniometer.

(4) When an orientation matrix has been obtained, the setting angles of any reflection can be calculated from its (hkl) indices. These can be used to find new reflections which after centering are introduced in a least squares procedure in order to obtain a refined orientation matrix.
(5) Alternative procedures can be very useful in some cases.
  (a) Reflections can be input from a diffraction photograph taken with an X-ray polaroid cassette placed on the front of the detector.
  (b) Special procedures may be used when the crystal has been pre-oriented and the unit cell parameters determined before being put on the diffractometer. The search for reflections will then be accelerated and indexing done unambiguously.

3.2.2 Automatic data collection

When an accurate matrix A has been determined, data collection can begin. For each reflection to be measured, the precision with which the goniometer axes will be positioned will depend on the quality of the orientation matrix. For intensity measurement an angular precision of a few per cent of a degree is required and cannot be achieved with a crude matrix.

The computers which control diffractometers have been programmed to collect data in many different ways and an experimenter can choose the method which best suits the measurement to be performed. For instance, the operator can choose the type of scan (ω or θ: 2θ scan) and the limits of the data collection. The indices of the reflections are generally given by a routine which generates the reflections in a systematic way minimizing symmetry redundancy and the time spent in driving the axes between two reflections.

All the parameters of the data collection cannot be examined here. It is more interesting to see how the computer is capable of making decisions in the course of the collection. As a first example, the computer must set an attenuator filter on the front of the detector to keep the counting rate at a safe level when a strong reflection is encountered.

By performing a fast pre-scan across the Bragg position of a reflection, it is possible to set rapidly a rough estimate of its intensity. Thereafter the computer can:

— decide not to measure this reflection if the intensity is too weak, i.e., less than a preset threshold;
— Adjust the scan speed for a second scan so as to reach a defined statistical accuracy.

Reference reflections should be monitored at regular intervals of the data collection and checks made that any changes are within acceptable limits. This facility is particularly important for samples having properties such that their structure slowly changes during prolonged exposure in the X-ray beam. This is true for photosensitive products as well as for oxygen or water sensitive molecules or for crystals containing solvent molecules. In such cases, it is necessary to mount the crystal in a capillary filled with a suitable protective atmosphere as is done for protein crystals [46].

By recording the intensity of reference reflections at regular X-ray exposure time periods the routine:

— controls the decay intensity and eventually stops the data collection if the loss of intensity is too large, say more than 20–30 %;
— stores the decay variation curve which will be used if a correction is necessary.

During data collection it is possible that the crystal can move, especially when the samples are mounted in capillaries. Reference reflections are monitored by the routine to provide an angular stability control. As soon as an angular shift greater than an acceptable value is detected the computer can make the decision to re-center the crystal and to calculate a new orientation matrix. However, only orientation changes can be corrected in this way; translations are ignored. Therefore it is better to warn the operator before continuing.

VI.4   PROGRAM SYSTEMS IN X-RAY STRUCTURE ANALYSIS

4.1   The different steps in structure determination

The main characteristic of X-ray structure analysis which distinguishes it from other methods used in chemistry, e.g., nmr or mass spectroscopy analysis, is that there is no direct interpretation of the experimental data. As indicated on Fig. 11, the path from the diffractometer data file to the study of the molecular conformation is far from being a direct way. The different steps which are involved are examined now.

4.2   Data reduction

The conversion of the diffractometer intensity measurement data file into a file of a consistent set of observed structure factors F_o and estimated standard deviations σ_o is performed by a group of programs.

Some of the factors which are involved in this process are briefly indicated below.

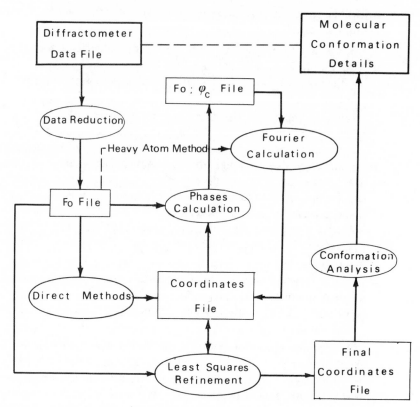

Fig. 11   Flow-chart of the different steps in the determination of a crystal structure.

For the extraction of the I_{NET} from measured I_{TOT}, it is necessary:

(a)   to subtract the background;

(b)   to apply coefficients taking into account the scan speed and eventually the attenuator filter factor.

For each reflection the intensity must also be corrected for the classical Lorentz-polarization factor [47], for decay and absorption [48, 49] if necessary and rarely for multiple reflection and extinction [50].

In the case of several measurements of the same or symmetry related reflection, a valid set of unique data must be formed by averaging the related reflections.

4.3   Fourier calculation

A program for calculation of Fourier series is one of the most important tools in structure determination analysis. This program is employed in nearly all stages of the analysis. It is used to compute:

(1)   The Patterson function, the interpretation of which leads to the location of the heavy atoms;

(2)   The E-maps in direct methods [32];
(3)   The electron-density maps to locate missing atoms;

It is important that these calculations be performed correctly and as efficiently as possible. The calculations are carried out by sampling, within the unit cell, the continuous Fourier function at the nodes of a regular network with spacing Δ of the order of 0.2–0.4 Å in the three directions.

As a practical case let us consider a crystal of unit cell volume $V = 1000$ Å^3 (three linear parameters near 10 Å, three angles of about 90°). In the triclinic system, in the absence of symmetry, there are about 3000 reflections for a d_{hkl} minimum of 0.9 Å and 64 000 points on the network for $\Delta = 0.25$ Å. Thus, even with the use of high-speed computers, the implementation of Fourier calculations in crystal diffraction analysis remains a fairly arduous task involving time-consuming procedures.

The general method used to evaluate the expression of $\rho(xyz)$ (10 and 11) is to factor the cosine function into its components, which is very cumbersome. Design specifications for a computer program can be found elsewhere [51, 52]. Development of new methods for accelerating these calculations is of great interest, particularly the application of the fast fourier transform algorithm [53, 54] as it is used to calculate the E-maps.

The values of the Fourier series at the grid points can be output on a line printer. In order to increase their readability, it is advantageous to contour them by drawing closed curves through points of constant value. The results are similar to a topographic map with atoms appearing as maxima surrounded by several concentric lines. However, this approach requires access to a plotter and the use of a contour program, which is also time consuming and requires the storage of the grid point values.

A more rapid and very convenient method consists of reading the temporary grid point file in order to determine the coordinates and the density values of the highest peaks and to store them in a peak file. A drawing of the spatial distribution of the peaks in the unit cell can be obtained by a program which calculates different projections of the peaks using a suitable scale. These projections are output by the line printer and, in conjunction with the corresponding peak–peak distances are easily interpretable in terms of molecular structure. This procedure is employed for the representation of the E-maps and is also very useful to complete a phasing model by allowing the connection of missing atoms, as shown in Fig. 12.

Fig. 12   Drawing of the projection of a structure as given by a line-printer. The highest peaks P1, P2 and P3 appearing on a difference-Fourier map are combined with the 16 Atoms of the model to complete the $C_{18}H_{19}N$ molecule [55].

4.4   Direct methods

4.4.1   General considerations

The number of structures solved by direct methods is rapidly increasing and this can be correlated to the development of very efficient and fairly automatic routines. Nowadays, most of the structures are studied by the Multisolution Tangent Formula Method and the corresponding MULTAN program. However, a second method still deserves attention: the Symbolic Addition Method. Historically it was the first direct method and as such it opened new perspectives in crystallography. Perhaps more important, symbolic addition may have success where Multan may fail merely because phases are generated in a different way in both methods. Finally, from an educational point of view it is an elegant method which can be easily developed even by hand.

The computer applications of the general principles given in (2-3-2) may be divided in five main steps for both methods:

(1)   The calculation of the normalized structure factors E [17] from the experimental data file;

(2)   The generation of the Sigma-2 triples is the repetition of the following operation: two reflections H and K of the data set being given, is there a third $H-K$ reflection which satisfies the Σ_2 relationships (13 and 14)? This search is in essence repetitive and strategies and computer oriented considerations that can improve the efficiency have been developed, but will not be considered here.

(3)   The selection of the initial set of phases already mentioned [33];

(4)   The phase determination to be considered now;

(5)   The phase extension and the phase refinement by utilization of the tangent formula (15). The tangent formula provides a general computationally efficient method of estimating phase values and thus serves a dual purpose:

   (a)   to refine a set of approximately known phases;

   (b)   to extend the phase determination to unknown phases.

The whole process—extension and refinement—will be repeated in an iterative procedure until the phases converge to an self-consistent set.

4.4.2   The symbolic addition method

The use of symbols in phase determination has been introduced by Zachariasen [22] in the study of the metabolic acid structure and developed by Karle and Karle [26, 56]. The procedure has been described in great detail [57] and the best way to understand the method is to deal with a practical case. Many examples can be found in the literature [2, 4, 58].

The basic principle of the method is suggested by the following remark: if the phase allotted to a reflection of the starting set is incorrect, then it will be necessary to repeat the entire phase determination. Therefore, instead of trying to evaluate the phases at the beginning of the process, an improved strategy will result in the specification of certain suitable phases by means of symbols. The phase determination is then carried out in terms of these symbols. If only a few are required, then the alternative possibilities arising from assigning numerical values to the symbols are readily considered.

Symbolic addition procedures have been automated in a great number of programs, particularly for the centrosymmetric case. In non-centrosymmetry, the programming problems are much greater and the number of successful automatic programs is smaller [59]. The iterative system SIMPEL [60] is probably the best adapted program for centro and non centrosymmetric structures.

4.4.3 *Multisolution tangent formula method and the MULTAN program*

In this method, explicit values are deliberately given to some reflections of the starting set in order to initiate a continuous phase determination process. This idea led to the development of MULTAN [61] which is considered now.

In non-centrosymmetric cases a general reflection is given one or another of the 'quadrant' values: $+$ or $-\pi/4$ and $+$ or $-3\pi/4$. Fixing reflections within quadrants is found to be satisfactory, yielding a maximum error of $45°$. All combinations of phases will be introduced in the tangent formula. Thus, if three general reflections are used in the starting set, then 64 solutions are to be generated.

The reflections that form the initial set are selected by the subroutine Converge [33, 62]. This subroutine has the function of looking for the strongly interlinked reflections giving rise to multiple phase indications and therefore corresponding to reliable phase determinations. Actually, Converge performs an elimination process of the worst reflections and should converge to that group of reflections which are linked together best of all.

Two other interesting features of MULTAN are the use of a weighted tangent formula [63] and the Magic Integer representation [64]. Each derivation of a phase is accompanied by the derivation of an associate weight which is included in the tangent formula. The Magic Integer concept, first suggested by White and Woolfson [65] as a phase determining procedure, has been introduced in MULTAN [66] in terms of efficiently producing a starting set.

The method involved in MULTAN is by definition a multisolution method: MULTAN is in position to generate a great number of solutions, 64 in the above example. Each set of phases produced by the program could be used to calculate an E-map which would then be examined.

However, it is much better to test the phases and to try to predict which maps are worth examining before proceeding with calculations. For that purpose, some figures of merit (FOM) are associated with each set of phases in order to judge their relative plausibilities. Several FOM have been defined and MULTAN uses a combined FOM [67] based on the combination of three individual FOM and which should have maximum values of 3.0. It is found that this combined FOM is a better FOM than three individual indications.

From the user's point of view the MULTAN package is very easy to use and the manual edited by the authors of the latest version of the program [68] is remarkably clear and complete. All decisions necessary for the solution of a crystal structure have been built into the program. Hence, for a not too complex structure, it happens frequently that a large fragment of the molecule appears on the first calculated E-map, if not the whole molecule. On the other hand, facilities are incorporated in the program allowing user intervention in case of difficulties.

Recent developments with MULTAN have proven to be very useful for solving fairly large crystal structures. MAGIC [69] and MAGEX [70] are procedures obtained by an extension of the magic integer representation [65]. YZARC [71, 72] is a fully automatic procedure which refines initially-random sets of phases

4.5   The method of least-squares refinement

Least squares refinement programs calculate structure factors, refine the atomic parameters, and evaluate the standard deviations of refine parameters.

4.5.1   Formulation

The method of least squares [73] can be applied to the general case of fitting a set of data points to a continuous function. If the second derivatives of the function with respect to the variables to be fit are zero, then the system is linear and the solution will be given by a single application of the method. If the second derivatives are non-zero, the method must be applied in an iterative fashion until convergence is achieved.

In the case of crystal structure analysis there are more observations than parameters available and the experimental observations do not depend linearly on the parameters. Typically, we have for the structure factor the expression:

$$F_c(H) = \sum_{j=1}^{n} f_j \exp[-2\pi i(H, X_j)] \tag{16}$$

where f_j is the scattering factor for the jth atom, suitably corrected for vibration effects; H is the order of the reflection, e.g., the Miller indices (hkl); X_j the position $(x_j y_j z_j)$ of atom j.

We observe $F_0(H)$ and compare it with $F_c(H)$. The least squares method is most generally applied by minimization of:

$$M = \sum_{r=1}^{m} \omega_r(F_{o_r} - k|F_{c_r}|)^2 = \sum_{r=1}^{m} \omega_r \Delta_r^2 \tag{17}$$

when the sum is over the m independent (hkl) reflections; ω_r is the weight of the reflection r, and k is an overall scale factor.

The n parameters $(n < m)$ are $p_1, \ldots, p_n$ and may include:

— the $x_j y_j z_j$ positional parameters of atom j;
— the thermal parameters, one to six parameters for each atom owing to an isotropic or anisotropic vibration [74];
— site occupation factor, scale factor . . . .

For the molecule under study we know n values $p_{1_0}, \ldots, p_{n_0}$ of the trial structure such that: $|F_c|$ is almost equal to $|F_{co}|$ for each reflection. For any of the $p_1, \ldots, p_n$ parameters we can write a Taylor series:

$$|F_c|_r = |F_{co}|_r + \sum_{j=1}^{n} (p_j - p_{j_0}) \frac{\partial F_{c_r}}{\delta p_j} \tag{18}$$

provided that the $\varepsilon_j = p_j - p_{j_0}$ are small and the higher derivatives of F_{co} are not too large.

It can readily be shown that the minimization of M is achieved by solving n normal equations of the form:

$$\sum_{i=1}^{n} c_{ij} \varepsilon_i = d_i$$

where

$$c_{ij} = \sum_{r=1}^{m} \omega_r \frac{\delta F_{c_r}}{\delta p_i} \frac{\delta F_{c_r}}{\delta p_j}$$

$$d_i = \sum_{r=1}^{m} \omega_r \Delta_r \frac{\delta F_{c_r}}{\delta p_i} \tag{19}$$

The normal equations may be written in matrix notation:

$$C\varepsilon = d$$

with C_{ij} and d_i the elements of the matrices.

Then the solution is given by:

$$\varepsilon = Bd,$$

where B is the inverse matrix of C.

The least squares method allows us to estimate the accuracy of the derived parameters. The estimated standard deviation σ_{p_j} (or the variance $1/\sigma_{p_j}^2$) of any parameter p_j is given by:

$$\sigma_{p_j} = \left(\frac{b_{jj} \left(\sum_{r=1}^{m} \omega_r \Delta_r^2 \right)}{(m - n)} \right)^{1/2} \tag{20}$$

where b_{jj} is the jth diagonal element of the inverse matrice B; m and n the number of reflections and parameters, respectively.

4.5.2 Comments

Since the normal equations have been derived by ignoring the terms in higher powers of ε_i and higher derivatives of F_{c_r} than the first, they do not generally give an exact result and may actually diverge. Fortunately direct methods plus Fourier analysis yield atomic positions which are accurate enough and the least-squares method generally converges but does not give exact adjustments. Therefore, it is necessary to repeat the calculation until it is clear that a further cycle of refinement would not give any significant shifts. It is a common experience that the refinement process is successful when the positions for the original trial structure have a root-mean square deviation from the final positions of less than 0.2 Å.

However, some false minima surprisingly close to the true minimum have been reported. A classical example is given by adamantane [75]. For greater deviations, refinement has frequently produced models of structures which have been proven to be 'false' in the sense that structural features were unacceptable. A method has been developed [76] which enlarges the convergence radius to about 0.75 Å by choosing a different minimization function and an unusual weighting scheme.

Programs for solving the normal equations can use the system library subroutines for solving simultaneous equations and inverting matrices or can be specifically written for crystallographic purposes using, for instance, Gauss –Seidel [77] or Cholesky [78] routines. The most important computing difficulties in forming the normal equations arise in a need for large storage capacity and the time taken for large problems.

Example: a full set of n equations for m structure factors needs about $\frac{1}{2}m \times n$ multiplications, e.g., 4.10^8 operations for the refinement of 50 atoms (anisotropic thermal parameters) and 4000 reflections.

If time must be saved or if there is a lack of random access storage, one can take advantage of the fact that the off-diagonal elements in the normal matrix are mostly small compared with those of the main diagonal. Some approximations which are usually used omit the calculation of most of these off-diagonal elements by construction of a block-diagonal submatrix [4].

Several least-squares programs are given in references [79] to [83].

4.5.3 Choice of weights

The estimation of the weights ω_r which are associated with each reflection in the least-squares method should be based on a detailed analysis of the errors. A good choice of weights is an important step in structure refinement because bad weighting schemes can yield misleading atomic parameters [84].

ω_r may reflect the precision of F_0:

$$\omega_r = 1/\sigma_r^2 \tag{21}$$

where σ_r^2 is the variance of F_o.

A typical weighting function is given by [85]:

$$\omega_r = 1/(C_1 + F_o + C_2 F_o^2) \tag{22}$$

where

$$C_1 = 2F_{min} \text{ and } C_2 = 2/F_{max}.$$

A method for optimizing relative weights is given in [86] and more complicated weighting schemes in [87] and [88(a) and (b)].

4.5.4   Rigid body refinement

Least-squares refinement programs should also incorporate other facilities such as extinction corrections [89(a) and (b)] and rigid body refinement.

In a group or rigid body refinement, certain parts of a crystal structure are constrained to a fixed geometry and refined as units. The motivation is often the desire to minimize the number of parameters, thus reducing computer time and storage requirements and giving a more favorable data to parameters ratio [90]. The most frequent application of constrained refinement is in complex structure analyses containing aromatic rings, the detailed structure of which is not of interest to the investigators. An early example is given by RhH (CO) $[P(C_6H_5)_3]$ [91]. Constrained refinement is of great interest for macromolecule structure refinement [92, 93].

A further application of group technique arises in refinement of disordered structures. Refinement of individual atomic parameters with partial multiplicity factors is a standard procedure in case of disorder. Such a refinement may not be possible if atoms are in very close proximity. Some problems of this sort have been successfully solved by group refinement methods [94(a), (b) and (c)].

In cases where high accuracy is necessary, for instance for charge density purposes, special refinement techniques are required and are given in [95].

4.6   Description of the molecular conformation

The information available from X-ray study, suitably analyzed, provides a valuable data base for conformational conclusions and predictions. For structure analysis the conformational study begins after the structure is determined. For that purpose the molecule under study is described by a set of conformational parameters and by some illustrations.

4.6.1   The conformational parameters

Bond lengths, bond angles and torsional angles, as well as their corresponding e.s.d.'s, can be directly calculated from the coordinates of the various atoms in the unit cell. These conformational parameters, being invariant with respect to any coordinate system, represent the molecular features and are particularly useful in comparing conformations under different environments. Some other interesting features are also obtained from the structure:

— the planarity of a group of atoms by calculation of the best (in the least-squares sense) mean plane and of the associated χ^2 text. Dihedral angles between different atomic planes are also to be considered.
— Intramolecular distances are calculated in order to detect any short contact or hydrogen bond.

PARST [96] is a convenient FORTRAN program for calculating these molecular structure parameters.

4.6.2   Graphic illustrations

The computer graphics technique is one of the best methods for analyzing and communicating structural information results. A common technique used in crystallography is to make the graphic representation as nearly like the familiar mechanical models as possible. Rather elaborate graphic details may be included, such as stereoscopic pairs of views and hidden-lines elimination. The goal of technical graphics is not to display everything but rather to select the important features of the subject and to display them in such a way that they become obvious to the reader. The artistic aspects of the illustration are also not to be neglected.

Several computer programs for plotter illustration of a crystal structure are available, such as PLUTO [97], SPACEFIL [98, 99], VANDER [100], and ORTEP [101] which are used routinely in many places.

ORTEP is a general plot package for the production of crystal and molecular illustrations in mono or stereoscopic presentation. These illustrations are of ball-and-stick type, an atom being represented by a circle, a sphere, or a thermal ellipsoid. The hidden line elimination is an essential step for illustration. This is performed by an analytical calculation of atom overlaps and bond intersections, the visible portion being only drawn. ORTEP is controlled through instructions which are used to program an illustration. Thus the user can define the subject of the drawing by choosing the atoms to be included, the orientation of the plot, the bonded atoms, etc.

With a minimum of modifications, ORTEP can produce pictures on a graphic CRT in an interactive manner. However, the generation and the manipulation in real time of complex molecular models need the utilization of high performance hardware displays and of sopisticated program systems such as BILDER [102], FRODO [103], GRIP [104].

4.7   Computers and crystallographic systems

4.7.1   Mini or large computers?

An increasing tendency towards the use of generalized program systems in crystallography has been observed for many years. The sophistication in software and general advances in crystallographic techniques have meant that few laboratories can afford to develop and maintain a complete set of their own crystallographic computer programs. This function is largely assumed by a small number of laboratories that specialize in developing and distributing program systems for other scientists to use.

On the other hand, recent developments in microcomputers have brought about a revolution in the organization of a crystallographic laboratory. Minicomputers with characteristics similar to those indicated in Table 1 have performances that compete well with bigger and more powerful computers available to crystallographers. Almost all crystallographic programs now running on a large computer can be modified to run a dedicated minicomputer. Moreover, a minicomputer is able to perform machine control

Table 1   Present day minicomputer

Hardware	Software
16–32 bit word	Operating System: monitor + service
0.5–1.0 μS cycle	routines (control operations, utilities,
CPU direct memory access	editors etc.)
Multi-level interrupt	
Memory 64 to 2 Moctets	Multitask–Multiuser time-sharing
Floating point processor	
Disk cartridge	Assembler
5 to 20 MB or more	
moving heads (fast)	Compiler for high level languages
Peripherals: mag tape;	
Line printer; plotter; visual display units	Overlay facility
(1 graphic).	

functions and can therefore be employed simultaneously as a diffractometer controller and an an 'in-house' computer for all calculations required for the study of crystal structures.

As long as the efficiency and the facilities of the local computer center are good, the need for an 'in-house' machine is minimized. However, many laboratories are opting for the acquisition of a minicomputer as an alternative to the central computer facility. Many factors are to be considered before making the decision and these include the cost of such machines. Characteristics and performances of the system to be run on the computer are also of importance and are examined in the next section.

4.7.2   System development in crystallography

A considerable range of program systems are currently available to crystallographers. Most of them have been reported in detail at computing schools of the International Union of Crystallography [105 to 108]. Problems such as generality, efficiency, transportability and flexibility of a system often demand opposite strategies in system design and have found different solutions which are briefly summarized here.

XRAY76 [109] and XTAL [110] are quite universal program systems where the problem of transportability is a priority. In XRAY76, a neutral subset of FORTRAN referred to as Pidgin Fortran is employed and RATMAC is a preprocessor language which was written for XTAL. As a result these systems have been implemented on almost all computer types (CDC, UNIVAC, IBM, ICL, HONEYWELL etc.), but generally on large computers.

The NRC program system [111] as well as its latest time-sharing version [112] is devoted to an IBM computer; UNICS [113] is the crystallographic system running on the Japanese computers HITAC 5020E and FACOM M-200.

The SHELX system [114] consists of several thousands of FORTRAN statements which are compatible with most of the large computers. This system is now installed on more than 200 computers, mainly in Europe; SHELXTL [115] is a new version written for a D.G. ECLIPSE S-140 which is a 16-bit minicomputer.

The advantage of an 'in-house' minicomputer is that it can be designed as a dedicated computer for crystallograhic applications. In the strategy of adapting medium or large programs to a minicomputer, some points are to be considered:

— First, one has to take advantage of the particular OS as much as possible. As a consequence, the system loses its transportable character and its usefulness for general distribtion is limited.
— Second, because of the limited amount of core, the programs are more disk oriented than those on large computers. The problem is to avoid the computer spending most of its time doing disk transfers.
— Third, large programs must be divided into many overlap segments so that each part will fit in 24 to 32 K of core memory. For instance, MULTAN requires 15 segments and full-matrix least-squares refinement requires 5. However, an efficient division is necessary in order to minimize the number of disk transfers.
— Fourth, a big advantage of mini's over remote computer centers is the greater amount of user interaction. Programs are conversational, detect input errors and allow the users to correct these errors without stopping. This is excellent for inexperienced users and for programs which require the user to look at the result of one calculation before deciding what to do next, but is slow and rather boring for specialists. A good compromise is to have the possibility to process long running programs in batches.

Two systems for minicomputers are distributed by private companies:

— XTL/XTLE by Nicolet-Syntex for a Data General Nova 800 and 1200 [116];
— SDP by Enraf–Nonius [117] is limited to the DEC PDP11 series of computers; a recent version is available for the 32-bit VAX series.

The Syntex routines on Nova have been adapted by the NCR on a PDP-8e 12-bit word and 32 K core memory [118]; the first version of CRYSTAN was running on a PDP 11–45 [119] and RONTGEN 75 [120] is a USSR crystallographic system running on a 48-bit word computer of 32 K core.

A complete list of programs and systems can be found in the World List of Crystallographic Computer Programs [121], and more recent systems are given in references [122] to [134].

Whatever the dedicated computer, mini or large, any system approach to crystallography must be motivated by and oriented towards the user's requirements. Because X-ray crystallography is now considered an essential tool in many other scientific fields (chemistry, mineralogy, biology), crystallographic software is often utilized by scientists who have little or no formal training in crystallography. Therefore these systems should attempt to minimize the mandatory input requirement. As many parameters as possible should be defined implicitly or calculated but at the same time, systems must provide the facility for a complete explicit specification.

The rule of minimum input is also valuable in the interactive mode, for it minimizes the risk of error in data entry.

The use of data files stored on disk or magnetic tapes is a standard feature in crystallography for data input. But the user must not be concerned by the database management.

The output on a line-printer, on a data file or on a video screen in the interactive mode should also be kept to a minimum. A limited output emphasizes the essential results necessary for a non-crystallographer to understand the calculations. However, as for the input data, full provision for obtaining more data should be available when the user so requires.

Documentation and error checking are two essential components of any system and deserve a lengthy discussion such as is found in [105(a)] pp. 430.

The documentation is the essential interface between the user and the system. The need for good documentation becomes more critical as the programs become larger and more sophisticated. Moreover, it is important to realize that the documentation must be aimed at the novice user and not at the expert. The parameters input by the user should be tested to see first if they are in a reasonable range and also that there are no internal conflicts between parameter combinations. Program constraints, like array dimensions and available core storage, must also be checked from the input data. In case of error detection, the message sent to the operator must clearly state where in the calculations the error occurred, what it was, and what to do to avoid it.

There are a number of system considerations that are specific to crystallographic computing and have to be discussed now. The first point is concerned with the treatment of space group symmetry. Programs must be independent of all symmetry considerations. Space group information should only have to be entered once by the user and be provided for automatically in subsequent calculations. However, the space group generality should not degrade the overall efficiency of the system.

The second point is related to the size of the problem. The range and amount of data that must be handled in crystal structure analysis is very large. A program system must clearly specify the limits of program dimensions and eventually should provide means to increase the limits. Traditionally crystallographers have measured the size of their computers by the number of variables that can be put into core by a full matrix least-squares program. Thus, Ibers [135] states that no more than 240 variables (about 25 anisotropic atoms) can be refined on CDC 6400. Nowadays, thanks to a judicious utilization of peripheral storage, full matrix refinement on a minicomputer can handle many more parameters: up to 500 in Syntex XTL and 800 in the last version of the Enraf-Nonius SDP.

VI.5 CONCLUSION

The development of X-ray diffraction analysis is a fascinating story which began with the elucidation of the structure of monoatomic metals and mineral salts and which is now able to undertake the study of very complex molecules such as proteins and viruses. Today the number of organic and organometallic compounds studied by X-ray analysis approaches 50 000. The results are collected in structural databases [136] which support a wealth of information [137]. The goal of this chapter was to examine the factors which are responsible for this development, namely: automatic diffract-ometers, digital computers and crystallographic programming systems and packages. Progress in crystallography is easy to understand, as it is related to progress in computer and diffractometry technology, e.g., area detectors [138], and to the development of new methods for the resolution and the refinement of structures as well. Because of the simplicity of use and of approach of the programs, crystallography is now the method of choice. Research projects which are developed by non-crystallographers in which precise structural data is required yield results that could not be obtained by other methods. We have not attempted to review here the enormous literature dealing with the multitude of studies of chemical compounds. In fact, a question arises at this point: can the computer do everything or, in other words, is human intelligence still needed for the study of small compound structures?

The computer is an irreplaceable tool for driving a diffractometer. How-ever, some problems cannot at present be solved by the computer alone, for instance, the unambiguous determination of the space group. Therefore the computer must be placed under the control of a qualified operator. The role of the operator is essential in order to avoid faults in the unit cell determina-tion, to detect some systematic errors and to reach the optimum accuracy in the intensity measurements.

For medium size molecules, e.g., 50 to 100 atoms, it is true that the solution of the structure is mostly given by direct methods. Thus the computer is able to calculate phases for normalized structure factors and to compute Fourier-series, and it can output a diagram as a representation of the largest peaks of the E-maps, but it is unable to translate this diagram into terms of molecular structure. This interpretation requires some imagination and intuition which are still not included in any software. This must still be done by the chemist, who tries to build up the molecule by correlating each peak to an atomic species in such way that the emerging molecule is chemically consistent and could eventually be completed by Fourier analysis and properly refined. This key role becomes essential as the molecule becomes more complex and also in some particular circumstances, for example when the solution must be extracted from the Patterson function or in case of disordered molecules, etc.

Finally an important consequence of the efficiency of modern crystallogra-phy is that more investigations can be dedicated to the preparation and to the significance of the results rather than to the means used to obtain them. Thus, scientists involved in structural studies can devote their time to the best choice of molecules to be investigated and can design experiments in order to obtain crystals suitable for an X-ray study.

VI.6 REFERENCES

[1] J. D. Dunitz, *X-ray Analysis and the Structure of Organic Molecules*, Cornell University Press, Ithaca, New York (1979).

[2] M. F. C. Ladd and R. A. Palmer, *Structure Determination by X-ray Crystallography*, Plenum Press, New York and London (1977).

[3] *International Tables for X-ray Crystallography*, 2nd ed., Kynoch Press, Birmingham (1968).

[4] G. H. Stout and L. H. Jensen, *X-Ray Structure Determination*, The McMillan Company, New York (1968).

[5] A. Guinier, *Theorie et Technique de la Radiocristallographie*, Dunod, Paris (1956).

[6] L. V. Azaroff, R. Kaplor, R. J. Kato, R. J. Weiss, A. J. C. Wilson and R. A. Young, *X-ray Diffraction*, McGraw-Hill, New York (1974).

[7] G. N. Ramachandran and R. Sunivasan, *Fourier Methods in Crystallography*, Wiley-Interscience, New York (1970).

[8] A. L. Patterson, *Phys. Rev.*, **46**, 372 (1935).

[9] H. Krebs, *Fundamentals of Inorganic Crystal Chemistry*, McGraw-Hill, New York (1968).

[10] F. A. Cotton and G. Wilkinson, *Advanced Inorganic Chemistry*, 4th ed., John Wiley and Sons, New York (1980).

[11] D. W. Green, V. M. Ingram and M. F. Perutz, *Proc. Roy. Soc.*, **A225**, 287 (1954).

[12] D. M. Blow and F. H. C. Crick, *Acta Cryst.*, **12**, 794 (1959).

[13] R. E. Dickerson, J. C. Kendrew and B. E. Standberg, *Acta Cryst.*, **14**, 1188 (1961).

[14] D. C. Phillips, *Advances in Structure Research by Diffraction Methods* (R. Brill and R. Mason, eds), **2**, pp. 75–117, John Wiley and Sons, New York (1966).

[15] J. Karle and H. Hauptman, *Acta Cryst.*, **3**, 181 (1950).

[16] D. Harker and J. S. Kasper, *Acta Cryst.*, **1**, 70 (1948).

[17] H. Hauptman and J. Karle, *Solution of the Phase Problem, I, The Centrosymmetric Crystal*, A.C.A. Monograph No. 3, Polycrystal Book Service, Pittsburgh (1953).

[18] J. Karle and H. Hauptman, *Acta Cryst.*, **9**, 635 (1956).

[19] A. J. C. Wilson, *Nature*, **150**, 151 (1942).

[20] I. L. Karle and J. Karle, *Acta Cryst.*, **17**, 835 (1964).

[21] D. Sayre, *Acta Cryst.*, **5**, 60 (1952).

[22] W. H. Zachariasen, *Acta Cryst.*, **5**, 68 (1952).

[23] W. Cochran, *Acta Cryst.*, **6**, 810 (1953).

[24] E. W. Hughes, *Acta Cryst.*, **6**, 871 (1953).

[25] M. N. Woolfson, *Acta Cryst.*, **7**, 61 (1954).

[26] J. Karle and I. L. Karle, *Acta Cryst.*, **21**, 849 (1966).

[27] W. Cochran, *Acta Cryst.*, **8**, 473 (1955).

[28] H. Hauptman and J. Karle, *Acta Cryst.*, **9**, 45 (1956).

[29] H. Hauptman and J. Karle, *Acta Cryst.*, **12**, 93 (1959),

[30] J. Karle and H. Hauptman, *Acta Cryst.*, **11**, 264 (1958).

[31] J. Karle, *Acta Cryst.*, **B26**, 1614 (1970).

[32] C. Weeks and H. Hauptman, *Z. Krist*, **131**, 437 (1970).

[33] G. Germain, P. Main and M. N. Woolfson, *Acta Cryst.*, **B26**, 274 (1970).

[34] I. L. Karle, H. Hauptman, J. Karle and A. B. Wing, *Acta Cryst.*, **11**, 257 (1958).

[35] U. W. Arndt and G. T. N. Willis, *Single Crystal Diffractometry*, Cambridge University Press (1966).

[36] CAD4 Operators Manual, Enraf Nonius, Scientific Instruments Division, Delft, The Netherlands (1984).

[37] H. Cole, Y. Okaya and F. W. Chambers, *Rev. Sci. Instr.*, **34**, 872 (1963).

[38] W. R. Busing, R. D. Ellison and H. A. Levy, *Abstracts of the A.C.A.*, pp. 59, Gatlingburg, Tennessee (1965).

[39] W. R. Busing and H. A. Levy, *ORNL-4143* (1968).

[40] R. A. Sparks, Syntex P 21, Operations Manual (1975).

[41] W. R. Busing and H. A. Levy, *Acta Cryst.*, **22**, 457 (1967).

[42] W. R. Busing, *Crystallographic Computing*, pp. 321–330 (F. R. Ahmed, S. R. Hall, C. P. Huber, eds), Munksgaard, Copenhagen (1967).

[43] *Philips Serving Science and Industrie*, **18**(2), 22–28, N. V. Philips, Eindhoven, The Netherlands (1972).

[44] R. A. Sparks, *Crystallographic Computing Techniques*, pp. 452–467 (F. R. Ahmed, K. Huml and B. Sedlacek, eds), Munksgaard, Copenhagen (1976).

[45] A. D. Mighell and J. R. Rodgers, *Acta Cryst.*, **A36**, 321 (1980).

[46] T. L. Blundel and L. N. Johnson, *Protein Crystallography*, Academic Press, New York (1976).

[47] C. K. Prout, *Computing Methods in Crystallography*, pp. 89–95 (J. S. Rollet, ed.), Pergamon Press, Oxford (1965).

[48] P. Coppens, *Crystallographic Computing*, pp. 255–270 (F. R. Ahmed, S. R. Hall and C. P. Huber, eds), Munksgaard, Copenhagen (1970).

[49] H. D. Flack, *J. Appl. Cryst.*, **8**, 520 (1975).

[50] (a) B. Post, *Acta Cryst.*, **A25**, 94 (1969).
      (b) W. H. Zachariasen, *Acta Cryst.*, **23**, 558 (1967).

[51] F. R. Ahmed, *Crystallographic Computing*, pp. 355–359 (F. R. Ahmed, S. R. Hall and C. P. Huber, eds), Munksgaard, Copenhagen (1970).

[52] A. C. Larson, *Crystallographic Computing Techniques*, pp. 392–395 (F. R. Ahmed, K. Huml and B. Sedlacek, eds), Munksgaard, Copenhagen (1976).

[53] J. W. Cooley and T. W. Tukey, *Math. Comput.*, **12**, 297 (1965).

[54] A. Immirzi, *Crystallographic Computing Techniques*, pp. 399–412 (F. R. Ahmed, S. R. Hall and C. P. Huber, eds), Munksgaard, Copenhagen (1976).

[55] P. Brouant, M. Pierrot, A. Baldy, J. C. Soyfer and J. Barbe, *Acta Cryst.*, **C40**, 1590 (1984).

[56] I. L. Karle and J. Karle, *Acta Cryst.*, **16**, 969 (1963).

[57] H. Schenk, *Computing in Crystallography* (R. Diamond, S. Ramaseshan and K. Venkatesian, eds), pp. 7.01–7.15, Indian Acad. Sciences, Bangalore and I.U.Cr (1980).

[58] I. L. Karle, *Crystallographic Computing Techniques*, pp. 27–70 F. R. Ahmed, K. Huml and B. Sedlacek, eds), Munksgaard, Copenhagen (1976).

[59] R. B. K. Dewar, *Crystallographic Computing*, pp. 63–65 (F. R. Ahmed, S. R. Hall and C. P. Huber, eds), Munksgaard, Copenhagen (1970).

[60] (a) A. R. Overbeek, Thesis, Univ. Amsterdam (1979).
      (b) H. Schenk, *Proc. Kon. Ned. Akad. Wet.*, **B79**, 341–343 (1976).

[61] G. Germain and M. N. Woolfson, *Acta Cryst.*, **B24**, 91 (1968).

[62] G. Germain, P. Main and M. N. Woolfson, *Acta Cryst.*, **A27**, 368 (1971).

[63] S. E. Hull and M. J. Irwin, *Acta Cryst.*, **A34**, 863 (1978).

[64] P. Main,. *Acta Cryst.*, **A33**, 750 (1977).

[65] P. S. White and M. N. Woolfson, *Acta Cryst.*, **A31**, 53 (1975).

[66] P. Main, *Acta Cryst.*, **A34**, 31 (1978).

[67] M. N. Woolfson, *Crystallographic Computing Techniques*, pp. 85–96 [F. R. Ahmed, K. Huml and B. Sedlacek, eds), Munksgaard, Copenhagen (1976).

[68] P. Main, S. J. Fiske, S. E. Hull, L. Lessinger, G. Germain, J. P. Declercq and M. N. Woolfson, 'Multan 80', A System of Computer Programs for the Automatic Solution of Crystal Structures from X-ray Diffraction Data, Univ. of York, England and Louvain, Belgium (1980).

[69] J. P. Declercq, G. Germain and M. N. Woolfson, *Acta Cryst.*, **A31**, 367 (1975).

[70] S. E. Hull, D. Viterbio, M. N. Woolfson and Zhang Shao-Hui, *Acta Cryst.*, **A37**, 566 (1981).

[71] R. Baggio, M. N. Woolfson, J. P. Declercq and G. Germain, *Acta Cryst.*, **A34**, 883 (1978).

[72] J. P. Declercq, G. Germain and M. N. Woolfson, *Acta Cryst.*, **A35**, 622 (1979).

[73] J. S. Rollett, *Crystallographic Computing*, pp. 168–181 (F. R. Ahmed, S. R. Hall, C. P. Huber, eds), Munksgaard, Copenhagen (1970).

[74] B. T. M. Willis and A. W. Pryor, *Thermal Vibrations in Crystallography*, Univ. Press, Cambridge (1975).

[75] J. Donohue and S. H. Goodman, *Acta Cryst.*, **22**, 352 (1967).

[76] J. S. Rollett, T. G. McKinlay and N. P. H. Haigh, *Crystallographic Computing Techniques*, pp. 413–419 (F. R. Ahmed, K. Huml and B. Sedlacek, eds), Munksgaard, Copenhagen (1976).

[77] R. A. Sparks, *Computing Methods and the Phase Problem in X-ray Crystal Analysis*, pp. 175–182 (R. Pepinski, J. M. Robertson and J. C. Speakman, eds), Pergamon Press, New York, Oxford, London, Paris (1961).

[78] J. S. Rollett, *Computing Methods in Crystallography*, pp. 16–21 (J. S. Rollett, ed.), Pergamon Press, Oxford (1965).

[79] W. R. Busing, K. O. Martin and H. A. Levy, 'ORFLS, a FORTRAN Crystallographic Least Squares Program.' ORNL-TM305, Oak Ridge National Laboratory (1962) and ORXFLS3 (1971) for an extensively modified version.

[80] L. W. Finger and E. Prince, *RFIN2*, Natl. Bur. Stand. (US), Tech. Note, pp. 854–981 (1975).

[81] P. Coppens, T. N. Guru Row, P. Leung, E. D. Stevens, P. D. Becker and Y. M. Yang, *Acta Cryst.*, **A35**, 63 (1979).

[82] R. L. Lapp and R. A. Jacobson, 'ALLS. A Generalized Crystallographic Least Squares Program.' US Department of Energy, Report IS 4708, Iowa State University (1979).

[83] W. Furey Jr, B. C. Wang and M. Sax, 'QWKREF.' *J. Appl. Cryst.*, **15**, 160 (1982).

[84] J. D. Dunitz and R. R. Ryan, *Acta Cryst.*, **21**, 617 (1966).

[85] D. W. J. Cruickshank, *Computing Methods in Crystallography*, pp. 112–115 (J. S. Rollett, ed.), Pergamon Press, Oxford (1965).

[86] K. Nielsen, *Acta Cryst.*, **A33**, 1009 (1977).

[87] J. D. Dunitz and P. Seiler, *Acta Cryst.*, **B29**, 589 (1973).

[88] (a) P. F. Lindley and M. M. Mahmoud, *Acta Cryst.*, **B34**, 445 (1978).
      (b) H. Borkakoti and R. A. Palmer, *Acta Cryst.*, **B34**, 482 (1978).

[89] (a) P. Coppens and W. C. Hamilton, *Acta Cryst.*, **A26**, 71 (1970).
      (b) P. Becker and P. Coppens, *Acta Cryst.*, **A30**, 129 and 148 (1974).

[90] C. Scheringer, *Acta Cryst.*, **16**, 546 (1963).

[91] S. J. La Placa and J. A. Ibers, *Acta Cryst.*, **18**, 511 (1965).

[92] K. D. Watenpaugh, L. C. Sieker, J. R. Herriott and L. H. Jensen, *Acta Cryst.*, **B29**, 943 (1973).

[93] W. A. Hendrickson and J. H. Konnert, *Biomolecular Structure, Function, Conformation and Evolution*, Vol. 1, pp. 43–57 (R. Srinivasan, ed.), Pergamon, Oxford (1980).

[94] (a) I. Bar and J. Bernstein, *Acta Cryst.*, **A39**, 266 (1983).
      (b) A. Armagan, *Acta Cryst.*, **A39**, 647 (1983).
      (c) K. J. Haller, *Acta Cryst.*, **A40**, C427 (1984).

[95] P. Becker, *Computational Crystallography*, pp. 354 (D. Sayre, ed.), University Press, Oxford (1982).

[96] N. Nardelli, 'PARST', *Computers and Chemistry*, **7** (3), 95 (1983).

[97] W. D. S. Motherwell and W. Clegg, 'PLUTO. A Program for Plotting Molecular Crystal Structures', Univ. of Cambridge, England (1978).

[98] G. M. Smith and P. J. Gund, *J. Chem. Inf. Comp. Sci.*, **18**, 207 (1978).

[99] D. R. Henry, *Computers and Chemistry*, **7** (3), 119 (1983).

[100] R. Norrestam, 'VANDER. Program for Raster Plot Representation of Crystal and Molecular Structures', Tech. Univ. of Denmark, Lyngby (1981).

[101] C. K. Johnson, 'ORTEP II. A Fortran Thermal Ellipsoid Plot Program for Crystal Structure Illustrations', ORNL-5138, (Third Rev. of ORNL-3794), Oak Ridge National Laboratory, Oak Ridge (1976).

[102] R. Diamond, *Biomolecular Structure, Conformation, Function and Evolution*, Vol. 1, pp. 63 (R. Srinivasan, ed.), Pergamon Press (1980).

[103] T. A. Jones, *J. Appl. Cryst.*, **11**, 268–272 (1978).

[104] E. G. Britton, A Methodology for the Ergonomic Design of Interactive Computer Graphic System and its Application to Crystallography', Univ. of North Carolina, Chapel Hill, NC (1977).

[105] (a) S. R. Hall, *Crystallographic Computing Techniques*, pp. 424–432 (F. R. Ahmed, K. Huml and B. Sedlacek, eds), Munksgaard, Copenhagen (1976).

(b) J. M. Stewart, Ibid., pp. 433–443.

(c) R. A. Sparks, Ibid., pp. 452–477.

[106] J. M. Stewart, *Computing in Crystalography*, pp. 3–16 (H. Schenk, R. Olthof-Hazekamp, H. Van Koningsveld and G. C. Bassi, eds), Delft University Press, Delft (1978).

[107] (a) S. R. Hall and J. M. Stewart, *Computing in Crystallography*, pp. 22.01–22.26 (R. Diamond, S. Ramaseshan and K. Venkatessan, eds), Indian Academy of Sciences and IUCr., Bangalore (1980).

(b) E. T. Gabe, Ibid., pp. 23.01–23.16.

[108] (a) J. M. Stewart. *Computational Crystallography*, pp. 497–505 (D. Sayre, ed.), Clarendon Press, Oxford (1982).

(b) G. M. Sheldrick, Ibid., pp. 506–514.

[109] J. M. Stewart, P. A. Machin, C. Dickinson, H. Heck and H. Flack, 'The X-RAY76 System', Techn. Report 446, Computer Science Center, Univ. Maryland, College Park, Maryland (1976).

[110] S. R. Hall, J. M. Stewart, A. P. Norden, R. J. Munn and S. Freer, 'The XTAL System for Crystallographic Programs'. Programmer's Manual, Report TR-873, Computer Science Center, Univ. of Maryland, College Park (1980).

[111] F. R. Ahmed, S. R. Hall, M. E. Pippy and C. P. Huber, 'MRC Crystallographic Programs for IBM 360 System,' National Research Council, Ottawa (1970).

[112] F. R. Ahmed, *Computing in Crystallography*, pp. 17–33 (H. Schenk, R. Olthof-Hazekamp, H. Van Koningsveld, G. C. Bassi, eds), University Press, Delft (1978).

[113] T. Sukurai, H. Iwasaki, Y. Watanabe, K. Kobayaski, Y. Bando and Y. Nakamichi, *Rep. Inst. Phys. Chem. Research*, **50**, 75–91 (1979).

[114] G. N. Sheldrick, 'SHELX76, Program for Crystal Structure Determination', Univ. of Cambridge, England (1976).

[115] G. N. Sheldrick, 'Integrated System for Solving Refining and Dis-

playing Crystal Structures from Diffraction Data', Univ. Gottingen, Fed. Rep. Germany (1981).

[116]   'XTL/XTLE, Structure Determination System', Syntex Analytical Instrument, Cupertino, California (1976).

[117]   B. Frenz, *Computing in Crystallography*, pp. 64–71 (H. Schenk, R. Olthof-Hazekamp, H. van Koningsveld and G. C. Bassi, eds), Delft University Press, Delft (1978).

[118]   E. J. Gabe, A. C. Larsen, F. L. Lee and Y. Wang, 'NRC–PDP8e Crystal Structure System', MRC, Ottawa (1979).

[119]   J. Burzlaff, V. Bohme and M. Gomm, 'CRYSTAN Crystallographic Program System for Mini Computer', Univ. of Erlangen, Fed. Rep., Germany (1975).

[120]   V. I. Andrianov, *Computing in Crystallography*, pp. 45–51 (H. Schenk, R. Olthof-Hazekamp, H. Van Koningsveld and G. C. Bassi, eds), Delft University, Delft (1978).

[121]   World List of Crystallographic Computer Programs, 3rd Ed., *J. Appl. Cryst.*, **6**, 309–346 (1973).

[122]   CRYSNET Manual, Brookhaven National Laboratory, Chemistry Department, Informal Report BNL 21714 (H. M. Berman, F. C. Bernstein, H. J. Bernstein, J. F. Koetzle and G. J. B. Williams, eds), BNL, Upton New York (1976).

[123]   J. A. Waters and J. A. Ibers, 'Northwestern University Crystallographic Library', *Inorg. Chem.*, **16** (2), 3273 (1977).

[124]   M. R. Churchill, R. A. Laschewycz and F. J. Rotella,'SUNY, Buffalo modified Syntex XTL', *Inorg. Chem.*, **16**, 265 (1977).

[125]   A. Karapides, *Inorg. Chem.*, **18**, 3034 (1979).

[126]   G. M. McLaughlin, M. Taylor and P. O. Whimp, 'ANUCRYS. Structure Determination Package,' Research School of Chemistry, Australian National University, Camberra.

[127]   C. T. Fritchie Jr., B. L. Trus, J. L. Wells, C. A. Langhoff Jr., M. Guise, W. Lamia, M. Krieger, J. T. Mague and R. Jacobs, 'LOKI, Crystallographic Computing System', Dpt of Chemistry, Tulane Univ., New Orleans, LA.

[128]   H. L. Carrell, H. S. Shieh and F. Takusagawa, 'The Crystallographic Program Library of the Institute for Cancer Research'. Fox Chase Cancer Center, Philadelphia, PA (1981).

[129]   D. H. Farras and N. C. Payne, *Inorg. Chem.*, **20**, 821 (1981).

[130]   J. C. Calabrese, Central Research and Development Department, E. I. du Pont de Nemours and Co, Wilmington, Delaware.

[131]   J. R. Carruthers and D. J. Watkin, 'CRYSTALS User Manual', Oxford University, Computing Laboratory, Oxford (1981).

[132]   R. Norrestam, 'XTAPL, Interactive APL Program for Crystal Structure Calculations', Technical Univ. of Denmark, Lynby (1982).

[133]   A. L. Spek, *Computational Crystallography*, p. 528 (D. Sayre, ed.), Clarendon Press, Oxford (1980).

[134]   G. N. Reeke Jr., *J. Appl. Cryst.*, **17**, 125 (1984).

[135]   J. A. Ibers, *Computational Needs and Resources in Crystallography*, pp. 18–27, National Academy of Sciences, Washington DC (1973).

[136]   F. H. Allen and O. Kennard, *Perspectives in Computing*, **3** (3), 28 (1983).

[137]   F. H. Allen, O. Kennard and R. Taylor, *Accts Chem. Res.*, **16**, 146 (1983).

[138]   U. W. Arndt, *Computing in Crystallography*, pp. 134–146 (H. Schenk, R. Olthof-Hazekamp, H. van Koningsveld and G. C. Bassi, eds), Delft University Press (1978).

CHAPTER VII

Mass Spectra and Kovats' Indices Databank of Volatile Aroma Compounds

Gaston Vernin, Michel Petitjean, Jean-Claude Poite and Jacques Metzger

Daniel Fraisse and Kim-Nuor Suon

C.N.R.S. Analysis Central Services, Vernaison, France

VII.1 INTRODUCTION

The reliable identification of the components of a complex mixture is always a tedious task for the analyst. Until recently it was necessary to proceed through a chemical or physical fractionation of the mixture, then to derivatization methods and successive purifications and finally to numerous varied identification and characterization methods. This methodology was time-consuming and required great technical support.

The increasing improvement of coupled gas chromatography–mass spectrometry has thoroughly changed this situation. But to handle the large amount of data now available from mass spectra, upon both electron impact and chemical ionization, the analyst must turn to the computer. Various computerized methods have been developed during the last two decades allowing one to deduce the formula of a compound from its mass spectrum. But this formula is not always sufficient for identification purposes, and one needs other data, such as Kovats' indices, to distinguish different products having very similar mass spectra.

The goal of this chapter is to review the general principles and mutual advantages and disadvantages of the three main types of computerized methods which allow the structural elucidation of an unknown product. Special attention will be given to comparative methods or library searches which appear to be more appropriate to the resolution of this problem [1]. We will give as an example of our own development of a system dealing with mass spectra and the Kovats' indices databank, called SPECMA, which is specially suited to the identification of volatile compounds in flavours and fragrances [2, 3].

Owing to the extensive nature of this topic, it as not been possible to discuss it in complete detail. Therefore, we also refer the reader to additional books and reviews given as references in the following discussion.

VII.2   VARIOUS APPROACHES FOR INTERPRETING MASS SPECTROMETRY DATA

There are three main groups of methods which deal with structural determination by mass spectrometry.

(1)   In theoretical methods, structure is predicted on the basis of fragmentation rules.

(2)   In statistical methods, also called *Pattern Recognition*, the identification of structural features is made by means of statistical analysis which provides classification of 'unknowns'. (These non-parametric methods constitute the great majority of published works in pattern recognition.)

(3)   Comparative methods or 'Library searches' compare an unknown spectrum with a file or library of reference spectra to find the best match.

There are a great many works devoted to these methods, and which have been well reviewed by Heller [4], Jurs [5], Smith [6], Chapman [7], and Mum and McLafferty [8].

2.1   Theoretical spectrum interpretation

Several programs have been described which interpret mass spectra recorded at low resolution [9-13]. They have been applied to the following families of compounds:

— saturated hydrocarbons and fatty methyl esters [9]
— aromatic esters, ethers, acids, ketones, aldehydes, alkenes, alkanes, alcohols, cycloalkanes, dienes and amines [10]
— amines and alcohols [11]
— esters, alcohols and phthalates [12, 13].

In this latter case, the main program selects the following spectral features associated with each family: molecular ion, key fragment ions and losses from the molecular ion. However, the presence of impurities sometimes makes impossible the molecular ion determination.

Most of theoretical methods require the knowledge of the molecular formula. When this is the case a program, such as heuristic DENDRAL algorithm, calculates the isomers which match the spectrum [6, 14]. *The term 'heuristic' refers to the judgmental rules which are used in the attempt to guide the search to a single correct solution* [6].

The list processing DENDRAL system which requires considerably greater computing power than is currently available with a microcomputer, incorporates some simple heuristics for inferring the presence of monofunctional groups in acyclic molecules such as ketones, ethers, amines, alcohols, thioethers and thiols. Much more powerful heuristics, related to those used by mass spectrometrists, can also be incorporated to explain the mass spectral pattern of molecules of known structures. The efficiency of the method is highly improved by the use of spectroscopic (^{1}H-NMR, IR) or other physical data.

The case of the tetradecylamine (MW 185) is a case in point. While 10 115 possible solutions are obtained from the mass spectrum alone, only one is listed when the ^{1}H-NMR spectrum is added to the data. The DENDRAL system effects a much more complete analysis of spectra than that used in the classification method described by Gray and Grönneberg [13] for unseparated compounds in GC. In this latter method, it is necessary to know the empirical formula and the family of the compound from which the user must select the corresponding file.

Other programs, such as INTSUM [15], CIFAS [16], and FAMOUS [17] are primarily used for the analysis of high resolution mass spectral data of particular classes of compounds. Similar programs working at low resolution (INTACT-INTERP) require a reference library [12]. A more sophisticated file search method, allowing the identification of compounds not contained in the reference files, has been perfected by McLafferty et al. [18–22]. The McLafferty Self Training Interpretive and Retrieval System (STIRS) calculates a set of indices from the following spectral features:

— ion series
— low, medium and high mass characteristic ions (most abundant odd and even mass ions),
— small and large neutral losses and secondary neutral losses,
— fingerprint ions,
— overall match factor derived as a linear combination of the above first three classes.

The STIRS system should be able to characterize the unknown by selecting best matches containing structural features that are the same as, or similar to, those in the unknown [7].

The program also calculates a threshold value above which the 'unknown' is selected. Three threshold values have been obtained from a reference library containing 13 000 spectra. However, the method is restricted, as is seen in the example of 4-picolyl-tert-butyl sulphide, where the program lists 23 possible compounds [23].

Several programs of spectra generation, such as ION GENERATOR [24], MAS, and REACT [25] have also been developed. In addition, completely theoretical methods allow ion intensities to be calculated from internal energy in the case of small molecules [26].

In short the theoretical methods present the same limitations as those of the chemist alone because they do not allow one to identify a compound from only its mass spectrum at low resolution, except in isolated cases.

2.2 Pattern recognition [5, 7, 27–32]

The pattern recognition approach in mass spectrometry fundamentally relies on the following concept: a spectrum is considered as a set of p-bits of information which can be represented by a point in a p-dimensional space. As an example [31], consider a set of compounds with known mass spectra where each compound is represented by only three fragments. Then one could plot a particular compound as a point in three-dimensional space, and a different compound in the data set with the same three fragments, but different values for them, would be represented by a different point in the same space. And if there were 100 compounds in a data set and each mass was represented by the same three fragments, they would be represented by a swarm of 100 points in this three-dimensional space. But there is no reason to restrict a mass spectrum to only three fragments; it is necessary to use many more. Then, each compound may be represented by p fragments and thereby form a p-dimensional space. The values of the measurements become the coordinates of each point in d-space so that the more similar the compounds, the nearer they cluster in d-space.

Pattern recognition methods then are simply a set of computer-based methods which allow one to investigate clustering of data represented according to this formalism.

According to Chapman [7] more or less sophisticated variants of pattern recognition methods, called supervised learning, have been applied to mass spectrometery. They are summarized below:

(i) *Methods leading to a classification by minimum distance from an average spectrum* [33–37]. These involve displays where one attempts to map the points in a multidimensional space into two-dimensions for direct viewing. Each class of compounds is represented by a single prototype point P about which all other points in that class tend to cluster. The coordinates of each point are the intensity values of the fragments in the mass spectrum. The Euclidean distance between the unknown spectrum and the pattern P is measured through the application of different functions [7]. A basic deficiency of this method is poorly developed clustering in many of the systems studied [37].

(ii) *Classification by learning machines* [38–41]. According to Jurs [5] any spectrum (at low resolution) of d elements can be represented in a high-dimensional space (hyperspace) as a pattern vector, $X \equiv (x_1 + x_2, \ldots, + x_d)$, where each x_j value is an intensity suitably encoded. An extra component, the $(d + 1)$th, whose value is unity, is added to each pattern vector so that a decision surface, passing through the origin, can be constructed to separate the clusters from each other. The decision surface is a hyperplane which is analogous to an ordinary two-dimensional plane in three space. The hyperplane has a normal vector at the origin, $W = (w_1, w_2, \ldots, w_{d+1})$, which is orthogonal to the decision surface. This vector is also called the 'weight vector'. The discriminant function, $g(x) = W \cdot X$, which is the dot product between the normal vector and the pattern vector, allows one to classify any pattern point in a d-dimensional space with respect to a hyperplane decision surface. It is represented as follows,

$$g(x) = w_1 x_1 + w_2 x_2 + \cdots w_d x_d + w_{d+1} = |W||X| \cos \theta \qquad (1)$$

in which θ is the angle between the two vectors.

For those patterns for which $\theta < 90°$ and thereby are on one side of the plane, the dot product is positive, and for those on the other ($i > 90°$) the dot product is always negative. The method called 'training' consists of finding effective decision surfaces (i.e., a set of values for the weight vector) which allow one to correctly classify the patterns (spectra).

When the training process does not converge (inseparable data), one may use a rejection zone or dead zone that has a finite width [42]. Several factors can influence the degree of success of learning machines [7] namely,

(a)   the character of the data
(b)   the choice of categories to be separated
(c)   the choice of the training set.

These learning machines have been applied not only to a wide range of classification problems using mass spectral data [43, 44] but also to a variety of chemical and structural analysis problems [45, 46]. They present certain disadvantages, such as the use of the largest, most representative training set available [39, 47] and the necessity of having a sufficient number of structural features present in order to access a rational formula for the required compound. Furthermore, every feature should give a separation frontier to the d-dimensional space. Finally, these methods (and the following ones) make considerable demands on computer time and storage.

(iii)   *K.nearest neighbor method* [30, 48–51]. The KNN procedure allows one to classify an unknown spectrum according to the class of the nearest K spectra in the data set. When $k \geq 1$ the compound is classified within the corresponding class, while when $K < 1$ the compound is rejected.

The intensity values of the d fragments are the coordinates of the pattern X representing the unknown spectrum. As in the first method, the usual measure of distance between the unknown and the nearest K spectra is the Euclidean distance. Since this method generates a unique classification, the method provides better results with inseparable data than do the learning machines. A parallel between the KNN and the learning machines methods has been made by Varmuza *et al.* [50, 51]. They determined the presence or absence of a particular chemical substructure among 31 categories (i.e., functional group, aliphatic, aromatic, alkyl group, oxygen or nitrogen presence, etc.). With $K = 1$, the two methods gave similar results, but when the number of neighbors is increased ($K = 6$) and if the category is determined by a majority vote, the results with KNN are highly improved. For example, 94% correct predictions are obtained for the presence of nitrogen in a molecule with KNN.

Other variants of the KNN method have been discussed by Kowalski *et al.* [49]. Using only three optimum variables and moment data rather than intensities, they obtained an increased classification performance. The method was applied to three classes of randomly selected hydrocarbon spectra. Another application of the KNN method deals with the mass spectra of deoxyoligonucleotides, whose features were selected to distinguish various sequences [52].

The KNN method gives, in most cases, better results than other non-parametric classification procedures such as the learning machines, but as already mentioned, they are computer time-consuming.

(iv) *Cluster analysis* [4]. Cluster analysis is usually considered a part of pattern recognition. It is concerned with finding homogeneous sets of data containing similar data. This approach to the recognition and classification of functional groups differs from the pattern recognition techniques described by Jurs [28, 29] in that it is usually non-supervised learning and is not limited (i.e., the number of clusters are not known). There are a number of approaches which have been developed in several books and reviews [53, 54].

(v) *Miscellaneous.* Other classification methods have been described [55, 56]. Some utilize factor analysis [57, 58]. But, as previously mentioned, they show the same restrictions as those mentioned and are inferior to the theoretical methods [59].

2.3 Library searches

The identification of low resolution mass spectra by means of computer matching techniques was first attempted by Abrahamson and Stenhagen [60]. Since 1964, this procedure has been widely studied [1–3, 10, 61–67]. Considering the difficulties encountered with theoretical methods and the drawbacks of statistical methods (which, furthermore, require a reference library) most authors have noted that the comparison of an unknown with a library of known compounds seems to be the most successful way to identify compounds. It is clear that this method presents two major advantages: it allows us to identify a totally unknown compound using only its low resolution mass spectrum, and it is a simple method, needing no special theoretical knowledge in mass spectrometry. However, comparative methods offer some disadvantages, such as the necessity of a reference library; but this point is also a handicap for other methods, except for certain theoretical methods. These latter, however, require a good knowledge of fragmentation rules.

Another problem arises when a compound is absent from the library. The problem may be circumvented with a reference library containing a great number of reference compounds. Based on the following implication:

Neighbor spectra → Neighbor compounds

It is possible to find a mass spectrum sufficiently similar to the unknown to detect a structural analogy. Nevertheless, as with other methods, the identification problem is not resolved in this case.

Comparative methods have been successfully applied in the case of unresolved mixtures (i.e., several compounds under a chromatographic peak) [68–71].

In all cases, comparative methods offer the same advantages as other methods, and in addition their effectiveness is greatly increased by the use of additional chromatographic data. Indeed, comparative methods allow one to identify compounds having similar mass spectra but different retention times or Kovats' indices (KI). But the choice of the most appropriate method is not only a function of the problem under consideration, but also of the anticipated applications. In our case, analysis of volatile aroma compounds by GC–MS coupling was dictated, hence the identification is made by comparing unknown data (i.e., MS and KI) with known data in the reference library. Accordingly, the databank will be specific for a given field dealing with the flavor industry.

In the two following sections we examine the major problems encountered in the construction of a computerized databank using comparative methods. The flowchart of the MS–KI bank design is shown in Fig. 1.

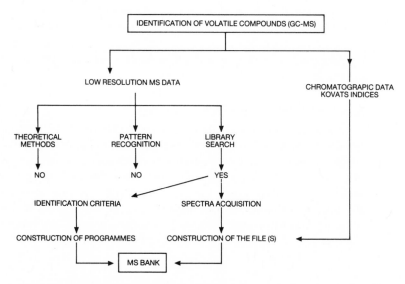

Fig. 1   The flowchart of the MS–KI bank design (reprinted from ref. 3, by courtesy of Academic Press, Ltd).

VII.3   MASS SPECTRA DATABANK CONSTRUCTION

When one wishes to construct a mass spectra databank, several problems must be solved (see Fig. 2):

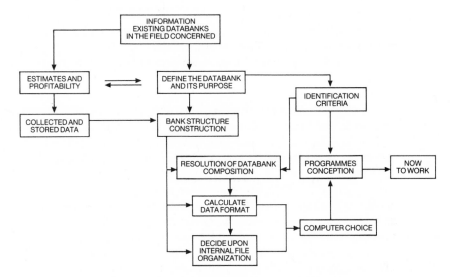

Fig. 2   Flowchart of SPECMA databank conception.

(1)   data acquisition and validation
(2)   the amount of data required to be stored in the reference library
(3)   the spectra search algorithm leading to quick retrieval of the stored compounds whose spectrum is supplied. These problems are closely related but the most important of them is the choice of a decision criterion (i.e., identification or non-identification) used in the research program. For this reason, all possible search algorithms will be examined first.

3.1   Various search algorithms

A mass spectrum is equivalent to a collection of N pairs of data (mass, intensity). The list of masses from one spectrum to another (whether or not of different compounds) is mathematically the same if zero intensity is assigned to an absent peak in a spectrum. Furthermore, since spectral intensities in both are expressed using the same unit, the decision criterion can be simplified to a simple comparison of intensities. If $\vec{S}_1$ are the intensities of an unknown spectrum, and $\vec{S}_2$ those of a reference spectrum, the comparison criterion can be reduced to a function $F(\vec{S}_1, \vec{S}_2)$ such that:

$$\vec{S}_1 = \vec{S}_2 \rightarrow F(\vec{S}_1, \vec{S}_2) = 0 \tag{2}$$

The most simple bijection arises from attributing the ratio $m/z = K$, the *Kth* coordinate of S (provided that m/z values are integer) [1].

In practice, the function F is always positive, and the closer to zero F becomes, the more similar the spectra are.

A threshold value F_S is defined such that:

$$F \leq F_S \rightarrow \text{identical compounds} \tag{3}$$

$$F > F_S \rightarrow \text{different compounds} \tag{4}$$

According to its form and properties (symmetrical, continuous or Boolean), the function F allows one to classify the search method. This is determined either empirically or from theoretical considerations (such as probability calculations).

Three main comparison methods have been described in the literature; some of them are based on binary encoding of intensities while others are based on intensity ranking or use stored intensity data.

Methods based on one bit encoding of peak height [61, 64, 72–75]

At each nominal mass a peak either exists, or does not exist, depending on the discrimination level (I_S)

$$I \geq I_S \rightarrow I = 1 \text{ (peak present)} \tag{5}$$

$$I < I_S \rightarrow I = 0 \text{ (peak absent)} \tag{6}$$

The binary encoding method, also called 'logical operator comparisons' or (XOR) instruction, has been widely used with Boolean functions or their linear combinations.

As example, the following function has been used by Grotch [76]

$$F = \mu N + \sum_{i=1}^{M} [(XOR)_i - \mu(AND)_i] \tag{7}$$

where μ is a weighting factor for logical AND (μ was found to have an optimum value of two); N is the number of peaks (ones) in the unknown; M is the number of masses to be compared; XOR is the logical exclusive OR instruction; and AND is the logical exclusive AND instruction.

If rapid searches are undertaken, the performance of the method is low, owing to encoding errors due to variations of intensities and/or the presence of impurities. There also exist intermediate cases between the binary encoding (two intensity levels) and classical encoding (101 levels, I ranging from 1 to 100%), i.e., four levels [76] or height levels [77].

Methods based on intensity ranking

In the method described by Knock *et al.* [62] the m/z values of the N most intense peaks in reference and unknown spectra are again compared, but an allowance is made for the relative positions of peaks with equal m/z values. The matching factor is given by the following equation

$$P_2 = \frac{1}{N^2} \sum_{k=1}^{A} (N - |i - j|_k) \tag{8}$$

where A is the number of agreements irrespective of order, and i and j are the positions in the respective sets of the kth pair of equal m/z values.

Later, the 'mass matching' system was improved [62] in specificity and performance through the use of a function having a finite number of values:

$$F = 1 - (1/R) \sum_{i=1}^{R} (N_c/N) \qquad (9)$$

R is the number of equal mass ranges, requiring that the N most intense peaks would have to be compared. N_c is the number of peaks common to the two spectra.

This function may be used with binary encoding and is well suited for spectra not requiring the storage of intensity data.

A reverse search system using the equation (9) has also been used [78]. In this case, the choice of the N most significant peaks has been adopted.

Methods requiring stored intensity data

Most comparison methods requiring stored intensity data use absolute intensity differences as a matching criterion [33, 76, 79–82]. These functions are summarized in Table 1.

Rather than use the absolute intensity difference, some authors, such as Biemann *et al.*, [63] have preferred to calculate the ratios of intensities of peaks at the same mass in the known and unknown spectra.

In this case, the mass spectra were abbreviated such that only two masses every 14 a.m.u. are considered, and the authors begin at the lowest mass common to both complete spectra. Other similar approaches have been adopted [83, 84].

The Mass Spectra Search System (MSSS) described by Heller [4, 85–87] is a conversational retrieval system based on an inverted file for the peak and search intensity. The program provides a wide choice of search options, for example:

1 Peak and intensity search
2 Molecular weight search
3 Molecular formula search (a) complete (b) imbedded
4 Molecular weight and peak search
5 Molecular formula and peak search
6 Molecular weight and molecular formula search
7 Dissimilarity comparison
8 Spectrum print-out
9 Microfiche display of spectrum
10 Display of spectrum
11 Plotting of spectrum
12 Crab-comments and complaints
13 Harvest-entering new data
14 News-news of the system
15 MSDC classification code list

Table 1   Various symmetrical functions used for the comparison of mass spectra[a]

Functions	Authors and Reference
$$F = \sum_{n=1}^{N} |I_u - I_r|_n$$	Abrahamson et al. [79] Dromey [80] Crawford and Morrisson [33]
$$F = \sum_{n=1}^{N} |I_r^{1/2} - I_u^{1/2}|_n$$	Crawford and Morrisson [33]
for normalized spectra such that: $$\sum_{n=1}^{N} I_n^{1/2} = 1$$	
$$F = \sum_{n=1}^{N} (I_r - I_u)_n^2$$	Crawford and Morrisson [33]
for normalized spectra such that: $$\sum_{n=1}^{N} I_n^2 = 1$$	
$$F = \frac{1}{N} \sum |I_u - I_r|$$	Abrahamson [81]
$$F = \sum \frac{|I_u - I_r|}{I_u + I_r}$$	Grotch [76] Dromey [80]
$$F = \sum \frac{(I_u - I_r)^2}{I_u + I_r}$$	Farbman et al. [66]
$$F = \sum \frac{(I_u - I_r)^2}{I_u^2 + I_r^2}$$	Grotch [76]
$$F = \frac{1000 \sum (I_u + I_r)^2}{(\sum I_u^2) \times (\sum I_r^2)}$$	VG Data System[b] (82)

[a] I_u and I_r are peak intensities in the unknown and the reference spectra.
[b] Two other functions $F(\text{Mix})$ and $F(\text{Rev})$ take into consideration only the peaks present in the unknown or only those present in the reference spectrum

The corresponding mass spectra data base is known as EPA/NIH (Environmental Protection Agency/National Institutes of Health) [88–93]. The STIRS program decribed by McLafferty et al. [18] also used comparison of intensity data, but it was mainly conceived to detect the presence of a structural feature rather than to provide an identification.

As a complement to the STIRS system, McLafferty et al. [94, 95] have devised Probability Based Matching (PBM). The function used in the method is:

$$F = 1/2^K \tag{10}$$

where $F = 0$, when $K > K_{\max}$ and $K = \sum K_i$. The K_i values (one for each peak) are interpreted knowing that 2^{K_i} is the average number of compounds, selected at random, whose mass spectra would have to be examined in order to explain the presence of peak i.

The probability of rendering an account of the peak i starting from $2^k i$ compounds chosen at random is $1/2^k i$. For the full spectrum, the corresponding value is $1/2^K$. Each value of K_i is the sum of four individual terms, empirically or statistically calculated from the reference library:

$$K_i = (U_i - A_i + W_i - D) \tag{11}$$

where $\frac{1}{2} I_u - A_i$ is the appearance frequency of peak i in the library (for I_u falling within a required limit in the unknown spectrum); U_i is the uniqueness of the m/z value of the ith peak; A_i is the modification of U_i based on the abundance of the peak in the target spectrum; D is a window tolerance; W_i and D are dependent on the ratios I_u/I_r; $K_i = 0$ if I_u/I_r fall outside of the predetermined tolerance limit. The independence of peaks inside of a spectrum-based hypothesis, is a disadvantage of the method [1].

A different approach was followed by Petitjean [1] for the SPECMA databank. The function used in this case allows one to work with variable tolerances with respect to the gaps in the spectra according to their origin.

In this function

$$F(\overrightarrow{S_u}, \overrightarrow{S_r}, X, R) = F(\overrightarrow{D_u}, \overrightarrow{D_r}) \tag{12}$$

X and R are the lists of parameters (other than the spectra themselves) for the unknown and reference compounds. These parameters take into account variations in experimental conditions and spectral quality.

Miscellaneous

Other functions based on information theory have been proposed, some of them are dissymmetrical [77, 96, 97] while others are symmetrical [66].

All the methods previously mentioned attempt to select and to compare reference compounds which are most consistent with the unknown spectrum. Some of their search systems use algorithms for this purpose which are more complicated than those described here.

Other data, such as intensity, molecular weight, IR, ^{1}H-NMR, ^{13}C-NMR, or retention indices are independent sets which can be chosen by the user (see Section VII.4).

3.2 Mass spectra abbreviation

Since the use of all peaks of a spectrum is impossible, it is necessary to select their number and size. Various devices used to abbreviate a mass spectrum are outlined below.

(i) N most intense peaks. Many authors choose the N most intense peaks ($N = 5$ to 10) in R ranges, each R containing m mass units [9, 10, 33, 62, 67, 79, 98, 99]. This method, and the following one can be criticized because it necessarily may omit interpretatively significant peaks in many instances. On the other hand, a lower number of peaks may still be, in some cases, sufficient for the purpose of identification. Conversely, a greater number of peaks has often been found necessary in the case of sesquiterpinic compounds [100–102].

(ii) N peaks in a limited interval. Other workers select the N most intense peaks in each M interval of mass spectrum ($N = 1$ or 2; $M = 7$ or 14 a.m.u.) [63, 80, 103–107]. Thus Hertz et al. [63] selected the two most intense peaks in each 14 a.m.u. interval throughout the complete spectrum.

Smith's method [108] relies on the widely known fact that mass spectrum of compounds within a given class are generally quite similar. This method consists of expressing as a set of numbers the percentage contribution of each peak to the total ion count within fourteen distinguishable ion series. Mass spectra are thus viewed as a set of fourteen ion types consisting of, for example, in the case of saturated alkanes, the possible ions series: $C_nH_{2n+2}]^+$, $C_nH_{2n+1}]^+$, $C_nH_{2n}]^+ \cdots C_nH_{2n-11}]^+$. Only fourteen series are possible because members of the homologous series $C_nH_{2n-12}]^+$ and $C_nH_{2n+2}]^+$ have the same nominal masses.

Using the Visually Interpretative Peaks method (VIP) [109], Alencar and Craveiro [110] select N peaks ($N = 8$ to 20 for most compounds), each peak m/z being more intense than its neighbors ($m + 1$) and also more intense than its homologues ($m + 14$) and ($m - 14$).

(iii)   In the case of a family of compounds showing similar characteristics and identical molecular masses, such as sesquiterpinic compounds (MW = 204), Fluckiger et al. [111] construct a microfile containing only the two most discriminant ion peaks of the mass spectrum. Unfortunately, the authors do not give a definite example, but they indicate that their method gives similar results to those which others have obtained.

The less intense ion peaks of a mass spectrum can also play an important role in distinguishing monoterpinic compounds that have similar spectra (MW = 136, base peak $m/z = 93$) [111].

In all cases, a reference file of very high quality is necessary.

The best method of spectrum encoding consists of selecting the N more intense peaks, with N sufficiently high so as to include all significant peaks (using a different number for N for each spectrum) [1].

In practice, it is necessary to distinguish:

(a)   the measured spectrum: all peaks within a certain mass range are recorded and the peaks at 28 a.m.u. and 32 a.m.u. are discarded to avoid air peaks;

(b)   the published literature spectrum: in most cases, this includes only a few fragments which are classified in order of decreasing intensity;

(c)   the formatted spectrum in memory: this is encoded according to one of the above-mentioned methods, involving a greater or lesser loss of information;

(d)   the part of the formatted spectrum useful to the search program: this is the only spectrum required for identification.

If the program does not use all encoded information, the result will be an inefficient use of memory and a lessening of search reliability. If the literature is the main source of spectra, the search algorithm must take into account the loss of information between the measured spectrum and the published spectrum.

3.3 Quality criterion for mass spectra

The specificity of a mass spectrum may be investigated, as well as established, when the following double implication is checked:

$$\text{IDENTICAL COMPOUNDS} \underset{2}{\overset{1}{\rightleftharpoons}} \text{IDENTICAL SPECTRA}$$

The first implication depends on the reproducibility of the mass spectrum in terms of experimental conditions, i.e.,

Ionization voltage and type of apparatus

Mass spectra are usually recorded at 70 or 80 eV. Few variations in relative intensities are observed between 80 and 20 eV, as shown for the case of furfural [1]. However, under 20 eV, this does not hold true. It is well-known that spectra of the same compound, recorded on different instruments, are not absolutely identical. Thus, the comparison of the mass spectra of α-pinene and citronellol in two single-focusing magnetic sector mass spectrometers (the Varian Mat 111 and the Atlas CH_4) show noticeable differences in relative intensities [67]. Camphor is another significant example [113]. The fragmentation pattern listed for this terpinic ketone in the Wiley and Sadtler files are strikingly different from its quadrupole mass spectrum. With sesquiterpinic compounds, different base peaks are often observed [62].

The quality of GC separation

With a low intensity peak, the background noise plays an important role, just as the presence of an undetected impurity under the chromatographic peak increases the variation of relative intensities.

Recording conditions

The shape of the mass spectrum can also depend on the position within the chromatographic peak at which the mass sample is taken. If the chromatographic peak is sampled at its maximum, saturation phenomenon may occur. Spectra of terpenoids such as α-pinene are very sensitive to these conditions [114].

Sample flow variation

This phenomenon occurs through the ion source during the course of scanning a spectrum. However, it can be corrected for by recording the total ion current and adjusting the peak intensities accordingly, or by averaging several rapidly-recorded spectra over a GC peak [108].

The above examples point out the precariousness of certain conclusions drawn from a mass spectrum which was recorded by GC–MS coupling. In the choice of the comparison criteria between spectra, it is necessary to take into account these variations in order to recognize whether two spectra of the same compound are truly identical.

The second implication cannot always be borne out because various compounds (homologues, stereoisomers, positional isomers) have strikingly similar spectra.

The comparison criteria have to be sufficiently discriminating to distinguish between various compounds having similar mass spectra.

3.4   Mass spectral acquisition and validation

The main source of mass spectra is the literature, where mass spectra are primarily compiled in data bases [115]. Data bases were first developed in the United States under the impetus of such organizations: the American Petroleum Institute (API) [116], Dow Chemical Company, the American Society for Testing and Materials (ASTM), the Environmental Protection Agency (EPA) and the National Institutes of Health (NIH). All the data compiled in the Mass Spectrometry Data Center (MSDC) [117] comprise the Mass Spectral Search System (MSSS) [88–93]. Including the data from Stenhagen *et al.* [118, 119], the MSSS contains more than 30 000 spectra. The Wiley/NBS Mass Spectral Base [120] contains 80 000 spectra arising from 68 000 different compounds. Mass spectra are also available from other data bases [99, 121–126] and in general texts [98, 127].

For a given category of compounds such as the volatile components of flavors and fragrances (i.e., heterocyclic, monoterpinic, sesquiterpinic, aliphatic, and aromatic compounds) mass spectra have been compiled in several specialized texts [128-131], atlases [132-134], and reviews or original works [100-102, 111, 112, 114, 135-138].

These data are of great value to flavor analysts, giving them the opportunity for the visual comparison of mass spectra.

Literature data are the least expensive acquisition form but they present some drawbacks, as summarized below.

Different representation methods for mass spectra

If the spectrum is submitted as a bar chart, it must be re-plotted. This is not always easy due to either a defective visualization or reproduction. In the best possible case, the full spectrum is published with a complete list of fragments, with their relative intensities.

Only a published mass list is available

Sometimes, the mass list is given in order of decreasing intensity. In this case, information is abridged and cannot be used completely. In other cases, only particular spectrum fragments are published.

Experimental conditions are not published

Ionization voltage, type of apparatus (quadrupole or sector), and limit damping mass are seldom indicated. The latter is often set at 40, but lower fragments, such as $m/z = 31$ $(CH_2OH)^+$, are characteristic of certain functions (alcohols, ethers, dioxolanes). On the other hand, the peaks at m/z 28 and 32, due to atmospheric air, can be omitted for very dilute compounds but this omission should be indicated.

Spectrum origin is unknown

Thus, it is impossible to proceed to a reference spectrum.

Several quite different spectra for the same compound

Unfortunately this is not a rare occurrence and it is most difficult to solve. As an example, we list in Table 2 the abbreviated mass spectra of several sesquiterpinic alcohols as reported by various authors.

The differences in the table cannot be ascribed simply to experimental procedures, but also to erroneous interpretation and sampling. In the present case, how can one determine which is the better spectrum? There are no definite rules but the authors' specialization, and our knowledge of fragmentation rules, and comparison with other spectra of the same products (when available) lead us to consider the results from Von Sydow *et al.* [132] and those from Hill *et al.* [112] as the better ones. The quality of certain publications, such as those dealing with coffee aromas [142] is another criterion to be considered. This problem of spectrum quality has been discussed by Speck *et al.* [141].

The drawbacks encountered with literature data have induced those search laboratories possessing a GC–MS coupled instrument to construct their own databank. Although this is the ideal case, it is also the most costly and time-consuming. Data must be obtained either from standard samples or from natural or synthetic mixtures whose composition is already known. In the first case, a sizeable collection of samples is needed. In the second, one must solve the identification problem in the absence of an MS reference databank.

In summary, the preliminary work required to construct a databank of mass spectra hinges on a threefold bibliographical search [141]:

(a) Acquisition of data;
(b) Evaluation of variations from one spectrum to another for the same compound;
(c) Evaluation of non-standardization of data, which is responsible for an ill-defined information reduction [3].

Furthermore, the decision to be made which is strongly dependent on these points, will be greatly improved by the use of selection criteria such as the Kovats' indices.

Table 2   Abbreviated mass spectra of some sesquiterpinic alcohols as reported by various authors[a]

Compounds	Ref.	m/z (I %)
Guaiol C15H26O (222)	[132]	181(100) 161(85) 59(62) 107(42) 105(40) 93(36) 81(33) 163(32)
	[139][b]	41(30) 43(29) 204(27) 189(25) 95(24) 57(100) 41(95) 43(90) 135(77) 55(70) 105(38) 77(24) 79(24)
	[132]	93(100) 81(96) 41(80) 107(60) 121(57) 161(40)
Elemol C15H26O (222)	[99]	59(100) 93(71) 81(47) 107(41) 161(40) 42(35) 67(34) 121(34)
	[139][b]	41(100) 59(84) 43(70) 67(64) 93(62) 39(58) 161(56) 55(50) 121(36) 135(31)
Ledol C15H26O (222)	[112]	43(100) 41(76) 69(74) 122(66) 109(66) 81(57) 107(47) 55(46)
	[140	161(100) 105(80) 93(70) 91 107 119 81 147 204 133 99 etc.
Bulnesol C15H26O (222)	[132]	135(100) 59(90) 107(70) 93(60) 161(50) 105(40) 81(37) 189(30) 204(30)
	[139][b]	43(100) 57(76) 41(68) 55(62) 56(43) 42(38) 70(31) 71(28)

[a] The first reference for each compound is considered as the better one.
[b] Also reported in *Eight Peak Index of Mass Spectra* [99].

VII.4   KOVATS' INDICES DATABANK CONSTRUCTION

Whatever the function chosen for the comparison of mass spectra, it will not always give the desired results, i.e., the infallible and reliable identification of an unknown. For this reason, and in order to reduce costs and search time, several authors have found it useful to insert into their search program other data which act as selection criteria. Among them, the molecular weight accessible from positive and negative chemical ionization mass spectrometry provides a useful additional piece of data, allowing the elimination of compounds which differ in molecular weight.

Another form of pre-search is the specification of a molecular weight range [143]. However, these criteria are not yet sufficient to identify an unknown. It is necessary to have recourse to other methods, including:

(a) accurate mass measurements (to provide the empirical formula);
(b) various spectroscopic data (Fourier transform IR, UV, ^{1}H-NMR, ^{13}C-NMR) [144–146].

But the chromatographic data are those to which the more attention has been paid. These data can be used to screen the compounds available in the MS databank before the search actually begins.

They can be also used after the completion of the search as reference data for the respective compounds. It is also possible to differentiate between compounds possessing similar mass spectra but different retention times.

4.1   Choice of chromatographic data

Several kinds of chromatographic data are available [147–151] (see Table 3). They are becoming more frequently used for the computer-assisted identification of complex mixtures by gas chromatography [152, 153].

Among these kinds of data, relative retention data [154–157] and Kovats' indices [158–165] are primarily recommended to assist mass spectra research for volatile aroma compounds. Results are further enhanced by the use of two columns with differing polarities [111, 152, 158] i.e., one polar (Carbowax 20 M or FFAP) and the other apolar (SE 30, OV 1, OV 101).

For GC programmed temperature the relation used to calculate Kovats' indices can be written thus: [1, 166]

$$KI(x) = 100\left(n + \frac{t'_{r(x)} - t'_{r(n)}}{t'_{r(n+1)} - t'_{r(n)}}\right) \qquad (13)$$

where $t'_{r(n+1)}$ is the reduced retention time of a linear alkane with $n + 1$ carbon atoms. Another reference scale, using the ethyl esters of linear fatty mono-acids as standards, has been described [166].

A simple relation permits us to convert from one scale (alkanes) to another (esters):

$$KI_0(x) = KI_0(n_1) + 100 \times D\left(\frac{KI_1(x)}{100}\right) \qquad (14)$$

where n_1 is the integer part of $KI_1(x)/100$ $[n_1 = INT(KI_1(x)/100)]$ and $D(KI_1(x)/100)$ is the decimal part of $KI_1(x)/100$:

$$D\frac{KI_1(x)}{100} = \frac{KI_1}{100} - n_1 = \frac{t'_{r(x)} - t'_{r(n1)}}{t'_{r(n1+1)} - t'_{r(n1)}} \qquad (15)$$

Accordingly, it is necessary to know n standard indices (or retention times) of esters $KI_1(n1)$ and $KI_1(n1 + 1)$ having $n1$ $[n1 = INT(KI_1(x)/100)]$ and $n1 + 1$

Table 3   Presentation of various retention data

Definitions	Formula
Retention data	
t_r: retention time	$t_r = d_r/u; \; V_r = D \cdot t_r$
V_r: retention volume	
d_r: retention distance	
u: chart speed	
D: carrier-gas flow rate	
Adjusted retention data	
t'_r and V'_r: adjusted retention time and volume, respectively	$t'_r = d'_r/u; \; V'_r = D \cdot t'_r = V_r - V_m$
d_m: atmospheric air retention distance	$d'_r = d_r - d_m$
V_m: 'gas hold-up' volume	
Limited retention data	
t^0_r and V^0_r: limited retention time and volume, respectively	$t^0_r = d^0_r/u; \; V^0_r = D \cdot t^0_r$
J: compressibility factor	$d^0_r = J \cdot d_r$
p_i: column inlet pressure	$J = \dfrac{3(p_i^2/p_0^2 - 1)}{2(p_i^3/p_0^3 - 1)}$
p_0: column outlet pressure	
Absolute retention data	
$t^0_{r'}$ and $V^0_{r'}$: absolute retention time and volume, respectively	$t^0_{r'} = d^0_{r'}/u; \; V^0_{r'} = D \cdot t^0_{r'}$
	$d^{0'}_r = J \cdot d'_r$
Specific retention volume (V_g)	
w: weight of stationary phase in the column	$V_g = V^0_{r'} \cdot \dfrac{273}{w \cdot T_c}$
T_c: column temperature (Kelvin degrees)	
Relative retention data (α_{AB})	
(between two solutes A and B)	$\alpha_{AB} = V_g(A)/V_g(B) = t'_r(A)/t'_r(B)$
Kovats' indices	
(Isothermal)	$KI = 100n + 200 \dfrac{\log t'_{r(x)} - \log t'_{r(n)}}{\log t'_{r(n+2)} - \log t'_{r(n)}}$
$t'_{r(x)}$, $t'_{r(n)}$ and $t'_{r(n+2)}$: reduced retention time of the unknown substance (x) and the linear alkanes with n and $n+2$ carbon atoms surrounding compound x	

carbon atoms. These values have been listed for SE 30 and Carbowax columns [166].

4.2   Properties of Kovats' indices

Kovats' indices exhibit a number of properties which can profitably be used as a route to identification [166-169];

(1)   Contrary to other chromatographic data, Kovats' indices are little influenced by temperature changes and may be used with programmed temperature;

(2)   By definition, the Kovats index of a linear alkane, C_nH_{2n+2} is equal to $100n$

(3)   The Kovats indices of higher members of a homologous series may be expressed as:

$$KI_n = KI + 100n \qquad \text{(for each liquid phase)} \qquad (16)$$

(4)   On a non-polar phase, the index difference (KID) between two isomers is related to the difference between their boiling points by:

$$KID = 5 \times \Delta E_b \qquad (17)$$

(5)   The Kovats' indices of non-polar compounds are usually independent of the nature of the stationary phase;

(6)   The Kovats' index of a solute is usually constant for all non-polar phases;

(7)   For unsymmetrically substituted compounds (R–X–R'), Kovats' indices may be calculated from symmetrical compound indices by:

$$KI(R\text{–}X\text{–}R') = [KI(R\text{–}X\text{–}R) + KI(R'\text{–}X\text{–}R')]/2 \qquad (18)$$

(8)   Similar substituents in similar compounds (i.e., belonging to the same family) give rise to the same index variations for a given liquid phase;

(9)   The more polar the column, the greater will be the index of a polar solute. The more polar the solute, the greater will be the retention increment for a given polar phase.

(10) If Kovats indices are measured on two columns of different polarity, the index difference (KID) is characteristic of the solute molecule:

$$KID = KIP - KIA$$

where KIP and KIA are retention indices on polar and non-polar columns, respectively. This difference is a measure of the additional effect of a polar column on the solute molecule when compared with the non-polar column. It increases with polarity of the liquid phase and also depends upon the polarity of the solute.

Another advantage of Kovats' indices is their additivity. On the basis of the Kovats' index of a parent molecule (PH) it is possible to calculate the indices of its derivatives by using the substituent increments:

$$KI(P\text{–}R) = KI(P\text{–}H) + \sum^{i}(\Delta KI)_R \qquad (20)$$

These increments have been listed in aliphatic [169] and in heterocyclic series [170–173], and they allow one to identify an unknown compound by comparing its experimental and theoretical KI values.

The increments for alkyl substituents not adjacent to the nitrogen atom in aromatic aza-heterocyclic compounds are not very different from those determined for the aromatic series. However, when an alkyl group is bonded in a position adjacent to the nitrogen atom, a decrease of the value of these increments, due to steric hindrance, is observed [3].

Retention data and Kovats' indices have also been used to determine the thermodynamics of the interaction of a volatile solute with a non-volatile solvent [174].

4.3   Choice of a comparison method [1]

As previously seen Kovats' indices are primarily dependent on the liquid phase polarity. Although this polarity varies continuously, in practice one must admit that only a finite number of phases is available. On the other hand, it has been established that Kovats' indices are well-defined by two factors only: the nature of the molecule and that of the stationary phase. As a result, for N phases, the comparison between indices is limited to a function F similar to the one used for mass spectra:

$$\overrightarrow{KI_{(x)}} = \overrightarrow{KI_{(y)}} \rightarrow F(\overrightarrow{KI_{(u)}}, \overrightarrow{KI_{(r)}}) = 0 \qquad (21)$$

The KI vectors as such that their N coordinates are the indices of a compound on N phases.

All the functions described in the previous paragraph for mass spectra may also be used here. A threshold value KI_S is determined at every stage such that:

$$|KI_{(u)} - KI_{(r)}| \leq KI_S \rightarrow \text{Compound non-rejected} \qquad (22)$$

$$|KI_{(u)} - KI_{(r)}| > KI_S \rightarrow \text{Compound rejected} \qquad (23)$$

The KI_S value may be varied according to the liquid phase. The difference index comparison method $|KI_{(u)} - KI_{(r)}|$ is more suitable than one comparing $\Delta(KI)/I$ ratios as these are not constant [1].

The Kovats' indices of more than a thousand aroma compounds on two different polarity phases (OV 1 and Carbowax 20 M) have been compiled [31]. They are of great value to a researcher wishing to construct a Kovats' indices databank.

VII.5   SPECMA DATABANK REALIZATION

Bearing in mind the foregoing discussion, the following points will also need to be considered:

(1)   Mass spectra will be selected from the literature and will be gradually supplanted by our own data;

(2) N intense peaks, with their relative intensities, will be selected, while rounding the lowest intensities to zero, and N will be variable according to the nature of the spectra;

(3)   Kovats' indices on polar and non-polar columns will be used as filters, and contrary to the conversational methods (MSSS–EPA–NIH or

DARC–CIDA databanks) where the responsibility for actually identifying an unknown is transferred to the user, the decision criterion will
primarily be made by the program, since the operator is considered a
non-specialist;

(4) Agreeing on a general solution containing a calculated number of
matching compounds, thanks to the use of threshold values, the search
program will find one compound as the best case solution and none if
the compound is not present in the library;

(5) A variable comparison function will be chosen, taking into account the
spectrum origin and that of parameters linked to it;

(6) Finally, other data such as legislative, organoleptic, toxicological, etc.,
will also be computerized for the specialist users.

The goal is now to store the acquired mass spectra in a computer and to
prepare a search program to compare the abbreviated spectra numerically
(unknown and reference) and to tell the user whether the spectra are identical
or not.

5.1 File Structures

The information was divided into two sets called SPECMA.DAT and
NOMREF.DAT files, respectively. Their content and structure are given in
Table 4.

The SPECMA.DAT file comprises all the elements needed for a spectrum
search. The empirical formula is formally restricted to carbon, hydrogen,
nitrogen, oxygen and sulphur atoms, but it can be extended to include all
other elements and isotopes represented collectively by Z (halogens, phosphorus, deuterium, etc.). The molecular weight is based on the most abundant isotopes. This option is particularly useful in chemical ionization, where
the molecular weight is easily deduced. The mass spectrum consists of the
largest peaks (originally up to 25 peaks). In the current version, this number
of peaks has been reduced to allow the introduction of three peaks in PCI and
in NCI, respectively.

The Kovats' indices on Carbowax 20 M and on Silicone SE 30 are
referenced to linear alkanes. A third value (KID), which represents the
difference between these two values may be useful in order to characterize a
given family. The limit damping mass is required to avoid hazards in
identification due to peaks below 40. Finally, the spectrum source, coded
according to Chemical Abstract reference is also given. Further extensions of
this file are also allowed for.

The NOMREF.DAT file contains all other information which is not
specifically useful for spectra searches, i.e., the name, flavor descriptor,
numbers on positive lists (FEMA GRAS, CoE, IOFI), toxicological data
(LD_{50}), olfactory threshold values, specific limits in foods and beverages, the
Registry Number. This last presents two advantages: it provides unequivocal
identification of a compound, even if its structure is ambiguous, and it allows

Table 4   File structures of the SPECMAIFF databank[a]

Files	Number of bytes
SPECMA.DAT	
Empirical formula (C, H, N, O, S, Z)	6(1 byte/element)
Molecular mass	2
Mass spectra (EI)	36(2 bytes/peak)
Mass spectra (PCI)	6(2 bytes/peak)
Mass spectra (NCI)	6(2 bytes/peak)
Kovats' indices (KIA, KIP, KID)	6
Lower mass limit	1
Subsequent extensions	65
Total	128
NOMREF.DAT	
Name	58
Descriptor[b]	28
Positive lists (FEMA GRAS, CoE, IOFI)[c]	5
Toxicological data (LD_{50}, ADI)[d]	4
Odor threshold values[e]	2
Specific limits of use (SL)[f]	4
Registry number	5
Reference	5
Subsequent extensions	17
Total	128

[a] The previous version of the databank called SPECMA did not contain toxicological and organoleptic data and recording required 128 bytes. The databank was realized on a North Star Horizon microcomputer of 64 Kbytes, equipped with a dual 8″ floppy disk unit. It is run by CP/M. The source program (PL1 80) was parameterized so as to optimize adjustments to various types of diskettes. The databank has also been loaded onto an Apple 2+ ($5\frac{1}{4}$ in diskettes) and contains some 2000 aroma compounds distributed between two diskettes.

[b] Compiled in part in Ref. [175].

[c] FEMA GRAS: Flavor and Extract Manufactures' Association—Generally Recognized as Safe, compiled in Ref. [176].

CoE (or CE): Council of Europe list compiled in Ref. [177].

IOFI: International Organization of Flavor Industries.

[d] LD_{50}: Acute lethal dose

ADI: Acceptable Daily Intake in mg/kg body-weight allocated by the Council of Europe.

[e] Compiled in part in Ref. [178].

[f] SL: in mg/kg food and beverages, recommended by the Council of Europe Committee of Experts.

us to forge a bond to other databanks using the Registry Number. Subsequent extensions of the NOMREF.DAT file are also allowed for.

These two files are first classified in increasing order of molecular weight and then in increasing number of atoms for each element, in the order C, H, N, O, S and Z. There also exists an access table of molecular weights stored in a third 1024 bits file called FDIR.DAT.

5.2   Program structure

The program is composed of a central procedure SM.COM, whose role is to print the 'Menu' of operations and to call one of the three following overlay subroutines:

(1)   PR.OVL: Mass spectral search program
(2)   AT.OVL: Modification of the databank content
(3)   UT.OVL: Utility programs not modifying the databank content.

In simplified form, the databank flowchart is shown in Fig. 3. Each subroutine provides for the correction of accidental errors occurring during the search (typing errors, exceeding limits, etc.).

The flowchart for the the PR.OVL search procedure is given in Fig. 4.

The file area to be selected is deduced from the molecular mass and empirical formula conditions. The filters used deal with the empirical

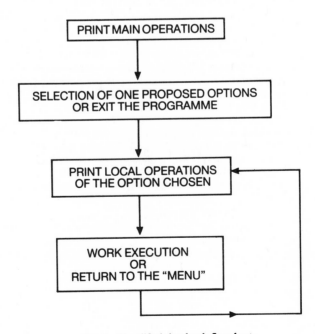

Fig. 3   Simplified databank flowchart.

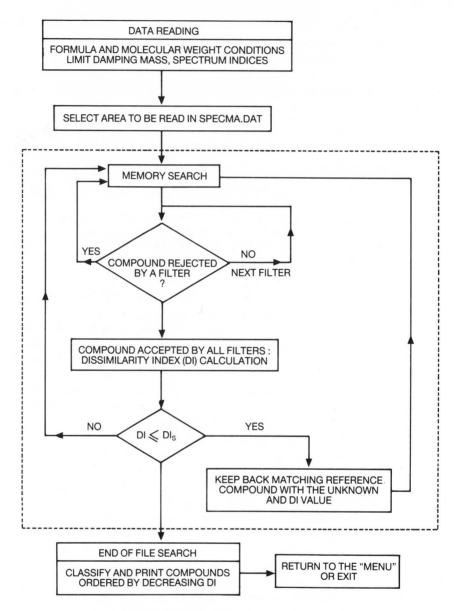

Fig. 4   PR.OVL search procedure flowchart.

formula, Kovats' indices and the spectrum, one after another. If the compound is passed by all filters, the DI calculation can be made. The compound is only printed if the DI value is less than or equal to a threshold value $DI_S = 2550$.

The first filter dealing with empirical formula conditions (i.e., nitrogen or sulfur atoms presence detected in GC using specific detectors) is primarily intended to reduce the number of compounds whose spectra will be compared with the unknown spectrum. Presently, the average performance time is a few seconds.

5.3 Databank utilization

All commands needed for the operations are printed at the console, thus lessening the task of the user. Databank control and checking is handled by the SM.COM program which prints the 'Menu' of the central procedure
A > SM

> †††††† SPECMA BANK 001 VERSION ††††††
>
> S : SEARCH OF UNKNOWN SPECTRA
> M : MODIFICATION OF BANK CONTENT
> U : UTILITY PROGRAMS
> E : END OF THE PROGRAM

Now the user must choose one of the options, typing the corresponding letter, followed by 'RETURN'.

The *S* option allows one to access the search program. The *C* and *S* options allow one to carry out a spectrum search or to return to the 'Menu'. In the first instance, the user must indicate empirical formula conditions. That being done, a range for each of the six elements (C, H, N, O, S, Z) is given.

In the following example (see Fig. 5), the user assigns the number of nitrogen atoms ($1 \leq N \leq 200$) and the sulfur atom presence ($S = 1$). A range of 0 to 200 indicates that nothing is known about the number of these atoms.

The program then asks for the molecular weight range. If known (from MS chemical ionization techniques) it is given; in this case it is 135. When unknown, it is sufficient to give an unrestricted range, such as $0 \leq M \leq 1000$. But it is usually more sufficient to determine the lower limit based upon the base peak. The upper limit rarely exceeds 300 among aroma compounds. Once these limits are determined, one must enter the lowest mass recorded by mass spectrometer. If that is unknown '0' is entered,

After the notice 'INPUT MS PEAKS' peaks with their intensities are entered, in any order. It is possible to constantly correct an error by re-entering the peak requiring correction.

The unknown spectrum entered by the keyboard must be as complete as possible, in order to diminish the number of possible answers. If Kovats' indices are known, they are also entered. (In the current version of the databank, the index difference (KID) has also been incorporated). The computer then lists the input data, and after a few seconds, the compound is printed. In the example, BENZOTHIAZOLE, which is certainly one of the most widespread heterocyclic compounds in food flavors, has been found with an excellent DI (DI = 510). When the DI is higher than 2550, the program prints 'UNKNOWN IN THE FILE'.

The following example (see Fig. 6) illustrates modification of the databank content (option M of the central procedure). After entering the letter 'M', the

††††† SPECMA BANK 001 VERSION †††††

S : SEARCH OF UNKNOWN SPECTRA
M : MODIFICATION OF BANK CONTENT
U : UTILITY PROGRAMMES
E : END OF THE PROGRAMME

CHOOSE : (S, M, U, E) ? S

SEARCH PROGRAMME : C)ontinue or S)top ?   C

EMPIRICAL FORMULA CONDITIONS : Y)es or N)o ?   Y

Input C Range : (Min Max) ? 0 200
Input H Range : (Min Max) ? 0 200
Input N Range : (Min Max) ? 1 200
Input O Range : (Min Max) ? 0 200
Input S Range : (Min Max) ? 1   1
Input Z Range : (Min Max) ? 0 200

Input Atomic Mass Range for Z : (Min Max) ? 2   300

Input Molar Mass Range (Min Max) ? 135   135

Input Limit Damping Mass of Scanning ?
(Give 0 if unknown)
? 20

INPUT MS PEAKS
    Integer or Half-Integer Mass
    Any Input Order for Peaks
    Any Units for Intensities
    Input a Second Time a Peak if Correction is Needed
    Input "?" if Correction of All Spectrum is Needed
    Input 0 0 to Stop

INPUT PEAKS
Peak 1 : Mass ? Intensity ? 135   100
Peak 2 : Mass ? Intensity ? 108   42
Peak 3 : Mass ? Intensity ?   82   12
Peak 4 : Mass ? Intensity ?   69   30
Peak 5 : Mass ? Intensity ?   63   12
Peak 6 : Mass ? Intensity ?   58   12
Peak 7 : Mass ? Intensity ?   45   10
Peak 8 : Mass ? Intensity ?   39   18
Peak 9 : Mass ? Intensity ?   0   0

Input Kovats' Indices (Ref. Alkanes)
(Give 0 if Index is Unknown)
KI apolar (SE 30)   = ?   1180
KI polar (Carb. 20M) = ?   1920

PROPOSED SPECTRUM
    M/E   I%
    135.0   100
    108.0   42
     82.0   12
     69.0   12
     58.0   12
     45.0   10
     39.0   18

KIA =   1180
KIP =   1920

POSSIBLE COMPOUNDS : 1

Compound Number : 393

BENZOTHIAZOLE   FM3250   CE0
  C7 H15 N S = 135   RN = 95-16-9
KIA = 1200   KIP = 1930
REF = 65 : 19456 a
Damping Mass. = 20; m/e 28 and 32 absent; Function 62
    135(100)
  136(9)   108(35)   82(12)   63(13)   54(9)
   45(12)   39(8)   38(8)

††††† DI = 510 †††††

Fig. 5   Main options of the SPECMA bank: search for benzothiazole (adapted from reference 179)

program gives the maximum storage bank capacity, 8192 compounds, and the actual capacity 1003 heterocyclic compounds.

We can choose between the four following options:

A : ADDING DATA
D : DATA REPLACEMENT
S : SUPPRESSION DATA
E : EXIT OF SUBROUTINE

If we want to suppress compound number 393 (option S), we type 393 and then the computer indicates the new bank capacity, 1002 compounds. Of course, this compound is again introduced to the file, using the option A. It will be registered under number 393.

MODIFYING DATA BANK
SPECMA BANK
† Maximum storage capacity : 8192 compounds
† Present capacity : 1003 compounds (heterocyclic compounds)

A : ADDING DATA
D : DATA REPLACEMENT
S : SUPPRESSION DATA
E : EXIT OF SUBROUTINE

Choose : (A, D, S, E) ?   S
Compound number ?   393

SPECMA BANK
† Maximum storage capacity : 8192 compounds
† Present capacity : 1002 compounds

The program send us back to the 'menu' (A, D, S, E) ?

Fig. 6   Various options for databank content modification.

UTILITY COMMANDS

Research of a compound from its Registry Number

CHOOSE : (S, M, U, E) ?　U

0　Listing of compounds having the same molecular mass
1　Listing of compounds having the same empirical formula
2　Listing of compounds having the same registry number
3　Listing of compounds having the same reference
4　Listing of compounds having the same Kovats' indices
5　Listing of compounds having the same name
6　Drawing spectrum
7　Memory search
8　Calculation of isotopic abundance
9　Exit of utility program

CHOOSE : (0, 1, 2, 3, 4, 5, 6, 7, 8, 9) ?　2

Fig. 7　Various options for utility programs.

In Fig. 7, the various possibilities of utility programs are illustrated.

Among them we select option 2, which means we wish to search for a compound from its registry number. In this case, we wish to verify if benzothiazole has been put in the file. After typing the registry number, i.e., 95–16–9, the program lists the information corresponding to benzothiazole and sends us back to the 'Menu' U : Utility programs.

Figure 8 shows how it is possible to calculate the isotopic abundance of

Input the Registry Number : (XXXXXX YY Z) ?　95–16–9

Compound number : 393

Benzothiazole　　　　　　　　　　FM 3256　　　CE 10
C7 H5 N S = 135　　　　　　　　　RN =　　95–16–9
KIA =　1200　　　KIP = 1930
REF = 65 : 19456 a

Damping Mass = 0　; m/e 28 and 32 absent; function 62
　　　　135(100)
136(9)　108(35)　82(12)　69(27)　63(13)　54(9)　45(12)
　38(8)　　38(8)

Send back to the menu 'U' (0–9)

CHOOSE : (0, 1, 2, 3, 4, 5, 6, 7, 8, 9) ?　8

Data from *CRC Handbook, 1980*
Input : (C, H, N, O, S)
(Type 0 0 0 0 0 to exit) ?　7 5 1 0 1

$(P + 1)/P = 9.104E　- 02 = 9.10\%$
$(P + 2)/P = 4.808E　- 02 = 4.81\%$

Fig. 8　Options 2 and 8 of utility programs.

nitrogen and sulfur atoms in benzothiazole (C7 H5 N S) from data given in the *CRC Handbook, 1980.*

VII.6   APPLICATIONS

The SPECMA databank has been tested with more than two hundred essential oils and model systems related to the Maillard reaction. Some of these studies, particularly those related to geranium [180], basil [181], garlic [182], Mirabelle [183(a)] and Reine-claude [183(b)] plums, tarragon [184], mentha crispa (213), Syrah wine (214) and heterocyclic compounds in model systems [1–3] have been published or presented on the occasion of Congresses. As space considerations make it impossible to summarize them here, we refer the reader to the references for more details.

VII.7   CONCLUSION

The low specificity of existing commercial MS banks, as well as their restricted information content and the absence of selectivity in results led us to conceive and realize our own databank specially adapted to the identification of volatile compounds in flavors and fragrances. The improvements realized are a greater specificity, further selectivity and the large amount of information useful to the flavor industry. However, we know that not all problems raised by comparative methods have been solved, especially the one of a compound 'UNKNOWN IN THE FILE'.

In that case, the trained spectroscopist can usually find a solution based upon fragmentation rules and pattern recognition. Beyond that, the combination of the library search method with theoretical and/or statistical methods can also bring useful structural information to bear on a product absent in the reference file. This possibility should not be neglected by the analyst who has a multidimensional view of things.

In the near future, it is quite evident that these combined computerized techniques, as well as the development of the high resolution capillary GC–MS coupling and the combination of chromatographic and other spectroscopic data in search programs will greatly facilitate the difficult work of organic analysists, not only in flavors and fragrances, but also in other fields such as pesticides, geochemistry, environmental studies, etc.

VII.8   ADDENDUM

Since the beginning of this work, advances in computer-assisted approaches to structure elucidation and applications of computers and microprocessors in mass spectrometry have been published. References to these works are summarized below. Computer-aided structure elucidation methods have been reviewed by several authors [185–187]. Several computerized search systems, such as SPEKTREN [188], CHEMICS [189], and CASE [190] have been developed for the structure elucidation of organic compounds [188–193]. Some of them allow one to remove redundant candidates by including spectroscopic data (^{1}H-NMR, ^{13}C-NMR, IR) [194, 195].

Computer analysis of high resolution mass spectral data [196] has been used either for assignment of a formula to experimental masses [197, 198] or for the determination of masses [199].

Pattern recognition [200] enables conclusions to be drawn concerning the presence or absence of certain structural features of an unknown compound and to attribute the compound to a definite chemical class.

Several artificial intelligence systems for computer-aided mass spectral interpretation have been developed [201, 202].

Recent research within the DENDRAL project into computer-assisted methods for chemical structure elucidation have been reviewed by Smith *et al.* [203].

Among several applications of computerized systems, one also can cite the characterization of natural and synthetic pyrethrins [204], alkylporphyrins [205], environmental spills [206], polymers [207] and clinical and medicinal agents (anti-inflammatory, analgesics, phenothiazine and analogous neuroleptics, antidepressants and opioids and their metabolites [208, 209].

In *Computer Methods of Molecular Structure Elucidation from Unknown Mass Spectra* Mum and McLafferty [210] reviewed interpretive programs (STIRS, CONGEN, GENOA) used to obtain partial or complete structure information. These algorithms are used when a retrieval program cannot give a satisfactorily matching reference spectrum. Kovats' indices as a preselection routine in mass spectra library searches of volatiles have been used by Alencar *et al.* [211]. This inclusion of Kovats' indices is greatly justified because it reduces computer processing time and increases the precision of the identification.

A computer pattern recognition method for the rapid identification of components in a complex mixture such as flavors and fragrances has been described by Chien [212]. The central part of this method is a comparison algorithm which compares the components in two mixtures using the K nearest neighbor classification rule. Essential oils in perfume samples can be detected by this way.

ACKNOWLEDGEMENT

We are grateful to the M.I.D.I.S.T. for financial support (Agreement No. 84 294 0222) which enable us to carry out this work and to CNRS Analysis Central Service for assistance in GC-MS analyses.

VII.9   REFERENCES

[1]   M. Petitjean, PhD, '*Mass Spectra and Kovats' Indices Bank of Heterocyclic Flavouring Compounds*' (in French), Faculty of Sciences and Techniques, Marseilles (1982).

[2] G. Vernin, M. Petitjean and J. Metzger, in *Enzymes Technol. Aliment. Symp. Int.*, 25–33 (P. Dupuy, ed.), Tech. Doc. Lavoisier, France (1982).

[3] M. Petitjean, G. Vernin and J. Metzger, in *Instrumental Analysis of Foods, Recent Progress*, Vol. 1, 97–124 (G. Charalambous and G. E. Inglett, eds), Academic Press, New York (1983).

[4] S. R. Heller, in *Proceedings of the Nato-CNA ASI on Computer Representation and Manipulation of Chemical Information*, 175–202 (W. T. Wipke, S. R. Heller, R. J. Feldmann and E. Hyde, eds), John Wiley, New York (1974).

[5] P. C. Jurs, in *Computer Representation and Manipulation of Chemical Information*, 265–285 (W. T. Wipke, S. R. Heller, R. J. Feldmann and E. Hyde, eds), John Wiley, New York (1974).

[6] D. H. Smith, L. M. Masinter and N. S. Sridharan, in *Computer Representation and Manipulation of Chemical Information*, 287–315 (W. T. Wipke, S. R. Heller, R. J. Feldmann and E. Hyde, eds), John Wiley, New York (1974).

[7] J. R. Chapman, in *Computers in Mass Spectrometry*, 101–219, Academic Press, New York (1978).

[8] I. K. Mun and F. W. McLafferty, in *Supercomputers in Chemistry*, Advances in Chemistry Series, Vol. 173, 117–124 (P. Lykos and I. Shavitt, eds) American Chemical Society, Washington (1981).

[9] B. Petersson and R. Rhyage, *Anal. Chem.*, **39**, 970 (1967); idem., *Ark. Kemi.*, **26**, 293 (1966).

[10] R. L. Crawford and J. D. Morrison, *Anal. Chem.*, **40**, 1464 (1968); ibid., **43**, 1790 (1971).

[11] J. F. O'Brien and J. D. Morrison, *Aust. J. Chem.*, **26**, 785 (1973).

[12] N. A. B. Gray, *Org. Mass. Spectrom.*, **10**, 507–514 (1975).

[13] N. A. B. Gray and T. O. Grönneberg, *Anal. Chem.*, **47**, 419–424 (1975).

[14] A. Buchs, A. B. Delfino, A. M. Duffied, C. Djerassi, B. G. Buchanan, E. A. Feigenbaum and J. Lederberg, *Helv. Chim. Acta.*, **53**, 1394 (1970).

[15] D. H. Smith, B. G. Buchanan, W. C. White, E. A. Feigenbaum, J. Lederberg and C. Djerassi, *Tetrahedron*, **29**, 3117 (1973).

[16] R. M. Hilmer and J. W. Taylor, *Anal. Chem.*, **51**(9), 1361–1368 (1979).

[17] L. W. McKeen and J. W. Taylor, *Anal. Chem.*, **51**(9), 1368–1374 (1979).

[18] F. W. McLafferty, *Pure Appl. Chem.*, **7**, 61 (1971).

[19] I. K. Mun, D. R. Bartholomew, D. R. Stauffer and F. W. McLafferty, *Anal. Chem.*, **53**(12), 1938 (1981).

[20] I. K. Mun, R. Venkataraghavan and F. W. McLafferty, *Anal. Chem.*, **53**(2), 179–182 (1981).

[21] F. W. Mclafferty, M. A. Bush, K-S. Kwok, B. A. Meyer, G. Pesina, R. C. Platt, I. Sakai, J. W. Serum, A. Tatematsu, R. Venkataraghavan and R. G. Werth, in *Mass Spectrometry and NMR Spectroscopy in Pesticide Chemistry* (F. J. Biros and R. Hague, eds), Plenum Press, New York (1974).

[22] F. W. McLafferty and D. B. Stauffer, *Int. J. Mass Spectrom. Ion Processes*, **58**, 139–149 (1984).

[23] R. Venkataraghavan, H. E. Dayringer, B. L. Atwater, G. M. Pesyna and F. W. McLafferty, *Adv. Mass. Spectrom.*, **7B**(10), 989–992 (1978).

[24] A. B. Delfino and A. Buchs, *Helv. Chim. Acta*, **55**(6), 2017–2029 (1972).

[25] N. A. B. Gray, R. E. Carhart, A. Lavanchy, P. H. Smith, T. Varkony, B. G. Buchanan, W. C. White and L. Creary, *Anal. Chem.*, **52**(7), 1095–1102 (1980).

[26] '*Mass Spectrometry*' *Biennial Review Bibliography*, *Anal. Chem.*, Vol. 5 (1982) and before.

[27] M. G. Kendall and A. Stuart, in *The Advanced Theory of Statistics*, C. Griffin Co Ltd, London (1961, 1963, 1966).

[28] P. C. Jurs, *Anal. Chem.*, **43**(13), 1812–1815 (1971).

[29] T. L. Isenhour and P. C. Jurs, in *Computers in Chemistry and Instrumentation* (J. S. Mattson, H. B. Mark Jr., and H. C. MacDonald, eds), Marcel Dekker Inc, New York, NY (1973).

[30] J. B. Justice and T. L. Isenhour, *Anal. Chem.*, **46**, 223 (1974); ibid., **47**(13), 2286–2288 (1975).

[31] P. C. Jurs, *Drug Informations J.*, **17**, 29–229 (1981).

[32] B. R. Kowalski and F. C. Bender, *J. Amer. Chem. Soc.*, **95**, 686 (1973).

[33] L. R. Crawford and J. D. Morrison, *Anal. Chem.*, **40**(10), 1464–1469 (1968); ibid., **40**(10), 1469–1474 (1968).

[34] R. J. Mathews, *Int. J. Mass Spectrom. Ion Phys.*, **17**, 217 (1975).

[35] D. H. Smith and G. Eglinton, *Nature*, **235**, 325 (1972).

[36] D. H. Smith, *Anal. Chem.*, **44**, 536 (1972).

[37] K. Varmuza, H. Totter and P. Krenmayr, *Chromatographia*, **7**, 522 (1974).

[38] D. D. Tunnicliff and P. A. Wadsworth, *Anal. Chem.*, **45**, 12 (1973).

[39] B. R. Kowalski, F. C. Bender and H. D. Shepherd, *Anal. Chem.*, **45**, 617 (1973).

[40] R. J. Mathews, *Aust. J. Chem.*, **26**, 1955 (1973).

[41] P. C. Jurs, B. R. Kowalski, T. L. Isenhour and C. N. Reilley, *Anal. Chem.*, **42**, 1387 (1970).

[42] P. C. Jurs, *Anal. Chem.*, **43**, 22 (1971).

[43] W. L. Felty and P. C. Jurs, *Anal. Chem.*, **45**, 885 (1973).

[44] P. Kent and T. Gaumann, *Helv. Chim. Acta*, **58**, 787 (1975).

[45] B. R. Kowalski and F. C. Bender, *J. Amer. Chem. Soc.*, **94**, 5632 (1972).

[46] K. L. H. Ting, R. C. T. Lee, G. W. A. Milne, M. Shapiro and A. M. Guarino, *Science*, **180**, 417 (1973).

[47] J. T. Clerc, P. Naegli and J. Seibl, *Chimia*, **27**, 639 (1973).

[48] B. R. Kowalski and C. F. Bender, *Anal. Chem.*, **44**, 1405 (1972).

[49] C. F. Bender and B. R. Kowalski, *Anal. Chem.*, **45**, 591 (1973).

[50] K. Varmuza, *Z. Analyt. Chem.*, **268**, 352 (1974).

[51] K. Varmuza and P. Krenmayr, *Z. Analyt. Chem.*, **266**, 274 (1973).

[52] D. R. Burgard, S. P. Perone and J. L. Wiebers, *Biochem.*, **16**, 1051 (1977).

[53] R. Duda and P. Hart, in *Pattern Classification and Science Analysis*, John Wiley (1972).

[54] W. S. Meisel, in *Computer-Oriented Approaches to Pattern Recognition*, Ch. 8; Academic Press, New York (1972).

[55] G. L. Ritter, S. R. Lowry, C. L. Wilkins and T. L. Isenhour, *Anal. Chem.*, **47**(12), 1951–1956 (1975).

[56] D. W. Fausett and J. H. Weber, *Anal. Chem.*, **50**(6), 722–731 (1978).

[57] R. W. Rozett and P. E. McLaughlin, *Anal. Chem.*, **47**, 1301 (1975); ibid., **47**, 2377 (1975); ibid., **48**, 817 (1976).

[58] F. J. Knorr and J. H. Futrel, *Anal. Chem.*, **51**(8), 1236–1241 (1979).

[59] J. A. Richards and A. G. Griffiths, *Anal. Chem.*, **51**(9), 1358–1361 (1979).

[60] S. Abrahamson, S. Stenhagen and E. Stenhagen, *Biochem. J.*, 92 (1964); S. Abrahamson, *Science Tob.*, **14**, 129 (1967).

[61] S. L. Grotch, *Anal. Chem.*, **42**, 1214 (1970).

[62] B. A. Knock, I. C. Smith, D. E. Wright, R. G. Ridley and W. Kelly, *Anal. Chem.*, **42**, 1516 (1970).

[63] H. S. Hertz, R. A. Hites and K. Bieman, *Anal. Chem.*, **43**, 681 (1971).

[64] L. E. Wangen, W. S. Woodward and T. L. Isenhour, *Anal. Chem.*, **43**, 1605 (1971).

[65] R. G. Ridley, in *Biochemical Applications of mass Spectrometry* (G. R. Waller ed.), Ch. 6 and references therein; John Wiley, New York (1972).

[66] S. Farbman, R. I. Reed, D. H. Robertson and M. E. F. Silva., *Int. J. Mass Sprectrom. Ion Phys.*, **12**, 123 (1973).

[67] R. J. Mathews and J. D. Morrison, *Aust. J. Chem.*, **27**, 2167–2173 (1974).

[68] D. D. Tunnicliff and P. A. Wadsworth, *Anal. Chem.*, **37**(9), 1082–1085 (1965).

[69] F. W. McLafferty, R. H. Hertel and R. D. Villwock, *Org. Mass Spectrom.*, **9**, 690–702 (1974).

[70] G. T. Rasmussen, B. A. Hohne, R. C. Wiebolt and T. L. Isenhour, *Anal, Chim. Acta*, **112**, 151–164 (1979).

[71] B. E. Blaisdell and C. C. Sweely, *Anal. Chim. Acta*, **117**, 17–33 (1980).

[72] J. Hardy and I. Jardine, in *Advances in Mass Spectrometry* (A. R. West, ed.), Vol. 6, 1061; Applied Science Pub., Barking (1974).

[73] S. L. Grotch, *Anal. Chem.*, **43**, 1362 (1971); ibid., **47**(8), 1825–1829 (1975).

[74] G. Van Marlen and J. H. Van den Hende, *Anal. Chim. Acta.*, **112**(2), 143–150 (1979).

[75] J. Kwiatkowski and W. Riepe, *Anal. Chim. Acta*, **112**(3), 219–231 (1979).

[76] S. L. Grotch, *Anal. Chem.*, **45**(1), 2–6 (1973).

[77] E. G. De Jony, J. Van Bekkun and H. A. Van't Klooster, *Adv. Mass Spectrom.*, **7B**(10), 1091–1098 (1978).

[78] T. O. Grönneberg, N. A. B. Gray and G. Eglinton, *Anal. Chem.*, **47**, 415 (1975).

[79] S. Abrahamson, G. Hagystrom and E. Stenhagen, 14th Annual Conference on Mass Spectrometry and Allied Topics, Paper No. 105, Dallas (1966).

[80] R. G. Dromey, *Anal. Chem.*, **51**, 229 (1979).

[81] F. P. Abramson, *Anal. Chem.*, **47**(1), 45–49 (1975).

[82] Documentation V. G. Data Systems; V. G. Instruments, 3 rue du Maréchal de Lattre de Tassigny, 78150 Le Chesney, France

[83] N. W. Bell, 20th Annual Conference on Mass Spectrometry and Allied Topics; Dallas, Paper R5, 364 (1972).

[84] V. L. De Gragnano and H. P. Hotz, in *Advances in Mass Spectrometry*, (A. R. West ed.), Vol. 6, 445, Applied Science Pub., Barking (1974).

[85] S. R. Heller, *Anal. Chem.*, **44**, 1951 (1972).

[86] S. R. Heller, H. M. Fales and G. W. A. Milne, *Org. Mass Spectrom.*, **7**, 107 (1973).

[87] S. R. Heller, R. J. Feldman, H. M. Fales and G. W. A. Milne, *J. Chem. Soc.*, **13**(3), 130–133 (1973).

[88] S. R. Heller and G. W. A. Milne, *EPA/NIH Mass Spectral Data Base*, Vol. 1–4, National Bureau of Standards, US Government Printing Office, Washington, DC 20402 (1978).

[89] S. R. Heller and G. W. A. Milne, *EPA/NIH Mass Specral Data Base* (1978), Molecular Weights and Indexes; Report NSRDS—NBS—68, order No. PB 290661 4654 (1978); Available from NTIS.

[90] *EPA/NIH Mass Spectral Data Base* Suppl. 1., US Environmental Protection Agency, USA; 849; Pub. Bernan Assoc. Lanham (1980).

[91] S. R. HELLER and G. W. A. Milne, *Anal. Chim. Acta*, **122**(2), 117–138 (1980).

[92] S. R. Heller, G. W. Milne and L. H. Gevantman, *EPA/NIH Mass Spectral Data Base*, Suppl. 2, 211; Pub. National Standards, Ref. Data System (1983); ibid., Suppl. 2, 1100, Pub. GPO (1983).

[93] S. R. Heller, W. L. Budde, D. P. Martinsen and G. W. A. Milne, *Int. J. Mass Spectrom. Ion Phys.*, **47**, 313–316 (1983).

[94] F. W. McLafferty, R. H. Hertel and R. D. Villwock, *Org. Mass Spectrom.*, **9**, 690 (1974).

[95] G. M. Pesyna, R. Venkataraghavan, H. E. Dayringer and F. W. McLafferty, *Anal Chem.*, **48**(9), 1362–1368 (1976).

[96] G. Van Marlen and A. Dijkstra, *Anal. Chem.*, **48**(3), 595–598 (1976).

[97] G. Van Marlen, A. Dijkstra and H. A. Van't Kloster, *Anal. Chim. Acta*, **112**(3), 233–243 (1979).

[98] A. Cornu and R. Massot, *Compilation of Mass Spectral Data*, 2nd Ed., Heyden and Sons Ltd, London (1975).

[99] *Eight Peak Index of Mass Spectra* MSDC 2nd Ed., MSDC AWRE, Aldermaston, Reading, RG7 4PR, UK (1974); idem., Vol. 1–4, compiled by the MSDC in coll. with ICI Ltd; Her Majesty's Stationery Office, P. O. Box 569, London SE1 9NH, UK (1975).

[100] S. Hayashi, H. Sato, N. Hayashi, T. Okude and T. Matsuura, *J. Sci. Hiroshima*, **31**(3), 217–231 (1967).

[101] Y. Hirose, *Shitsuryo Bunseki*, **15**(314), 162–178 (1967).

[102] M. G. Moshonas and E. D. Lund, *Flavour Industry*, **1**, 375 (1970).

[103] R. A. Hites and K. Biemann, in *Advances in Mass Spectrometry* (E. Kendrick, ed.), Vol. 4, 37; Institute of Petroleum, London (1968); idem., *Anal. Chem.*, **40**, 1217 (1968).

[104] B. S. Finkler, D. M. Taylor and J. E. Bonelli, *J. Chromatog. Sci.*, **10**, 312–333 (1972).

[105] C. Merrit Jr., D. H. Robertson, J. F. Cavagnars, R. A. Graham and T. L. Nichols, *J. Agric. Food Chem.*, **22**, 750 (1974).

[106] B. E. Blaisdell, *Anal. Chem.*, **49**, 180–186 (1977).

[107] D. H. Smith, M. A. Achenback, W. J. Yeager, P. J. Anderson, W. L. Fitch and T. C. Rindfleisch, *Anal. Chem.*, **49**, 1623–1632 (1977).

[108] D. H. Smith, *Anal. Chem.*, **44**(3), 536–547 (1972).

[109] M. C. Hamming and N. G. Foster, *Interpretation of Mass Spectra of Organic Compounds*, Academic Press, New York, NY (1972).

[110] A. A. Craveiro and J. Alencar, 178th ACS National Meeting, Paper 70, AGFD Div., Washington D.C, USA (1979).

[111] R. Flückiger, Y. Kato and S. Hishida, in *Mass Spectrometry in Biochemistry and Medicine*, 277–280 (A. Frigerio and N. Castagnoli, eds) Raven Press, New York (1974).

[112] H. C. Hill, R. I. Reed and M. T. R. Lopes, *J. Chem. Soc. (C)*, 93–101 (1968).

[113] J. Kumamoto and R. W. Scora, *J. Agric. Food Chem.*, **32**, 418–420 (1984).

[114] R. P. Adams, M. Granat, L. R. Hogge and E. Von Rudloff, *J. Chromatog Sci.*, **17**, 75–81 (1979).

[115] B. S. Middleditch (ed.), *A Guide to Collections of Mass Spectral Data* (1974), American Society for Mass Spectrometry.

[116] *Catalog of Selected Mass Spectral Data*, American Petroleum Institute Research, Project 44.

[117] Mass Spectrometry Data centre (MSDC) AWRE, Aldermaston, Berks, England (1968).

[118] E. Stenhagen, *Registry of Mass Spectral Data*, Wiley-Interscience, New York (1970).

[119] *Wiley's Registry of Mass Spectral Data Base* (E. Stenhagen, S. Abrahamson and F. W. McLafferty eds) Vol. 1–4, John Wiley and Sons, New York (1969–1974).

[120] *Wiley/NBS Mass Spectral Data Base*, Data Base Marketing, John Wiley and Sons Inc., New York (1984).

[121] J. G. Grasseli and W. M. Ritchey, *Atlas of Spectral Data and Physical for Organic Compounds*, Vol. 1–4, Ohio CRS Press Inc. (1975).

[122] H. Birkenfeld and G. Haase, *Possibilities of Data Analysis in the Data Storage System HMETH for Mass Spectrometry*, Zfi-Mitt. (Zimido), V. B. Vortr. Fachtag. Fakten Dokumenteninformationsverarb. Eser,. Anlagen Ist., 144–122 (1977).

[123]   L. H. Keith, Report EPA/600/4-83-036; order No. PB 83-255844, 35 (1983); available from NTIS.

[124]   S. Saeki, O. Yamamoto, K. Someno, U. Hiraishi and H. Sawada, *Kagiken (Japan)* 63-66; *Tokyo Daigaku Rigakubu Gosei Kagakuka* (1980).

[125]   S. Salki and D. Yamamoto, *Codata Bull.*, **40**, 53-56 (1981); idem., *Proc. Int. Codata Conf. (Data Sci. Technol.)* 262-265 (1981).

[126]   Y. Koyama, T. Susuki and K. Maeda, *Shitsuryo Bunseki (Shibak)*, **30**(1) 37-53 (1982).

[127]   H. Budzikiewicz, C. Djerassi, A. H. Jackson, G. W. Kenner, D. H. Newman and J. M. Wilson, *Structure Elucidation of Natural Products by Mass Spectrometry* (1964).

[128]   Q. W. Porter and J. Baldas, in *Mass Spectrometry of Heterocyclic Compounds*, Wiley Interscience, New York (1971).

[129]   G. Vernin (ed.), *The Chemistry of Heterocyclic Flavouring and Aroma Compounds*, (Ellis Horwood Pub.), Chichester, England (1982).

[130]   Y. Masada, *Analysis of Essential Oils by Gas Chromatography and Mass Spectrometry*; John Wiley and Sons, New York (1976).

[131]   W. Jennings and T. Shibamoto, *Qualitative Analysis of Flavor and Fragrance Volatiles by Glass Capillary Gas Chromatography*; Academic Press, New York (1980).

[132]   E. Von Sydow, K. Anjou and G. Karisson, 'Archives of Mass Spectral Data' SIK—Report No. 279, Goteborg, Sweden, Vol. 1 (1970).

[133]   *Spectral Atlas of Terpenes and the Related Compounds* (Y. Yukawa and S. Ito, eds), Hirokawa Pub. Co, Tokyo (1973).

[134]   N. De Brauw, J. Bouwman, A. C. Tass and G. F. La Vos, *Compilation of Mass Spectra of Volatile Compounds in Food*, Central Institute for Nutrition and Food Research, TNO, The Netherlands (1981).

[135]   E. Von Sydov, *Acta Chem. Scand.*, **17**, 2504-2512 (1963).

[136]   R. Ryhage and E. Von Sydow, *Acta Chem. Scand.*, **17**, 2025-2035 (1963).

[137]   A. F. Thomas and B. Willhalm, *Helv. Chim. Acta*, **47**, 475 (1964).

[138]   B. Shieh and Y. Matsubara, *Shit Suryo Bunselei (Shi Bak)*, **29**, 97-111 (1981).

[139]   J. Hanuise, J. P. Puttermans and R. R. Smolders, *Tetrahedron*, **25**, 1757-1769 (1969).

[140]   R. Chennouli, J. P. Morizur, H. Richard and F. Sandrel, *Riv. Ital. EPPOS*, **LXII**(7), 353-357 (1980).

[141]   D. D. Speck, R. Venkataraghavan and F. W. McLafferty, *Org. Mass Spectrom.*, **13**(4), 209 -213 (1978).

[142]   M. Stoll, W. Winter, F. Gautschi, I. Flament and B. Willhalm, *Helv. Chim. Acta*, **50**(2), 628-694 (1967).

[143]   D. H. Robertson and C. Merritt, 22nd Annual Conference on Mass Spectrometry and Allied Topics, Philadelphia, Paper U7, 447 (1974).

[144]   J. Zupan, M. Penca, D. Hadzi and J. Marsel, *Anal. Chem.*, **49**, 2145 (1977).

[145] S. Sasaki, H. Abe, K. Saito and Y. Ishida, *Bull. Chem. Soc. Japan*, **51**(11), 3218-3222 (1978).

[146] H. Idstein, W. Herres and P. Schreier, *J. Agric. Food Chem.*, **32**, 383-389 (1984).

[147] E. Kovats, *Helv. Chim. Acta*, **41**(7), 1915-1932 (1958); idem., *Adv. Chrom.*, **1**, 229-247 (1965).

[148] I. Tranchant, in *Manuel Pratiaque de Chromatographie en Phase Gazeuse* (Masson and Co eds) (1968).

[149] G. Vernin, in *Chromatographie, Synthèse et Réactivité*, Manuel de Travaux Pratiques (Dunod Unoversité ed.) Paris (1970).

[150] M. A. Kaiser, in *Modern Practice of Gas Chromatography*, 153-158 (R. L. Grob, ed.), John Wiley, New York (1977).

[151] G. Vernin, *Parf. Cosm. Arômes*, **39**, 77-88 (1981).

[152] L. Huber and H. Obbens, in *Flavour 81* 339-343 (De Gruyter and Co eds), Berlin-New York (1981).

[153] J. C. Demirgian, *J. Chromatog. Sci.*, **22**(4), 153 (1984).

[154] N. H. Andersen and M. S. Falcone, *J. Chromatog.*, **44**, 52-59 (1969).

[155] B. J. Tyson, *J. Chromatog.*, **111**, 419-421 (1975).

[156] B. E. Blaisdell, *Anal. Chem.*, **49**(1), 180-186 (1977).

[157] K. R. Betty and F. W. Karasek, *J. Chromatog.*, **166**, 111 (1978).

[158] H. Nau and K. Biemann, *Anal. Letters*, **6**(12) 1071-1081 (1973); idem., *Anal. Chem.*, **46**(3), 426-434 (1974).

[159] C. C. Sweely, N. D. Young, J. F. Holland and S. C. Gates, *J. Chromatog.*, **99**, 507-517 (1974).

[160] D. H. Smith, M. A. Achenback, W. J. Yeager, P. J. Anderson, W. L. Fitch and T. C. Rindfleisch, *Anal. Chem.*, **49**, 1623-1632 (1977).

[161] S. C. Gates, M. J. Smisko, C. L. Ashendell, N. D. Young, J. F. Holland and C. C. Sweely, *Anal. Chem.*, **50**(3), 433-441 (1978).

[162] S. C. Gates, N. Dendramis and C. C. Sweely, *Clin. Chem.*, **24**(10), 1674-1679 (1978).

[163] B. E. Blaisdell and C. C. Sweely, *Anal. Chim. Acta*, **117**, 1-15 (1980): ibid., **117**, 17-33 (1980).

[164] A. A. Craveiro, A. S. Rodrigues, C. H. S. Andrade, F. J. A. Matos, J. W. Alencar and M. I. L. Machado, *Technical Data, Cannes-Grasse*, 241-254, 8th Int. Congress Essential Oils (October 1980).

[165] A. A. Craveiro, A. S. Rodrigues, C. H. S. Andrade, F. J. A. Matos, J. W. Alencar and M. I. L. Machado, *J. Nat. Products*, **44**(5), 602-608 (1981).

[166] H. Van den Dool and P. D. Kratz, *J. Chromatog.*, **11**, 436-471 (1963).

[167] L. S. Ettre, *Anal. Chem.*, **36**(8), 31A-41A (1964).

[168] P. H. Wiener and J. F. Porcher, *Anal. Chem.*, **45**(2), 302-307 (1973).

[169] G. I. Spivakovskii, A. I. Tishchenko, I. I. Zaslavskii and N. S. Wulfson, *J. Chromatog.*, **144**(1) 1-16 (1977).

[170] J. P. Aune, H. J. M. Dou and J. Crousier, in *The Chemistry of Heterocyclic Compounds:Thiazole and its Derivatives* (Vol. 1-3) Part I, 361 (J. Metzger, ed.) John Wiley and Sons, New York (1979).

[171]   M. Petitjean, G. Vernin and J. Metzger, *Ind. Aliment. Agric.*, **98**(9), 741-751 (1981).

[172]   R. Tressl, K. G. Grünewald and R. Silvar, *Chem. Mikrobiol. Technol. Lebensm.*, **7**, 28-32 (1982).

[173]   J. M. Memiaghe, 'Synthèse des alkyl-pyryliums par diacylation d'oléfines en catalyse protonique: influence de la structure des réactifs sur l'orientation de la réaction', PhD, Aix-Marseilles III (1981) in French.

[174]   P. H. Weiner and J. F. Parcheer, *Anal. Chem.*, **45**(2), 302-307 (1973).

[175]   S. Forss, in *The Maillard Reactions in Foods and Nutrition* (G. R. Waller and M. S. Faether, eds), ACS Symposium Series 215, Amer. Chem. Soc., Washington, DC (1983).

[176]   *Flavor and Fragrance Materials* 1981, Allured Pub. Co., Wheaton Illinois, USA.

[177]   *Flavouring Substances and Natural Sources of Flavourings*, Council of Europe, 3rd ed., Maisonneuve (1981).

[178]   L. J. Van Gemert and A. H. Nettenbreijer (eds.), *Compilation of Odour Threshold Values in Air and Water*, Central Institute for Nutrition and Food Resarch, TNO, Zeist, The Netherlands (1977).

[179]   G. Vernin, M. Petitjean and J. Metzger, *Parf. Cosm. Arômes*, **51**, 43-51 (1983).

[180]   G. Vernin, J. Metzger, D. Fraisse and C. Scharff, *Parf. Cosm. Arômes*, **52**, 51-61 (1983); 9th International Congress of Essential Oils (14-17 March 1983), Singapore, *Technical Paper*, **Book 3**, 100-120.

[181]   G. Vernin, J. Metzger, D. Fraisse, K-N. Suon and C. Scharff, *Perf. and Flavorist*, **9**(5), 71-86 (1984).

[182]   G. Vernin, J. Metzger, D. Fraisse and C. Scharff, *Planta. Med.*, 96-101 (1986).

[183](a)   G. Vernin, J. Metzger, K-N. Suon and D. Fraisse, *Parf. Cosm. Arómes*, **62**, 69-74 (1985).

[183](b)   G. Vernin, D. Bouin, J. Metzger, D. Fraisse and C. Scharff, in *The Shelflife of Foods and Beverages*, 4th International Flavor Conference, Rhodes, Greece; ACS Series (G. Charalambous, ed.), 255-283 (1986).

[184]   G. Vernin, M. Petitjean, J. Metzger, D. Fraisse, K-N. Suon and C. Scharff, in *Gas Chromatography in Essential Oils Analysis* (C. Bicchi and P. Sandra, eds), Huthig Verlag, Heidelberg, 1986 (in press).

[185]   R. D. Sedgwick, *Mass spectrometry*, **6**, 174 (1980).

[186]   H. Abe, T. Yamasaki, I. Fujiwara and S. Sasaki, *Anal. Chim. Acta*, **5**(4), 499 (1981).

[187]   M. Munk, *Fresenius Z. Anal. Chem.*, **311**(4), 317 (1982).

[188]   M. Zippel, J. Mowitz, I. Köhler and H. J. Opferkuch, *Anal. Chim. Acta*, **140**(1), 123 (1982).

[189]   I. Fujiwara, T. Okuyama, T. Yamasaki, H. Abe and S. Sasaki, *Anal. Chim. Acta*, **5**(4), 527 (1981).

[190]   C. A. Shelley and M. E. Munk, *Anal. Chim. Acta*, **5**(4), 507 (1981).

[191] K. S. Lebedev, V. M. Tormyshev, B. G. Derendysev and V. A. Koptyug, *Anal. Chim. Acta*, **5**(4), 517 (1981).

[192] S. P. Kirshansky and K. S. Lebedev, *Izv. Sibirsk. Otdelen. Akad. Nauk.*, **1**, 97 (1984).

[193] P. G. Oza, H. S. Mazumdar, A. K. Choudhary and K. Copalan, *Proc. Indian Acad. Sci. Chem. Sci.*, **93**(2), 189 (1984).

[194] A. P. Uthman, J. P. Koontz, J. Hinderlitresmith, W. S. Woodward and C. N. Reilley, *Anal. Chem.*, **54**(11), 1772 (1982).

[195] K. Maeda, Y. Koyama, K. Sato and S. Sasaki, *Anal. Chim. Acta.*, **5**(4), 561 (1981).

[196] R. A. Heppner and J. M. Knox, *Chem. Biomed. Environ. Instr.*, **11** (5-6), 425 (1981).

[197] S. E. Scheppele, C. S. Hwang and K. C. Chung, *Int. J. Mass Spectrom. Ion Physics*, **49**(2), 143 (1983).

[198] S. E. Scheppele, Q. G. Grindstaff, C. S. Hwang, R. D. Grigsby and S. R. McDonald, *Int. J. Mass Spectrom. Ion Physics*, **49**(2), 179 (1983).

[199] Y. Tandeur, J. R. Hass, D. J. Harvan and P. W. Albro, *Analyt. Chem.*, **56**(3), 373 (1984).

[200] D. C. Leegwater and J. A. Leegwater, *Trends in Analyt. Chem.*, **3**(3), 66 (1984).

[201] C. Damo, C. Dachum, K. Teshu and C. Shaoyu, *Anal. Chim. Acta.*, **5**(4), 575 (1981).

[202] B. Adler, B. Eitner, A. Herrmann and E. Sorkau, *Zeitschrift für Chemie*, **24**(7), 237 (1984).

[203] D. H. Smith, N. A. B. Gray, J. G. Nourse and C. W. Crandel, *Anal, Chim. Acta.*, **5**(4), 471 (1981).

[204] H. Kohlmann, *Int. J. Mass Spectrom Ion Phys.*, **48**, 141 (1983).

[205] P. J. Marriott, J. P. Gill, R. P. Evershed, C. S. Hein and G. Eglinton, *J. Chromatog.*, **301**(1), 107 (1984).

[206] M. H. Carter, *J. Chromatog.*, **235**(1), 165 (1982).

[207] S. Foti, D. Garozzo and G. Montaudo, *Chim. Ind (Milan)* **65**, 641-644 (1983).

[208] D. Laue, *Fresenius Z. Anal. Chem.*, **311**(4), 318 (1982).

[209] H. Maurer and K. Pfleger, *Fresenius Z. Anal. Chem.*, **314**(6) (1983); ibid., **317**(1), 42 (1984); idem., *J. Chromatog.*, **306**, 125 (1984); idem., *J. Chromatog.*, **305**(2), 309 (1984).

[210] In Ki Mum and F. W. McLafferty, in *Supercomputers in Chemistry*, pp. 117-124 (P. Lykos and I. Shavitt, eds), ACS Symposium Series 173, American Chem. Soc., Washington, DC, USA (1981).

[211] J. W. Alencar, A. A. Craveiro and F. J. A. Matos, *J. Nat. Products*, **47**(3), 890-892 (1984).

[212] Mingilien Chien, *Anal. Chem.*, **57**, 348-352 (1985).

[213] D. Fraisse, K-N. Suon, C. Scharff, G. Vernin, G. M. F. Vernin, R. M. Zamkotsian and J. Metzger, *Parf. Cosm. Arómes*, **65**, 71-75 (1985).

[214] G. Vernin, *Parf. Cosm. Arómes*, **68** (1986).

CHAPTER VIII

Online Access to Chemical Information

Parina Hassanaly and Henri Dou

University of Aix-Marseilles III, France

VIII.1 INTRODUCTION

1.1 Importance of chemical information

Chemical information is a large field proceeding from simple scientific journals to patents, reports, dissertations, conference proceedings, books, etc. The growth of chemical and biochemical information is tremendously rapid [1, 2].

This means that the modern chemist, if he really wants to cope with his discipline, must give a larger portion of his time to chemical information. The number of various publications is now so great that computer help becomes increasingly necessary. Moreover, if we want to deal with information concerned with structures in combination with properties, syntheses, etc., it amounts to more than ten million bits of information which must be taken into account. Furthermore, if we want to deal with substructures, it must be realized that over 6 million structures are indexed in computerized chemical files, and this means tens of millions of bits of structural information.

As in all the disciplines in the sciences, however, chemical information has its own vocabulary, rules, tools, etc. The progress made in this area with the introduction of online access to computerized information systems has become so important in recent years that it is a duty of the chemist to know at least what can be expected from such systems.

This is the goal of this chapter. It is beyond the scope of these few pages to present all the various aspects of this field, which normally takes several hundred hours to learn.

We will present what we think are the most important ideas: the development of large host computers and their access from telephone networks, electronic printing, and simple aspects of database structure. We will then develop the use of structural files, and close with patent files and specialized databases dealing with chemistry.

1.2   Growth of chemical information

If we add all the papers, patents, reports, theses etc., related to chemistry, every year, we will obtain a rate of several papers per minute. This figure which seems rather high can be easily verified:

Chemical abstracts takes into account about 500 000 papers each year. This already means about one paper per minute. If we add all the publications related to chemistry and not necessarily present in chemical abstracts, we can see that the rate is tremendously important. Of course, we will not need all the references, and we do not have to consider all the various fields of chemistry, but the spread of the required information makes it very difficult if we do not use computerized techniques to obtain, in a reasonable time, a complete view of a given problem. This is the reason online information retrieval is becoming, year by year, a more necessary tool in all the branches of chemistry.

VIII.2   FROM HARD COPY TO ONLINE ACCESS

During the past decade, the price of large computers has decreased and storage capacity has increased. The cost of memory has fallen along with the computer cost. At the same time, international data transmission networks were developed on a worldwide basis with an accuracy in data transmission of nearly one hundred per cent. These networks (Tymnet, Telenet, Euronet, Transpac etc.) are well known and they allow almost everyone to access a large computer using a telephone, a modem and a terminal.

Figure 1 represents schematically, the transmission of data from a remote terminal to a host computer, and vice versa.

During the same decade, computers began to be used to assist the printing industry. It is commonplace to use word processing with home computers. In almost the same way computers are used to print papers, books etc. The text is stored on a magnetic support, and the computer and the appropriate software help present the text in the desired format: type of presentation, fount, columns, etc. It is easy to see that the conjunction of both techniques can give rise to the creation of databases and also to their interrogation from places very remote from the location of the mainframe computer.

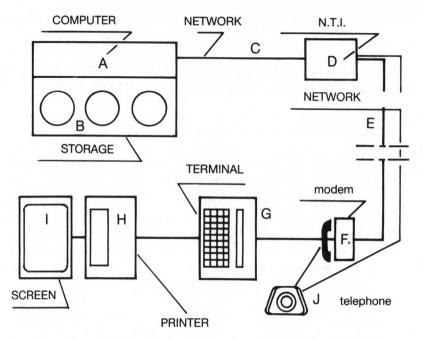

Fig. 1   Reversible transmission of data from a remote terminal to a host computer.

VIII.3   HOW TO MAKE INFORMATION AVAILABLE IN AN INTERACTIVE FORM

To understand how a database works, it is necessary to first have an idea of what a reference is, from a bibliographic point of view.

3.1   Reference

A reference is composed of various *items* called fields. Those fields are wide enough to describe almost anything. In a primer approach we will distinguish, for instance:

(1)   title   TI
(2)   authors   AU
(3)   address   OS (organization source)
(4)   journal   SO (source)
(5)   index term   IT
(6)   abstract   AB

Other fields can be present, for example: Language (LA), country (abbreviate), document type, registry number (RN), coden of journals or books, supplementary terms, etc.

A basic index (BI) exists in almost all systems. The field is by default and does not need to be indicated to the computer. It normally includes TI, IT, and AB as a minimum. Thus, a reference by itself, is a small piece of a chemical story in which the information is sliced into fields, each used in the same way according to rigid rules. This part of the work is called indexing and is *the most important task* when chemical information is abstracted.

According to the objectives of the various abstracting services, sets of references dealing with particular aspects of chemistry (general, food, corrosion, patents, environment, polymers, safety, energy, medical, biology etc.) will be created from journals, patents, reports, dissertations, conference proceedings, books. All this information may be stored on magnetic support and may be used to produce hard copy (chemical abstracts, biological abstracts, pollution abstracts, for example), or online databases, or both.

3.2   Batch access

At the very beginning it was not possible to ask a question of a database and to obtain an answer almost immediately. All the information (the references) were stored on a magnetic tape. To access the 988th reference for instance, it was necessary to go through all the preceding ones. This was called batch research. A profile (a set of questions using any combination of terms appearing in the fields), was made and the terms of the profile compared with the contents of the fields of each reference, one after another. The process was very long, and generally was carried out overnight; the useful references were printed when selected. This procedure is known as *sequential access* to information.

3.3   Interactive searches

(1)   Inverted files,
(2)   Boolean operators,
(3)   Example.

The progress in the power of computers and the storage capacity and speed of access now allow another way of dealing with the preceding information search. Even if the magnetic support and its organization remains the same (the content of a reference and the various fields), the content of a reference and the various fields), the computerized treatment can be quite different with:

(1)   Inverted files:

Each field gives rise to a dictionary of all significant words (excluded are the so-called stop words: an, so, the, they, an, at ... These dictionaries contain two types of information.

This sequence continues for all the words present in all the titles considered. The same treatment is performed for each of the fields needed for

Table 1   Inverted file of the title words: T1

Words in the dictionary	Frequency in the database	Reference locations on the magnetic support
Animal	750	Locations of the 750 references
Baby	234	Locations of the 234 references
Ketone	123 000	Locations of the 123 000 references
Wine	11 345	Locations of the 11 345 references

interactive searches. It is easy to see that for a database containing millions of references, the size of each inverted file must be considerable. Also we need to know that this treatment made by the computer is not definitive, and will continue when new references are added.

3.4   Boolean operators

The Boolean operators AND, OR and NOT are used to combine answers (the frequencies of words) between each other as shown in Fig. 2.

Using these operators the interactive searches will continue as follows (using, for example, the title words animal and ketone):

> question 1: animal   answer 1: 750
> question 2: ketone   answer 2: 123 000
> question 3: 1 OR 2   answer 3: 123 601

1 OR 2 means to search for answers containing *either* answer 1 or answer 2. Answer 3 may be less (as in our case) than the total of 1 plus 2 if some references already contain both words.

> question 4: 1 AND 2   answer 4: 149

those answers containing *both* answer 1 *and* answer 2.

> question 5: 1 NOT 2   answer 5: 601

those answers containing answer 1 *but not* answer 2.

> question 6: 2 NOT 1   answer 6: 122 851

those answers containing answer 2 *but not* answer 1; Note that questions 5 and 6 *are not the same.*

An identical treatment can be repeated as often as we like, on any number of fields. Combinations of answers coming from different fields are possible. When it is felt that the proper question has been treated, the order print, PRT or PRINT, or EDIT etc. (with indication of the format and the number of

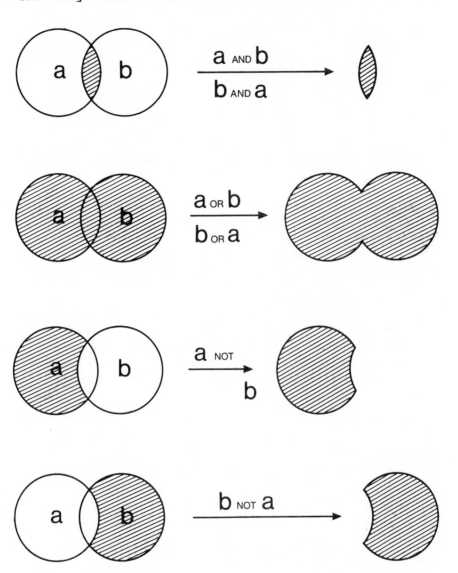

Fig. 2   Boolean operators used to combine answers.

references) will indicate to the computer that the references composing the answer can be printed. The computer thus finds the reference locations and prints them directly (i.e. *direct access*) without going through all the main file. It is always the most recent reference which is printed first.

This sequence, which will be more or less complicated depending upon the databases and the type of questions posed, is basically the one used for online searches.

3.5 Example

In this example, taken from Chemical Abstracts database, the host computer is System Development Corporation (SDC) using logical ORBIT IV (the orders used during the interrogation are part of a language, different for almost each host computer).

The questions will be SS 1, SS 2, etc. (SS = search statement).

The answers will be PSTG (PSTG = postings).

The Boolean operators are AND, OR, NOT.

The field used, which has to be indicated to the computer is: /AU (authors), the basic index does not need indication.

The print order is PRT FU INDENTED (to indicate the full reference and the various names of the fields).

Print can also be tailor-made: PRT TI AU SO, or they can be standard (PRT).

The question to be posed in computer terminology to the computer is:

Find the references having the authors P. Hassanaly and H. Dou which deal with catalysis.

A very simple search can be performed using the general truncation (:) after each surname. Since both are associated (AND) here, the probability of getting homonyms is almost nil. Simple association with the term catalysis (taken here from the basic index) will lead to all publications of the authors which have the term catalysis either in the title or in the index terms or in the supplementary terms.

This search is made with the Chemical Abstracts file dealing with the years 1977 to 1981.

Example 1*

SS 1 /C? USER: ALL HASSANALY, P:/AU AND ALL DOU, H:/AU

PROG:
OCCURS   TERM
18   ALL HASSANALY, P:/AU
35   ALL DOU, H:/AU
SS 1 PSTG (17) SS 2 /C? USER: 1 AND CATALYSIS

PROG:
OCCURS   TERM
15463   CATALYSIS
SS 2 PSTG (11) SS 2 /C? USER: PRT FU INDENTED 2

PROG:

Example 2

ACCESSION NUMBER	CA95-131765(15)

TITLE	Phase transfer catalysis in heterocyclic chemistry
AUTHORS	Gallo, Roger; Dou, Henri J. M. Hassanaly, Parina
ORGANIZATIONAL SOURCE	IPSOI, Fac. Sci. Tech. St Jerome, Marseille Fr., 13013

---

* In all examples refer to explicative note at the end of the chapter.

SOURCE	Bull. Soc. Chim. Belg. (BSCBAG), V90 (8), p. 849–79, 1981, ISSN 00379646
DOCUMENT TYPE	J (Journal)
LANGUAGE	Eng
CATEGORY CODES	SEC22-0
INDEX TERMS	Catalysts and Catalysis, phase-transfer: (for reactions of heterocyclic compds.)
INDEX TERMS	Heterocyclic compounds: (reactions of, phase-transfer catalysis in)
SUPPLEMENTARY TERMS	Review; heterocycle

Example 3

ACCESSION NUMBER	CA94-15639(3)
TITLE	Reactions of ambident dianions in phase transfer catalysis. Case of S- and N-alkylation
AUTHORS	Dou, Henri J. M.; Hassanaly, Parina; Metzger, J.
ORGANIZATIONAL SOURCE	Cent. St Jerome, Lab. Chim. Org. A., Marseille, Fr., 13397/4
SOURCE	Chim. Acta Turc. (CATUA9), V 7 (3), p. 291–295, 1979, ISSN 03795896
DOCUMENT TYPE	J (Journal)
LANGUAGE	Fr
CATEGORY CODES	SEC28-10
INDEX TERMS	74-96-4; 78-77-3; 106-93-4: (*N*- and *S*-alkylation of imidazolidinethione by, catalysts for)
INDEX TERMS	96-45-7: (N- and S-alkylation of, catalysts for)
INDEX TERMS	1643-19-2: (catalysts, for N- and S-alkylation of imidazolidinethione)
INDEX TERMS	29704-02-7; 76048-66-3; 76048-67-4; 76048-68-5; 76048-69-6: (prepn. of)
INDEX TERMS	Alkylation catalysts: (tetrabutylammonium bromide, for imidazolidinethione)
SUPPLEMENTARY TERMS	catalyst; imidazoline; alkyl; alkylthio; imidazothiazole; perhydro

VIII.4   CHEMICAL ABSTRACTS TEXT FILES

It was at about the beginning of the century that Chemical Abstracts started. Since then, rules in indexing, chemical nomenclature, keywords, indexes, have changed many times. Even if you used online retrieval (which goes back mainly to 1967) the use of the *Index Guide* is a prudent practice in order to attain the best searches. For those interested in a more thorough explanation of chemical abtracts, *Chemical Abstracts: An Introduction to its Effective Use* will give guidelines [3, 4].

4.1   General presentation

The CA (Chemical Abstracts) text files arise from a very complicated process, which can be summarized as follows:

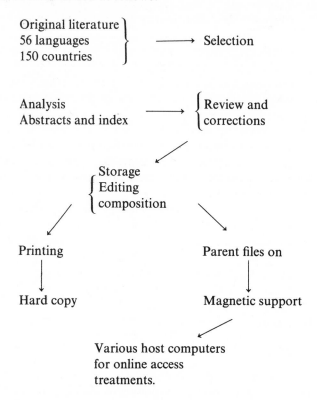

Original literature
56 languages $\longrightarrow$ Selection
150 countries

Analysis
Abstracts and index $\longrightarrow$ Review and
                                            corrections

Storage
Editing
composition

Printing   Parent files on

Hard copy   Magnetic support

Various host computers
for online access
treatments.

All the products from Chemical Abstracts, except 'Patent Concordance' are available online. Often, the abstracts are not present, except in the CAS ONLINE files. However, the presence of index terms with associated keywords and *Registry Numbers* (*RN*) gives useful information which is often more easily used than the abstracts themselves. The information accessible online dates from 1967 forward, and they are divided into collective index:

8th collection 1967–1971, 9th collection 1972–1976,

10th collection 1977–1981, 11th collection 1982–1986

Owing to space restrictions here, all the CA rules cannot be presented. Some will be presented, as appropriate in the section dealing with Chemical Abstracts dictionaries. As far as the textual file and online retrieval are concerned, the main rules are as follows:

(1) Titles are always presented in English, but the Original Title (OTI) can also be used for searching.
(2) Chemical names in the titles are as written by the authors,
(3) Authors are indexed up to a maximum of 10,

(4)　Special notation and rules are used for names such as Sister, De....
Generally indications are given by the host computer.

(5)　The address indicated (OS) is that of the first author,

(6)　Journals can be searched directly using either coden or issn,

(7)　Chemicals are never indicated in the Index Terms by their name, but
rather by their registry number (RN). These numbers have no chemical
significance: they are used only for a unique computer indexation.

(8)　Index terms contain the RN, keywords, abbreviations and general
chemical names (e.g. ketone, phenol, etc.) sometimes with small
sentences related to a chemical operation starting or leading to the
cited RN,

(9)　The keywords used in the tables of the weekly issues of CA are not the
same as those used in the semi-annual or collective index. They are
present as supplementary terms (ST) and reflect ideas and words used
freely by authors at the time of the abstraction of the article.

(10)　Document types are indicated by letters, such as P for patent, T for
thesis, R for report, B for books etc. Reviews are indicated by REVIEW
either added after the title, or in the supplementary terms.

(11)　Country names are abbreviated, a list is generally given,

(12)　Language names are also abbreviated,

(13)　Patents are not all abstracted by CA Only the *source patent* (the first
patent that CA indexes) is present in the text files available online.
Other related patents are not present.

(14)　Priority countries are also abbreviated in patents.

4.2　Structure of references

The Chemical Abstracts databases contain a number of references. The most
important are the articles originating in chemical journals. We have already
seen that only the source patents are present. The papers dealing with the
economics of chemistry are not in the main CA files, but rather are compiled
in CIN (Chemical Industry Notes). The main documents abstracted by CA
are: journal articles, proceedings from congresses and symposia, edited
collections, dissertations and theses, technical reports, new book announce-
ments and patents. For more information, see the section entitled, 'Use of the
Chemical Literature,' edited by the American Chemical Society.

4.3　Survey of the reference fields

In a reference issued from Chemical Abstracts through online retrieval, there
are a certain number of fields which can be used in combination. There are
also other fields which are not printed in the reference, but which are present
in the magnetic tape. They can be used in interactive searches: e.g. PN (patent
number), the Coden of journals or books, priority countries for patents etc.
　　There is also the possibility, when the fields to be searched are not present,
of using string searches (STRS or TEXT...) on the references retrieved. A
string search consists of a sequential search within a field of a sequence of
letters, figures, or words.

4.4 Strategies and languages

It is obvious that the more fields which exist and the more details which are available, the better the search will be. This is the reason that the same database issued by the same parent file, can be very different on various host computers. Also the search strategies will be different, since some data will not be directly available (e.g. direct access to coverage limitation dates, to Coden's etc.). But, the main limitation is very often the capacity of the language. The more powerful the language is, the faster and better the search will be.

Classically (and this will be the same for all other databases as well), masks and truncations can be used. For example:

Thiazol: is identical to thiazolyl, thiazolo and thiazole. and : is the symbol of a general truncation.

Thiazol £ *is identical to thiazole and thiazol.* £ is a mask for one character or space, or nothing.

Thiazol ££ *is identical to thiazole, thiazoles, thiazol.*

Other operators, such as proximity can also be used. For example: Ethyl-(W)acetate means that ethyl and acetate have to be present as ethyl acetate. The symbol xW can be used in brackets to determine the number of letters (spaces or words) which may be present between two words. Other operators exist, such as thiazole(L)nitration (L = link), indicate that thiazole and nitration have to be present in the same subfield, and so on.

File segments

To assist broad searches in CA, it is possible to make some divisions in the main file. Section groupings, or file segments (FS) can be used; these will limit the search, for instance, to ORGCHEM/FS for the organic chemistry, or PCHEM/FS for the physical chemistry, etc.

Selecting terms

From one or several references, it is also possible to select index terms and to automatically reintroduce them into the file, in order to retrieve the corresponding references. This is generally done with RNs or keywords.

Save, store searches

Most of the files present in a computer host are standardized in format (e.g. the fields used are all the same for all the references contained into the available databases). It is then possible to save (for one operation) or store (permanently) a search strategy, and to recall it when necessary to be executed on the database of your choice.

Other operations are possible, but they are beyond the scope of this chapter. What must be emphasized is that *almost all possible operations are feasible*; they have only to be adapted to the proper questions and strategies.

4.5 Hosts and formats

Depending on the host used, the format (presentation of the references, the fields present, their names, etc.) will be different. Also the print-out will be different. However, in CA text files, English will be the only language used (certain hosts do allow the use of original titles, OTI). The following example illustrates various references issued from several host computers.

4.6 Various examples

The following set of examples illustrates some of the capabilities of the CA textual file.

Example 1

'Search all papers published by French laboratories which deal with iodo-naphthalene.' We used the limitation FR./LO (FR. stands for France taken into the field location LO). Iodonaphthalene is searched not by its name, but in a more general and complete way using its RN. (RNs are all located in index terms IT, and are always used by CA to characterize a chemical. Names are only used as spelled in titles, if present there.) In this example three references are obtained: PERIODE 1977–1981.

```
SS 5 /C?  USER: 90-14-2/RN;

PROG:
SS 5 PSTG (43);  SS 6 /C?  USER:  5 AND FR./LO

PROG:
OCCURS    TERM
72105     FR./LO
SS 6 PSTG (3)  SS 7 /C?  USER:  PRT

PROG:
-1-
AN — CA93-113572(11)
TI — Current dips in polarography and cyclic voltammetry associated with electrochemical
inducement of chemical reactions.
Aromatic, nucleophilic substitution
AU — Amatore, C.; Pinson, J.; Saveant, J. M.; Thiebault, A.
OS — Univ. Paris VII, Lab. Electrochim., Paris, Fr., 75221/05
SO — J. Electroanal. Chem. Interfacial Electrochem.
(JEIEBC), V 107 (1), p. 75-86, 1980, ISSN 00220728
LA — Eng.  CC — SEC22-5
-2-
AN — CA93-83553(8)
TI — The solvent as hydrogen-atom donor in organic electrochemical
     reactions: Reduction of aromatic halides
AU — M'HALLA, F.; Pinson, J.; Saveant, J. M.
OS — Univ. Paris 7, Lab. Electrochim., Paris, Fr., 75221/05
SO — J. Am. Chem. Soc. (JACSAT), V 102 (12), p. 4120-7, 1980, ISSN 00027863
LA — Eng.  CC — SEC72-11; SEC22; SEC26; SEC27
-3-
AN — CA87-135849(17)
TI — Arylation of the Reformatsky reagent catalyzed by zerovalent: complexes of palla-
dium and nickel
AU — Fauvarque, J. F.; Jutand, A.
OS — Univ. Paris-Nord, Cent. Sci. Polytech., Villetaneuse, Fr.
SO — J. Organomet. Chem. (JORCAI), V 132 (2), p. C17-C19, 1977
LA — Eng.  CC – SEC29-13; SEC25
```

Example 2

'Find all references with thiolacetic acid, and find those dealing with NMR.' Again, the acid is characterized by its RN, and the association of the term NMR used in the basic index, leads to the answer (eight references, one is printed in normal mode: bibliographic reference).

The RN of thiolacetic acid can be obtained from the Index Guide or from various chemical catalogs, or chemical dictionaries, online, or either by purely structural searches.

In this example, the period considered is from 1977 to 1981.

```
SS 9 /C?  USER: 507-09-5/RN
PROG:
SS 9 PSTG (262) SS 10 /C? USER: 9 AND NMR
PROG:
OCCURS    TERM
  25075   NMR
SS 10 PSTG (8); SS 11 /C? USER: PRT 1
PROG:
–1–
AN — CA94-46411(7)
TI — NMR study of rapid proton exchange with the participation of thiocarboxylic acids
AU — Pogorelyi, V. K.; Turov, V. V.
OS — Inst. Fiz. Khim. im. Pisarzhevskogo, Kiev, USSR
SO — Teor. Eksp. Khim. (TEKHA4), V 16 (5), p. 643–648, 1980, ISSN 04972627
LA — Russ,  CC — SEC22-3
```

Example 3

'Find all references dealing with the analysis and the preparation of thiolacetic acid.' Here we used the neighboring function. The RN, present only in index terms is neighbored, and all the qualifiers used by CA are visualized, with a number for selection (select). The number following the selected term, is the frequency of the term in the database. We then select SEL 12 where ANAL = analysis, and we obtained seven references, with one printed. After we select SEL 16 where PREPN = preparation, and we obtained three which were printed. The period considered is from 1977 to 1981.

```
SS 11 /C?  USER: NBR 507-09-5/RN 20
PROG:
SELECT£  POSTINGS  TERM
    1         1     50698-53-8, CHARGE-TRANSFER COMPLEXES W/RN
    2         3     50698-68-5/RN
    3         4     50698-69-6/RN
    4         1     50698-75-4/RN
    5         1     50698-76-5/RN
    6         1     50698-77-6/RN
    7         1     50698-78-7/RN
    8         1     50699-50-8/RN
    9         4     50699-52-0/RN
   10        18     507-02-8/RN
   11       262     507-09-5/RN
   12         7     507-09-5, ANAL/RN
```

```
13        12        507-09-5, BIOL STUDY/RN
14         1        507-09-5, DERIVS./RN
15         1        507-09-5, ESTERS/RN
16         3        507-09-5, PREPN/RN
17         8        507-09-5, PROP/RN
18         1        507-09-5, REACTION PRODUCTS WITH ACRYLI/RN
19         1        507-09-5, REACTION PRODUCTS WITH ALKENE/RN
20         1        507-09-5, REACTION PRODUCTS WITH AMINES/RN
UP N OR DOWN N?   USER: SEL 12
```

PROG:
SS 11 PSTG (7); SS 12 /C? USER: PRT 1

PROG:

–1–
AN — CA94-162828(20)
TI — High-performance liquid chromatographic analyses of the antihypertensive drug captopril
AU — Perlman, Solomon; Kirschbaum, Joel
OS — Squibb Inst. Med. Res., New Brunswick (NJ), USA, 08903
SO — J. Chromatogr. (JOCRAM), V 206 (2), p. 311–17, 1981, ISSN 00219673
LA — ENG. CC — SEC64-3
SS 12 /C? USER: SEL 16

PROG:
SS 12 PSTG (3); SS 13 /C? USER: PRT

PROG:
–1–
AN — CA92-41116(5)
TI — CNDO/2 calculations on the thiocarbonyl stretching frequencies in thioformaldehyde and its derivatives
AU — Randhawa, H. S.
OS — Punjab Agric. Univ., Dep. Chem., Ludhiana
SO — J. Mol. Struct. (JMOSB4), V 56 (1), p. 143–14, 1979, ISSN 00222860
LA — Eng. CC — SEC22-8
–2–
AN — CA91-4869(1)
TI — Effect of solvation upon carbonyl substitution reactions
AU — Fukuda, Elaine K.; McIver, Robert T., Jr.
OS — Univ. California, Dep. Chem., Irvine
SO — J. Am. Chem. Soc. (JACSAT), V 101 (9), p. 2498–2499, 1979, ISSN 00027863
LA — Eng. CC — SEC22-8
–3–
AN — CA87-135154(17)
TI — A novel method for acetalization of the formyl group at the C3-position of the 2,3-dihydro-1H-pyrrolo°1,2-a§indole skeleton
AU — Kametani, Tetsuji; Kigawa, Yoshio; Takahashi, Kimio; Nemoto, Hideo; Fukumoto, Keiichiro
OS — Tohoku Univ., Pharm. Inst., Sendai, Japan
SO — Heterocycles (HTCYAM), V 6 (7), p. 887–93, 1977
LA — Eng. CC — SEC28-2

4.7 Access to the full abstracts (without structural formulas)

Only one database, the one created by Chemical Abstracts (e.g. CAS Online), gives access to full abstracts. An example, taken from the STN host in Europe, is reproduced here, with the courtesy of the CNIC. As we will see with the structural databases, the CAS Online allows one to search by structures or substructures.

STN INTERNATIONAL

ANSWER 2
AN   CA100(11):85688m
TI   1-Substituted indazoles with a substituent at the benzene ring
CS   Asahi Chemical Industry Co., Ltd.
LO   Japan
PI   Jpn. Kokai Tokkyo Koho JP 58/159473 A2 [83/159473], 21 Sep 1983, 16 pp.
AI   Appl. 82/40688, 17 Mar 1982
CL   C07D231/56, C07D401/06, A61K31/395-, A61K31/40-,
      A61K31/415-, A61K31/445-, A61K31/495-, A61K31/535-,
      C07D401/06J, C07D231/00J, C07D211/00J
SC   28-8 (Heterocyclic Compounds (More Than One Hetero Atom))
SX   1
DT   P
CO   JKXXAF
PY   1983
LA   Japan
AB   Indazoles I [Z = $(CH_2)_n$, alkylated $(CH_2)_n$; n = 1–6; R = Cl, Br, F, MeO, Me, Et,
      Pr, Me_2CHCH_2; R^1, R^2 = H, alkyl, or NR^1R^2 = heterocyclyl] and their HCl salts
      were prepd. I are antiinflammatory agents and also inhibit the stomach ulcer-
      forming side-effect of nonsteroidal acidic antiinflammatory agents. Thus, *N*-
      acylation of 3-amino-4-chloroindazole with phthalic anhydride gave the 3-
      phthalimido deriv., which was *N*-alkylated by $Et_2N(CH_2)_3Br.HBr$ and K_2CO_3 in
      DMF to give the 1-[(3-diethylamino)propyl] deriv. (II). Deprotection of II by
      NH_2NH_2 in EtOH gave the 3-amino deriv. (III). At 100 mg/kg orally, III.2HCl
      reduced carrageenan-induced edema in rats by 58%. For diagram(s), see
      printed CA Issue.
KW   indazole aminoalkyl; antiinflammatory aminoindazole prepn; ulcer inhibition
      ethylaminopropylchloroindazole prepn
IT   Inflammation inhibitors and Antiarthritics
          (aminoindazoles)
IT   Ulcer
          (inhibition of, by aminoindazoles)
IT   88805-84-9P   88805-85-0P   88805-86-1P   88805-87-2P   88805-88-
      3P
      88805-89-4P   88805-90-7P   88805-91-8P   88805-92-9P   88805-93-
      0P
      88806-09-1P   88806-10-4P   88806-11-5P   88806-12-6P   88806-13-
      7P
      88806-14-8P   88806-15-9P   88806-16-0P   88806-17-1P   88806-18-
      2P
      88806-19-3P   88806-20-6P   88806-21-7P   88806-22-8P   88806-23-
      9P
      88806-24-0P   88806-25-1P   88806-26-2P   88806-27-3P   88806-28-
      4P
      88806-29-5P   88806-30-8P   88806-31-9P   88806-32-0P   88806-33-
      1P
      88806-34-2P   88806-35-3P   88806-36-4P   88806-37-5P   88806-38-
      6P
      88806-39-7P   88806-40-0P   88806-41-1P   88806-42-2P
          (prepn. and antiinflammatory activity of)
IT   88805-83-8P
          (prepn. and deprotection of)
IT   88805-82-7P

Listing 1 Reproduced by courtesy of CNIC

VIII.5 CHEMICAL ABSTRACTS STRUCTURAL FILES

The structural files are of two main types. The first one uses the structural elements which are included by CA files such as the Registry Handbook, the Ring Index, etc. These elements are textual and lead to the selection of packages of RNs which share the same structural characteristics. These files are called chemical dictionaries. The second one is fully structural and uses as search key the structural formula or part of it. Again, it selects RNs and these RNs can be transferred to bibliographic files to obtain the references, in the same manner as the select option discussed in the preceding section. It is not possible in a short presentation to completely treat these structural aspects, but we will nevertheless present the possibilities of these files in three ways: chemical dictionaries, Telesystemes DARC and CAS Online [5–8].

5.1 Chemical dictionaries

Elements such as the preferred chemical abstracts name (N), the molecular formula (MF), the ring system descriptor (RSD), the number of ring system (xURS), various codes introduced by the host computer . . . are used as search elements. The synonyms (SYNM), the trade-names are also indicated. The chemical names are subdivided into parents (P), substituents (S), and fragments (PF, NF, MFF). It is not possible here to completely explain chemical dictionaries but the following examples will help the reader to understand how to use these tools to perform various searches.

Chemical dictionaries, which contain about 6 million structures, are very often divided (as are the bibliographic databases of CA) into several parts. A few lines, printed by the host computer before using the dictionary indicate the divisions. In other host computers, the files have no names, but a number which is used to indicate to the computer the file.

In this example, /NF means name fragment, /PF means parent fragment (in the sense used by CA), FROM 7-7 limits the number of carbon atoms to 7, and PRT FU 2 means: print 2 references, out of the 15 indicated, in full.

In the reference, RN is the registry number, MF is the molecular formula, N is the name, RSD is the ring system descriptor, (NCSC2) describes the thiazole ring, and SYNM means synonyms.

```
SS 1 /C?  USER: THIAZOLE/PF

PROG:
SS 1 PSTG (8026); SS 2 /C?  USER: 1 AND BUTYL/NF

PROG:
SS 2 PSTG (228); SS 3 /C?  USER: 2 AND FROM 7-7

PROG:
SS 3 PSTG (15); SS 4 /C?  USER: PRT FU 2

PROG:
-1-
RN — 53833-33-3
MF — C7H11NS
N — Thiazole, 4-butyl-
RSD — RING(S), C3NS (NCSC2)
```

```
SYNM  —  4-n-Butylthiazole; 4-Butylthiazole
RN  —  52414-90-1
MF  —  C7H11NS
N  —  Thiazole, 5-butyl-
RSD  —  1 RING(S), C3NS (NCSC2)
```

In the next example, we searched the above response (15 in SS 3) for the thiazoles which have a substituent in the 2 position. To achieve this, we used a string search (STRS) in the field name /N and the chain :THIAZOLE, 2: to obtain the compounds with a substituent in the 2 position:

```
SS 4 /C?  USER: STRS :THIAZOLE, 2: /N

PROG:
SS 4 PSTG (8); SS 5 /C?  USER: PRT FU

PROG:
–1–
RN  —  41731-67-3
MF  —  C7H10BrNO2S2
N  —  Thiazole, 2-bromo-5-(butylsulfonyl)-
RSD  —  1 RING(S), C3NS (NCSC2)
SYNM  —  2-Bromo-5-butylsulfonylthiazole
```

In the preceding example, the ring present was used to select compounds with a pyridine and a thiazole ring present (/RPR) in a unique ring system (1URS/ CC CC stands for category code). We obtained 389 structures, from which only the ones with a nitro group were selected (NITRO/NF nitro as name fragment). This leads to six compounds, and data from the first three were printed in the default mode:

```
SS 10 /C?  USER: NCSC2/RPR AND NCC5N/RPR AND 1URS/CC

PROG:
SS 10 PSTG (389); SS 11 /C?  USER: 10 AND NITRO/NF

PROG:
SS 10 PSTG (6); SS 11 /C?  USER : PRT 3

PROG:
–1–
RN  —  55393-58-3
MF  —  C13H13N3O4S
N  —  3H-Thiazolo(4,3-a)isoquinolin-3-imine,
        5,6-dihydro-8,9-dimethoxy-1-nitro-
–2–
RN  —  53952-54-8
MF  —  C7H6BrN2O2S.Br
N  —  Thiazolo(3,2-a)pyridinium,
        6-bromo-2,3-dihydro-8-nitro-, bromide
–3–
RN  —  53952-50-4
MF  —  C7H7N2O2S.Br
N  —  Thiazolo(3,2-a)pyridinium, 2,3-dihydro-6-nitro-, bromide
```

5.2 Telesystemes DARC

Telesystemes DARC allows one to search the chemical compounds present in the CAS files since 1967 by a purely structural approach. In the chemical dictionaries the structural elements are introduced by using descriptor such as name, name fragments, rings, molecular formula fragments, etc ... while in this file, only the structural formulas or part of them are used [9, 10].

The structural elements in the search (e.g. these to be present in the structure to be retrieved) are introduced into the computer by graphic input (for a graphic terminal only), or by a textual sequence which describes the backbone of the molecule or the substructure using a sequence of numbers (all the atoms of the structure are numbered; for instance, 1-2-3-4-5-6-1 stands for a benzene ring ...), lists the atoms present (by their atomic symbol and number in the backbone), the type of bonds (single, double, tautomeric, ...), and the substitutable positions in the molecule.

The answers may be seen and/or printed (graphic terminal and graphic printer), may be printed offline, or may be obtained via the RN of the family of compounds selected (if a non-graphic terminal is used).

In all the cases, the RNs, are selected and may be transferred to the bibliographic databases to obtain the references dealing with the chemicals selected.

It is also possible to limit the search to certain types of compounds, using special codes, or to polymers (using a special file), or to use the DARC software to search the Chemical Abstracts Markush formulae for patents.

This structural file may be used to select compounds with the same structural feature, or to cross two groups of compounds with a specific structure, select compounds with both, or to obtain the molecules which exist from an initial parent molecule, etc. The possibilities are almost endless, and will certainly change the way one looks at organic synthesis within a few years.

In the following example, we display the results obtained (four structures among several others) by searching for the 2-oxo-, (S-methyl)-4 δ-4 thiazolines substituted in the 5 position by any chemical group (see Fig. 3).

It is also possible, by using various chemical fragments coming from different molecules to screen the six million structures to select the one containing them. This leads to various RN packages and to references which can be crossed together.

With this purely structural file, it is also possible to access chemical catalogs to select reactives with various substructural elements (Janssen product by Janssen Chimica).

5.3 CAS ONLINE

CAS ONLINE is the name of a database developed by Chemical Abstracts Service. This database allows the combination of bibliographic searches with a possible access to the full abstract, chemical dictionary searches using part of the registry file and pure structural searches by using the structural part of the registry file [11-13].

This is the only database which allows such a large set of possibilities.

The structural searches, use the following items for the chemical dictionaries: names and names fragments, molecular formulas and fragments, formula weight, element count, identifiers for classes of substances, and periodic

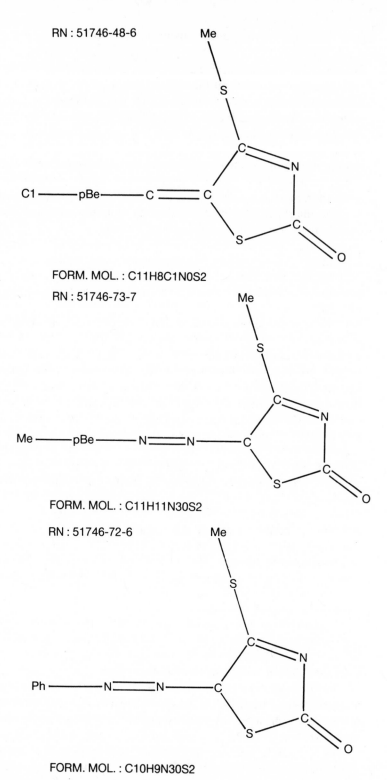

RN : 51746-48-6

FORM. MOL. : C11H8C1N0S2

RN : 51746-73-7

FORM. MOL. : C11H11N30S2

RN : 51746-72-6

FORM. MOL. : C10H9N30S2

Fig. 3   Search for 2-oxo-, (*S*-methyl)-4 δ-4 thiazolines variously substituted in the 5 position.

groups. It is worthwhile to note that these search elements can be combined with purely structural elements arising from a search for part of a molecular structural formula.

The purely structural part of the registry file also allows the search for Markush formulae in patents.

The searches and their results may be initiated and completed either with graphics or non-graphics terminals, in the same way as was described for Telesystemes DARC.

The following examples show a portion of an offline print-out showing a structure diagram, and the partial answer of a structural search concerning various pyrazinecarboxylic acids (short format).

VIII.6   THE PATENT FILES [14–17]

The importance of patents in industrial and commercial fields is so dramatic that many patent files, more or less exhaustive or specialized, exist. In chemistry, specialized data bases such as Paperchem, Ialine, Fsta, etc. deal with patents. Chemical Abstracts takes into account a large number of patents, but is not exhaustive, and only abstracts the source patent (e.g., the first patent abstracted by CA and which describes an orginal idea). This means that equivalent patents are not present (with abstracts etc.) in CA. These can only be obtained from the CA indexes by using the patent concordance which is only available in hard copy, and is not available online.

The Derwent files are perhaps the most used and the most reliable, patent index in chemistry. They are available online on several host computers (SDC, Telesystemes Questel, and LIS). They are searchable online as usual, but also with special codes (punch codes), and for the chemical patent by a chemical code. Unfortunately, it has no correspondence with either the RN (issued by CA), or the CA patent number format (which is not always the same, due to the codes of various countries).

The European patents are available through Telesystemes Questel, and various national host computers offer their national patent file online (INPI), PERGAMON INFOLINE, SDC, LIS, INKA, DIMDI, etc.).

However, legislation, which differs from country to country, the codes used, changes in titles, claims, the international classification, and the Markhus formulas have made it very difficult for a non-specialist to efficiently use patent files. Nevertheless, the patent literature is very rich, *and should be used in academic research more often.*

The following examples will successively show some print forms of various patents arising from different databases and host computers. They only show part of the possibilities since many other files have patents as part of their data.

6.1   Patents in Chemical Abstracts

Chemical Abstracts: The titles are changed if not meaningful enough and the index terms are as usual, with the RNs, followed by qualifiers and indexing

STN INTERNATIONAL

PRINT RESULTS –

REGISTRY NUMBER = 69301-61-7   ANSWER NUMBER = 2
INDEX NAME = 1,3-Dithiane, 2-(2,6-dimethyl-5-heptenyl)-2-(3-methyl-2-butenyl)- (9CI)
MOLECULAR FORMULA = $C_{18}H_{32}S_2$

$Me_2C:CHCH_2$—

$Me_2C:CHCH_2CH_2CHMe\ CH_2$

1 REFERENCES IN FILE CA (1967 TO DATE)
REFERENCE   1
AN   CA90(13):104155d
TI   Terpenes and terpene derivatives.   VII. Terpenes from C_5 and C_{10}
      building blocks through alkylation of 1,3-dithianes
AU   Hoppmann, A.; Weyerstahl, P.
CS   Inst. Org. Chem., Tech. Univ. Berlin
LO   Berlin, Ger.
SO   Tetrahedron, 34(11), 1723-8
SC   30-20 (Terpenoids)
SX   27, 28
DT   J
CO   TETRAB
IS   0040-4020
PY   1978
LA   Ger
AB   Dithiane derivs. I [R = $Me_2C:CH(CH_2)_2CHMeCH_2$, *E*-,
      Z-$Me_2C:CH(CH_2)_2CMe:CH$, $Me_2C:CH$, 4-isopropenylcyclohex-1-en-1-
      yl, 6,6-dimethylbicyclo[3.1]hept-2-ene-2-yl, 3-furanyl; R^1 =
      $Me_2C:CCH_2$, *E*-$Me_2C:CH(CH_2)_2CMe:CHCH_2$], prepd. by alkylation of
      the corresponding I (R as before, R^1 = H) with $Me_2C:CHCH_2Br$ and
      E-$Me_2C:CH(CH_2)_2CMe:CHCH_2Br$ (BuLi, THF, −78°), were converted
      to $RCOR^1$ and the corresponding RCH_2R^1 on hydrolysis (MeOH,
      HgO–$HgCl_2$, 25°, 2 h) and hydrogenolysis (Et_2O, Na/NH_3, 10 min), resp.
      E.g., 55 % $RCOR^1$ and 41 % RCH_2R^1 (R =
      6,6-dimethylbicyclo[3,1.1]hept-2-en-2-yl, R^1 = $Me_2C:CHCH_2$) were
      obtained from I (R, R^1 as before). For diagram(s), see printed CA Issue.
KW   dithiane alkylation prenyl bromide; furanyldithiane alkylation geranyl
      bromide; bicycloheptenyldithiane alkylation geranyl bromide;
      alkenyldithiane solvolysis hydrogenation; ketone diterpene sesquiterpene
IT   Alkylation
         (of dithianes, by geranyl and prenyl bromide)
IT   Diterpenes and Diterpenoids
         Sesquiterpenes and Sesquiterpenoids
         (prepn. of, by alkylation of dithianes)
IT   870-63-3   6138-90-5
         (alkylation by, of dithianes)
IT   69460-06-6
         (alkylation of, by geranyl and prenyl bromides)

Listing 2 Portion of an offline print showing structure diagram. Reproduced by courtesy of CNIC.

```
20 NOV 81 15:01:46          *** CAS ONLINE ***                    P0007
                                 WORKSHOP
                                                          ANS      1
REG  1154-82-1
FOR C13 H13 C1 N4 02
IND Pyrazinecarboxylic acid, 3-amino-5-(benzylamino)-6-chloro-, methyl ester (8CI)
```

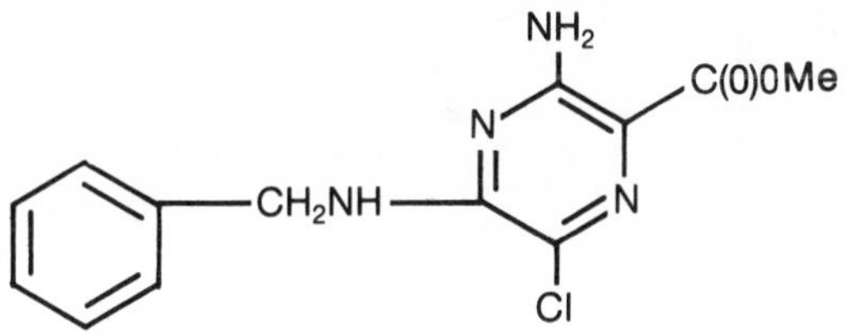

```
FULL FILE SEARCH COMPLETE
   SEARCHED   SEARCH TIME      ANSWERS
   100.000%      00.00.22          3
ENTER (DIS), STATUS, OR ?:
```

Fig. 4   Partial answer of a structural search dealing with various pyrazine carboxylic acids.

terms from CA. The list of priority countries (special code) is available in the information tools given by either CA or the host computers.

6.2   Patents in WPI, WPIL (Derwent)

WPI, WPIL: Derwent files, and WPIL (L = latest) deal with patents from 1980 to date, while WPI covers 1963 to 1979, but only in some areas (agriculture and pharmaceutical from 1963, all chemistry from 1970, mechanical from 1974). The Derwent patent files are the most complete, but they come directly from a manual (punched cards) system which has been transposed to a computerized system. This is the reason some of its codes have no correspondence with any other files on the market, and also the reason why in chemistry, the chemical codes are difficult to handle. All the equivalents of a patent are indexed. Abstracts are available on SDC.

6.3   Patents in USP

USP: this is the American US Patent Office files. They are available from 1970, exclusively from SDC, in full format (front page of the patent). The information given in this format is very large, rich and important. This format is very useful with regard to American legislation.

6.4   Patents in INPI

INPI produces files issued from the French Patent Office. They have also a special file for European granted patents. These files note the presence of all consulted sources when advice to grant patent has been given. (The same field exists in USP.)

Various examples:

File Chemical Abstracts (1977-1981):

```
-1-
AN — CA87-23259(3)
TI — Sulfoxides
AU — Durant, Graham John; Ganellin, Charon Robin
OS — Smith Kline and French Laboratories Ltd., Engl.
SO — Ger. Offen. 2634432, 22 pp. Addn. to Ger. Offen. 2,406,166.,
              2/10/77 Pat App/Prty = 7531969 (in Brit.), 7/31/75.
PCL CO7D-233/64 DT — P (Patent)
CC — SEC28-7
IT — 55884-24-7: (oxidn. of)
IT — 7790-28-5: (oxidn. by, of nitro bis° (thiazolylmethylthio)ethylamino ethylene)
IT — Antihistaminics; Inflammation inhibitors: (thiazole sulfoxide deriv.)
IT — 62941-24-6; 62941-25-7
ST — antihistaminic; antiinflammatory
```

The PRINT FU HIT, allows one to print only the index terms which contain the elements used during the search: THIAZOLE and NITRO
SS 3 /C? USER: PRT FU HIT

```
PROG:
-1-
AN CA87-23259(3)
TI — Sulfoxides
AU — Durant, Graham John; Ganellin, Charon Robin
OS — Smith Kline and French Laboratories Ltd., Engl.
SO — Ger. Offen. 2634432, 22 pp. Addn. to Ger. Offen. 2,406,166.,
              2/10/77 Pat App/Prty = 7531969 (in Brit.), 7/31/75.
PCL C07D-233/64; DT — P (Patent)
CC — SEC28-7
IT — 7790-28-5: (oxidn. by, of nitro bis° (thiazolylmethylthio)ethylamino ethylene)
IT — Antihistaminics; Inflammation inhibitors: (thiazole sulfoxide deriv.)
ST — NO HITS
```

```
File Derwent WPI:
-1-
AN — 80-90052C/50
TI — 3-Nitro-pyrazole-r-carboxylic acid nitro-furfurylidene hydrazide(s)—useful as
      antimicrobials, trypanocides and herbicides
DC — B03/C02
PA — (ELIL) ELI LILLY and CO
IN — JONES RG, TERANDO NH
PN — US4235995-A (8050)
PR — 78.09.25 78US-945170; 71.03.15 71US-124463; 71.12.23
      71US-211791;
      73.05.04 73US-357135; 74.06.06 74US-477118; 75.03.24
      75US-561139;
      76.06.18 76US-697516; 77.10.13 77US-842006
      AB — Nitrofurfurylidine hydrazides of formula (I) are new.
      In (I) R is:
H, 3-4C epoxyalkyl, 2-10C alkenyl, 3-10C alkynyl, 3-10C
alkynyl, 3-10C cyanoalkyl, cyclopropylmethyl, 3-10C cycloalkenyl,
2-10C alkyl mono- or di-substd. by halogen or 1-3C alkoxy, or
1-10C alkyl opt. monosubstd. by R2, or pH, thiazolyl or pyridyl,
all 3 being monosubstd. by NO2; R2 is SH, CONH2, = O, OH,
phthalimido, 1-3C alkylthio, 1-3C alkylsulphonyl, 1-3C alkanoyl
R1CO-O or Ph opt. monosubstd. by 1-3C alkyl, 1-3C haloalkyl, OH
or halo; and R1 is 1-3C alkyl, 1-3C haloalkyl, 3-6C cycloalkyl or
Ph opt. monosubstd. by 1-3C alkyl, 1-3C haloalkyl, halogen or OH.
Provided that when R is Me it is only substd. by Ph opt. substd.
:(I) are also animal growth promoting agents, and are broadly disclosed in US.
```

File USP: U.S. Patent Office:

–1–
PN — US4235995
TI — 3-Nitropyrazole derivatives
IN — Jones, Reuben G., Cedar City, UT; Terando, Norman H., Indianapolis, IN
PA — Eli Lilly and Company, Indianapolis, IN, US
PD — 80.11.25
AP — 945170, Filed 78.09.25, Division of 842006 Filed 77.10.13 (Now patented
US4145554, Issued 79.03.20), which is a division of 697516 Filed 76.06.18 (Now
patented US4066776, Issued 78.01.03), which is a continuation in part of 561139 Filed
75.03.24 (Abandoned), which is a continuation in part of 477118 Filed 74.06.06
(Abandoned), which is a continuation in part of 357135 Filed 73.05.04 (Abandoned),
which is a continuation in part of 211791 Filed 71.12.23 (Abandoned), Continuation in
part of 124463 Filed 71.03.15 (Abandoned)
NO — 7 Claims, Exemplary Claim 1, 0 DRAWINGS, 0 Figures Examiner: Rotman, Alan
L.; Harkaway, Natalia Atty/Agent: Page, Kathleen R. S.; Jones, Joseph A.
PCL — 542/408, Cross Refs: 424/273P X
IC — C07D2-231/16
FLD — 542/408
DT — INVENTION PATENT
FS — To U.S. Company or Corporation
CT — US3014916, 12/1961, Wright, 424/273P;
        US3102890, 9/1963, Wright, 548/375;
        US3121092, 2/1964, Geiszler, 548/375;
        US3234217, 2/1966, Schmidt et al., 542/408;
        US3658839, 4/1972, Kiehne et al., 548/378;
        US3719759, 3/1973, Sarett et al., 424/273R;
        BE780675, 9/1972; [Cheng, J. Hetero. Chem. 1968, vol. 5, pp. 195–197].
AB — Novel 1,4-disubstituted-3-nitropyrazoles having antimicrobial, parasiticidal, and
herbicidal activity are prepared by a reaction sequence of which the individual steps are
conventional. The new 3-nitropyrazoles are characterized by a 1-substituent and a
usually carbonyl-containing 4-substituent. The novel 3(5)-nitro-4-pyrazolecarbonitrile is
obtained as an intermediate in the preparation of the biologically-active compounds.
Preferred compounds are 1-alkyl or -alkenyl-4-pyrazolecarboxamides are carbonitriles.
The new compounds are particularly useful for the control of bacterial animal diseases.
MCLM–We claim: 1. A compound of the formula
HRNNCXCCNO/sub 2/wherein R represents (A) C/sub 3/-C/sub 4/epoxyalkyl, (B)
hydrogen (C) C/sub 1/-C/sub 10/alkl, C/sub 2/-C/sub 10/alkenyl,
    C/sub 3/-C/sub 10/alkynyl, C/sub 3/-C/sub 10/cyanoalkyl,
    cyclopropylmethyl, C/sub 3/-C/sub 10/cycloalkenyl,
C/sub 2/-C/sub 10/alkyl mono- or disubstituted with halo or C/sub
1/-C/sub 3/alkoxy, or C/sub 1/-C/sub 10/alkyl monosubstituted with (1) mercapto, (2)
carboxamido, (3) keto oxygen, (4) hydroxy, (5) phthalimido, (6) C/sub 1/-C/sub 3/
alkylthio, alkyl, (b) C/sub 1/-C/sub 3/haloalkyl, (c) hydroxy, or (d) halo, OOCR/sup 1/,
(11)
wherein R/sup 1/represents: (a) C/sub 1/-C/sub 3/alkyl, (b) C/sub 1/-C/sub 3/
haloalkyl, (c) C/sub 3/-C/sub 6/cycloalkyl, (d) phenyl, or (e) phenyl monosubstituted
with (1) C/sub 1/-C/sub 3/alkyl, (2) C/sub 1/-C sub 3/haloalkyl, (3) halo, or (4)
hydroxy, or (D) phenyl, thiazolyl, or pyridyl,
monosubstituted with nitro; and X represents OCHNNHCO o o o o NO/sub 2/provided
that a C/sub 1/alkyl R group is substituted only with phenyl or substituted phenyl.

The claims CLM 1 to 7 have not been printed to save space, but they are also
available.

Also, in the field CT (Cited), the cited patents, or *Journal papers* are
present.

Files INPI (French Patent Office). The search is presented with various
commands of the language used to communicate with the computer, to show
some of the differences with the one presented in the preceding chapters.

ETAPE FREQ
  1   21 MELEZE? OU PIN?
  2   21 1 OU LARIX
  3   1 2 ET ECORCE?
COMMANDE, OU ETAPE DE RECHERCHE 4
?..VI ET 3 MAX
-1- 528424 C.INPI
PUB : 2529509; EN : 8211521; NAT : B; DDP : 820630
AVD : RAP. RECH. – FR2148879(A)(Cat. X); BE542480(A)(Cat. Y);
GB840128(A)(Cat. X); CH510719(A)(Cat. Y); FR2072740(A)(Cat. A);
GB2028841(A)(Cat. X); FR998192(A)(Cat. Y); DE1404967(B)(Cat. A);
GB1028819(A)(Cat. X) – (JAPANESE PATENTS GAZETTE, section Ch: Chemical,
Week A18, 12 juin 1978, abr:g:32751A/18 Derwent Publications
Ltd. (LONDRES, GB) – JP – A – 53 030 663 (HITACHI CHEMICAL K.K.)
03-09-1976)
DPD : 840106
BPD : 84-01
DRR : 840106
BRR : 84:01
DEP : BOYER GUY
ADEP : FR-43
MND : M WEINSTEIN
TI : NOUVEAU MATERIAU AGGLOMERE A BASE D'ECORCE D'ARBRE
FORESTIER TEL QUE PIN ET SAPIN ET PROCEDE DE FABRICATION DE CE
MATERIAU AGGLOMERE
CIB : B29J-5/04; B29J-5/02; B29B-1/12; B32B-13/00; B32B-15/20; B32B-21/02;
B32B-27/04; B32B-27/40; E04B-1/74
COMMANDE, OU ETAPE DE RECHERCHE 7
?THIAZOL+; 58 TERME(S) RETENU(S); *7* RESULTAT 381
COMMANDE, OU ETAPE DE RECHERCHE 8
?7 ET MEDIC+; /BI MEDIC+: 26 TERME(S) RETENU(S) – RESULTAT 5036;
8 RESULTAT 95
COMMANDE, OU ETAPE DE RECHERCHE 9
?..VI MAX
-1- 550827 C.INPI
PUB : 2552089; EN : 8314680; NAT : A; DDP : 830915; DPD : 850322;
BPD : 85-12; DRR : 850322; BRR : 85-12; DEP : ROUSSEL UCLAF;
ADEP : FR-75; INV : ANDRE HEYMES, ALAIN BONNET ET DIDIER PRONINE;
MND : ROUSSEL UCLAF; TI : NOUVEAUX PRODUITS DERIVES DE L'ACIDE
2-AMINO
THIAZOLYL 2-OXYIMINO ACETAMIDO BICYCLO-OCTENE CARBOXYLIQUE,
LEUR PROCEDE DE PREPARATION ET LEUR APPLICATION COMME
MEDICAMENTS
CIB : C07D-501/22; A61K-31/545
-2- 546691 C.INPI
PUB : 2547729; EN : 8310456; NAT : B; DDP : 830624; AVD :
RAP. RECH. – FR3184(M)(Cat). X); FR2444459(A)(Cat. X) –
(HELVETICA CHIMICA ACTA, vol. 30, no. 1, 1 fevrier 1947, pages 432–440, Bale,
CH) (CHEMICAL ABSTRACTS, vol. 72, no. 3, 19 Janvier 1970, page 202, no. 11312r,
Columbus, Ohio, US)
DPD : 841228; BPD : 84-52; DRR : 841228; BRR : 84-52; DEP :
RIKER LABORATORIES INC; ADEP : US; INV : CLAUDE MOREAU,
RENE CHEVALIER ET DANIEL LAGAIN; MND : RINUY SANTARELLI;
TI : MEDICAMENT A BASE D'ACIDE DL THIAZOLIDINE CARBOXYLIQUE
CIB : A61K-31/425; C07D-277/06
?..BA INPI-2
  BASE EN COURS DE CONNEXION
  BASE CONNECTEE: INPI-2 27/03/85
  COMMANDE, OU ETAPE DE RECHERCHE 1
?..OP MS AU
  COMMANDE, OU ETAPE DE RECHERCHE 1
?THIAZOL+ ET MEDIC+
/BI THIAZOL+: 71 TERME(S) RETENU(S) – RESULTAT 131

/BI MEDIC+: 24 TERME(S) RETENU(S) – RESULTAT 1299 *1* RESULTAT 20
COMMANDE, OU ETAPE DE RECHERCHE 2
?..VI MAX 2
–1– 128355 C.OEB/INPI
PUB : 0124911; AN : 84105322.6; NPD : A2; LDP : De; LPD : De;
LPR : De; DDP : 840510; PR : – DE; 830510; 3317000; DSG : AT;
BE; CH; DE; FR; GB; IT; LI; LU; NL; SE; DPD : 841114;
BPD : 84-46; DRE : 840510; DEP : GPDECKE AKTIENGESELLSCHAFT
Salzufer / 16 D-1000
Berlin 10 / ADEP : DE; INV : Satzinger, Gerhard, Dr. Im Mattenbuhl / 7 D-7809
Denzlingen Herrmann, Manfred, Dr. Wolfweg / 25 D-7811 St.
Peter Fritschi, Edgar, Dr. Am Scheuerwald / 2 D-7811 St.
Peter Hartenstein, Johannes, Dr. Fohrenbuhl / 23 D-7801
Stegen-Wittental Bartoszyk, Gerd Buchenweg / 6 D-7808 Waldkirch / AINV:
DE; DE; DE; DE; DE
ET : 4-Oxo-thiazolidin-2-yliden-acetamide derivatives, process for their preparation, and
medicaments for the treatment of central nervous system illnesses containing them.
FT : Derives de 4-oxo-thiazolidine-2-ylidene-acetamide, procede pour les preparer et
medicaments pour le traitement des maladies du systeme nerveux central les contenant.
GT : 4-Oxo-thiazolidin-2-yliden-acetamid-Derivate, Verfahren zu deren Herstellung und
diese enthaltende Arzneimittel zur Bekampfung von Erkrankungen des
Zentralnervensystems.
CIB : C07D-277/34; CIB2 : A61K-31/425

VIII.7 OTHER FILES DEALING WITH CHEMISTRY OR RELATED TOPICS

Of the main files offered by various host computers, all of them accessible online, *there are at least a hundred files concerned with chemistry and related areas* of applications and sciences.

It is not possible here to give even an overall view of the information contained in these files concerning all the subjects treated, so it is better to give a general approach to this problem rather than presenting in detail a few topics which may only interest a few readers.

7.1 The dissemination of scientific information

If we are dealing with a huge amount of information (several million references, which is the normal case when using large databases), the dissemination of the information is very important. It may be divided in various areas:

— dissemination by subjects (number of useful references versus those not useful)
— dissemination by sources (e.g. journals, reports, theses, books etc.)
— dissemination by authors and laboratories.

A general rule of distribution can be applied to these areas:

— There is always a core to the dissemination where you are able to get the best compromise between what you want to obtain and the work necessary to obtain it. This has been called the law of least effort and we completely agree with it.

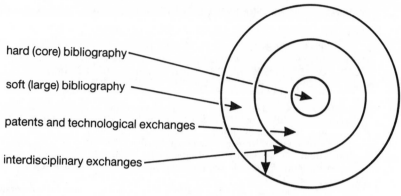

hard (core) bibliography

soft (large) bibliography

patents and technological exchanges

interdisciplinary exchanges

Fig. 5

In practice, if you want to obtain a hundred percent coverage of a subject, the number of sources to consult will increase exponentially, as you approach the total, but in the case of the sciences (academic research for instance), only a small number of journals cover fifty to seventy percent of the field. This is illustrated by the 15 000 journals covered by Chemical Abstracts for the chemistry only, and the 2 500 included in Current Contents in interdisciplinary research.

A naive view of the problem is represented on Fig. 5 (18).

7.2   How to deal with large information sources

All of the software on the market today allows one to search a database in almost all possible ways, from subjects to compounds, from concepts to techniques, from country to city and languages, from sources (journals, reports, reviews, patents, from author names to laboratory addresses, etc.). These unique capabilities allow one to use the databases as bibliographic, statistical, or decision-making tools. Nowadays, the bibliographic uses (hard or soft) are shifted to a different role: to provide special information to assist the decision-maker. However, it is not the purpose of this chapter to develop these ideas further [19].

Scientists are conservative and many people consider that databases will never replace the consultation of various journals or issues of CA in hard copy. This view is probably naive, arising from the fact that most scientists are not familiar with database searches. Using the browsing facilities they offer, they will be able to test ideas, chemical concepts, etc., thoroughly and not only for one issue, but hundreds of them. Indeed, the size is not comparable and these non-bibliographic searches, very often performed by the scientist himself will be a unique way to support, help, and to lead to innovations in everyday work [20].

7.3   Specialized databases available

As we stated at the outset of this section the number of databases available is very large, and we can only give a few examples: biology (Biosis), environment (Pollution, Environline etc.), biotechnologies, paper chemistry, (Paperchem) fabrics (Titus), polymers (Rapra, DARC) foods (FSTA, IALINE), patents (DERWENT, C.A., USP, INPI, Claim VETDOC, PESTDOC, JAPIO etc), pluridisciplinary databases (Pascal, NTIS ...), paints, softwares, river waters and oceans (Oceanic, BNDO, Waterlit...), agriculture (Tropag ...), metallurgy, alloys, corrosion (Corrosion), thermodynamic data (Thermodata), spectra (IR, NMR C13, Mass spec. ...), Congress, safety, physics, medical uses of chemicals, (Medline, Excerpta Medica...), petroleum (Apilit, Apipat), Commercial (Predicast, CIN...), engineering (Compendex, C.A., ...), education (Eric, C.A.), theses (CDI), chemical reactions (CRDS, ISI-IC, INDEX CHEMICUS...), energy (Energyline, EDFDOC, ...) and so on [21, 22].

The various repertoires of host computers and databases are available in almost all countries, and the best course is to consult them prior to making a decision. One must remember that the name of a database is nothing by itself. Rather, it is the contents which are important, and, at the same time, the available software. Also, it is obvious that databases overlap in certain fields, and that only through expert consultation and practice will you be able to choose the best core of databases to use.

Some specialized databases are given in Table 1.

7.3.1   Environline

This database is concerned with all studies dealing with environmental research. In this way, it is multidisciplinary, but the overlap with specialized files is sometimes great. Abstracts are available online. The search presented here deals with polyaromatic hydrocarbons present in coal, and their effects on health.

Table 1   Example of a few Specialized Databases

Field coverage	Database name
Environment	ENVIROLINE
	DOCODEAN
	OCEAN
	WATERLIT
US government contracts	NTIS
Chemical reactions	CRDS
	INDEX CHEMICUS
Bibliography and citation index	SCISEARCH
Food and related areas	FSTA—IALINE
Paper and wood	PAPERCHEM—FOREST

SS 1 /C? USER: PAH AND COAL

PROG:
OCCURS   TERM
   118   PAH
  6086   COAL
SS 1 PSTG (20); S 2 /C? USER: PRT FU

PROG:
-1-
AN - 84-05755
TI — DETERMINATION OF NITRATED POLYCYCLIC AROMATIC
HYDROCARBONS IN DIESEL PARTICULATES BY GAS CHROMATOGRAPHY
WITH CHEMILUMINESCENT DETECTION
AU — YU, WING C.; FINE, DAVID H.; BIEMANN, KLAUS; CHIU, KIN S.
SO — ANALYTICAL CHEMISTRY, JUN 84, V56, N7, P1158 (5)
OS — THERMO ELECTRON CORP, MA
CC — U.S. POLICY & PLANNING
AV — MICROFICHE AV. FROM EIC
IT — *POLYCYC AROMATIC HYDROCARBON; *PARTICULATES;
*CHROMATOGRAPHY, GAS; DIESEL ENGINES–TRANSPORT;
STACK EMISSIONS; CHEMILUMINESCENCE; AUTOMOBILE EMISSIONS
AB — AN ANALYTICAL TECHNIQUE FOR THE DETERMINATION OF NITRATED
POLYCYCLIC AROMATIC HYDROCARBONS (PAH) IN AUTOMOTIVE DIESEL
PARTICULATES IS DETAILED. THE APPROACH INVOLVES USE OF CAPILLARY
GAS CHROMATOGRAPHY FOR THE SEPARATION OF NITRO-PAH
AND SUBSEQUENT DETECTION BY CHEMILUMINESCENCE.
1-NITROAPTHALENE,2-NITROAPHTHALENE, 2-NITROFLUORENE, AND
1-NITROPYRENE WERE SELECTIVELY DETECTED IN COAL COMBUSTION
EMISSION AND DIESEL EMISSION SAMPLES. THE APPROACH IS USEFUL
FOR THE SCREENING OF NITROAROMATICS IN COMPLEX ENVIRONMENTAL
SAMPLES. (4 GRAPHS, 35 REFERENCES, 2 TABLES)
SS 2 /C? USER: 1 AND HEALTH

PROG:
OCCURS   TERM
  8664   HEALTH
SS 2 PSTG (5); SS 3 /C? USER: PRT 1

PROG:
-1-
AN — 82-02158
TI — HISTORICAL PERSPECTIVE OF POLYCYCLIC AROMATIC HYDROCARBON
CARCINOGENESIS: IMPLICATIONS FOR THE TRANSITION TO REGENERATIVE
ENERGY
AU — LLOYD, J. WILLIAM
SO - PRESENTED AT SOCIETY FOR OCCUPATIONAL AND ENV HEALTH (ANN
ARBOR) CONF, SALT LAKE CITY, APR 4-7, 79, P301(6)

7.3.2   NTIS (National Technical Information Service)

This file deals with all the reports of the contracts sponsored by the US
government. This is a multidisciplinary database, important abstracts are
available online, and, in the index terms field, the more important topics are
indicated by a *. This database is searchable by using weighted words.

Our example is 'Are there any studies of the bark of larix (a resinous
mountain tree)':

SS 3 /C? USER: LARIX

PROG: SS 3 PSTG (17); SS 4 /C? USER: 3 AND BARK

```
PROG:
OCCURS     TERM
   212     BARK
SS 4 PSTG (1); SS 5 /C? USER: PRT FU
```

```
PROG:
–1–
AN — PB84-131432
TI — Above-Ground Weights for Tamarack in Northeastern Minnesota.
AU — Carpenter, E. M.
OS — North Central Forest Experiment Station, St. Paul, MN.
SPN — 020332000
SO — 7 Dec 83; 13p
IS — u8406
PR — NTIS Prices: PC A02/MF A01
LA — English
DT — Forest Service Research paper (Final)
RP — FSRP-NC-245
CC — 2F (AGRICULTURE—Forestry); 48D (NATURAL RESOURCES and EARTH
SCIENCES—Forestry)
IT – *Biomass; *Forest trees; Weight(Mass; Mathematical prediction; Density(Mass/
volume); Moisture content; Height; Minnesota; *Larix laracana; Tamarack trees.
AB — The author used trees from natural, uneven-aged stands in northeastern
Minnesota to develop predictive equations to estimate total tree, stem, crown, live and
dead branch weight. Presented in the report are specific gravity and moisture content by
d.b.h. and height in tree, as well as bark, sapwood, and heartwood ratios.
```

7.3.3   CRDS

Three examples will be shown of this file, because this is a good example of chemical reaction files. This product is developed by Derwent (Chemical Reactions Documentation Service), from Theilheimer's Synthetic Method Journal. In this file, the titles are rewritten to be significant, and the searchable fields are as follows:

— Title /TT
— Starting material /SM
— Product /PR
— If part of the structure does not change during the reaction, the indexation is xxxx)/PR
— Thematic groups such as Q/TG (patent), O/TG (oxidation), W/TG (rearrangement), A/TG (Aromatic substitution and exchange, ring opening, ring closure, aromatization, dearomatization), V/TG (irradiation)
— Reaction symbols HC OC becomes HC-A-OC/SY
— Elements.

A list of all indexed terms is available. We give three examples which illustrate part of the potentiality of such chemical reaction files.

The first one shows a simple concept: what are the reactions indexed with hydrocarbon as starting material (but the concept of hydrocarbon is given by the list of indexed terms), and two references in full print are given.

The second example deals with the concept of oxidation by hydrogen peroxide with phase transfer catalysis in patents. We use reaction terms to define hydrogen peroxide, which is given by peroxide, inorganic and then we associate this concept to the one of quaternary ammonium to describe phase transfer catalysis. The restriction to patents is made by using a thematic group Q/TG. One reference in full is printed, note in the field CI the patent number, which makes the bridge with the derwent patent files.

The third one deals with a search of rearrangement of the thiazole ring. To do so we first select the term thiazole in the title /TT and we associate the concept of rearrangement by using the thematic group W/TG. Two references are printed in standard form.

SS 3 /C? USER: HYDROCARBON/SM

PROG:
SS 3 PSTG (8); SS 4 /C? USER: PRT FU 2

PROG:
–1–
AN — 77813V OYS
TI — CARBOXYLIC ACID ESTERS FROM HYDROCARBON GROUPS.
DEGRADATION WITH LOSS OF 3 C-ATOMS. SELECTIVE OXIDATION/ OC-X-C /
CI — J.ORG.CHEM. 45, NO.22 4385–87 (1980).
KW * /HYDROCARBON/ GIVES *C-ESTER* *(STEROID)* CLEAVAGE-3C
OXIDN, SELECTIVE PEROXY-OXIDN PEROXIDE, INORG (HYDROGEN-PEROXIDE)
ACID =
4 INORG-ACID, OXO (SULFURIC-ACID) C-ACID (TRIFLUOROACETIC-ACID
TEMP = 3 PROTIC, AQ TEMP = 5
–2–
AN — 78144U U
TI — /
REVIEW OF THE CHEMISTRY AND PREPN. OF HYDROCARBONS. 75019S/
HC-E-O /
CI — CHEM. ABSTR. 90, 21666 (1979), ULLMANNS ENCYKL. TECH. CHEM.
4 AUFL. 1977, 14, 663–92.
KW — /HYDROCARBON/ REVIEW *HYDROCARBON* REF = 247
****
SS 1 /C? USER: PEROXIDE, INORG/RT
PROG: SS 1 PSTG (444); SS 2 /C? USER: AMMONIUM, QUART/RT

PROG:
SS 2 PSTG (390); SS 3 /C? USER: 1 AND 2 AND Q/TG

PROG:
SS 3 PSTG (2); SS 4 /C? USER: PRT FU; S 4 /C? USER: PRT FU

PROG:
–1–
AN — 75552X AOQ
TI — /CATECHOLS FROM O-HYDROXYKETONES/ OC-X-C /
CI — J5-6147737 95894D-BCE.
KW — /KETONE/ /OXYCARBONYL-O/ GIVES *PHENOL* *DIOL* *DIOXY-O*
(PHENOL) PEROXY-OXIDN PEROXIDE, INORG (HYDROGEN-PEROXIDE)
CATALYSIS AMMONIUM, QUAT (TETRABUTYLAMMONIUM-HYDROXIDE)
CLEAVAGE-C PROTIC, AQ HYDROXIDE, ALKALI (SODIUM-HYDROXIDE)
TEMP = 3
****
SS 9 /C? USER: THIAZOLE/TT

PROG:
SS 9 PSTG (15); SS 10 /C? USER: 9 AND W/TG

PROG:
SS 10 PSTG (3); SS 11 /C? USER: PRT 2

PROG:
–1–
AN — 75041T W
TI — /4-ALCOXYTHIAZOLE FROM THIAZOLE 3-OXIDES/OC-R-ON/
CI — CHEM. PHARM. BULL. 25, NO. 12, 3270–76 (1977).
–2–
AN — 75400R HLW
TI — /THIAZOLE RING FROM PENICILLINS/SC-E-O/
CI — CAN. J. CHEM. NO. 22, 3425–38 (1975).
****
USER: A/TG AND O/TG AND V/TG

PROG:
SS 13 PSTG (38); SS 14 /C? USER: PRT 3

PROG:
–1–
AN — 76735Z AJVO
TI – /AROMATIZATION BY CYCLODEHYDROGENATION. NAPHTHO(1,2-B:
6,5-B')BISBENZO(B)THIOPHENES/CC-E-H/
CI — J. HETEROCYC. CHEM. 21, NO. 2, 321–5 (1984).
–2–
AN — 75688Y AVJO
TI — BENZOCYCLOBUTENES FROM 4,5-DIMETHYLENECYCLOHEXENES/CC-R-
CC, CC-E-H/
CI — J. ORG. CHEM. 48, NO. 2, 273–6 (1983).
–3–
AN — 75152Y AOQV
TI — /SYM. DISELENIDES FROM HALIDES. DIARYL DISELENIDES/MC-X-G/
CI — SU-891656 92135E-E.

7.3.4   Index Chemicus Online

The Index Chemicus can be accessed online. This is the version of Current Abstracts of Chemistry and Index Chemicus produced by ISI. The ISI Index Chemicus Online database includes new organic compounds and related bibliographic data from 1962 to 1965 and from 1977 to present. The ten missing years will soon be available.

One advantage of this database, as far as chemical reactions are concerned, is that it includes unisolated and theoretical intermediates, which is not the case in other databases such as Chemical Abstracts (*note*: a RN is given only for reaction products and isolated intermediates). This will lead to more flexible and versatile retrieval.

The database will contain approximately 200 000 new structures included in 20 000 new articles, each year. The retrieval will be possible via the structural formula using the TELESYSTEMES DARC software, a link being created to jump from the structural searches to the bibliographical records. The following examples by courtesy of ISI, show both types of information retrieved. To help the researcher, several alerts are indexed and included at the article level. These alerts will help achieve a more efficient retrieval focused on directly useful articles. These alerts are:

— Biological activities alert: a list of activities is given provided the term activity itself is also indexed.
— Instrumental methods alert: 24 methods are included.

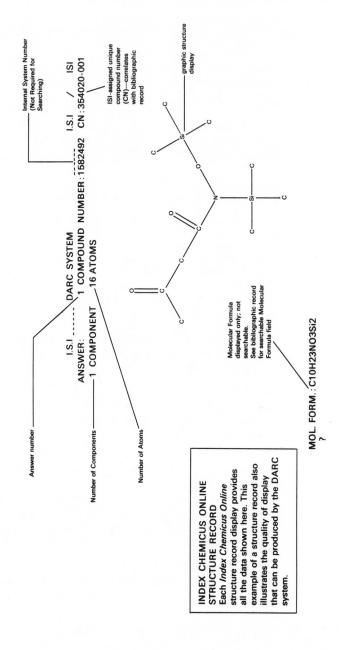

INDEX CHEMICUS ONLINE
STRUCTURE RECORD
Each *Index Chemicus Online*
structure record display provides
all the data shown here. This
example of a structure record also
illustrates the quality of display
that can be produced by the DARC
system.

Answer number

Number of Components

Number of Atoms

I.S.I ------ DARC SYSTEM
ANSWER: 1 COMPOUND NUMBER:1582492
1 COMPONENT 16 ATOMS

Internal System Number
(Not Required for
Searching)

I.S.I / ISI
CN:354020-001

ISI-assigned unique
compound number
(CN)—correlates
with bibliographic
record

graphic structure
display

Molecular Formula
displayed only; not
searchable.
See bibliographic record
for searchable Molecular
Formula field

MOL. FORM.:C10H23NO3Si2

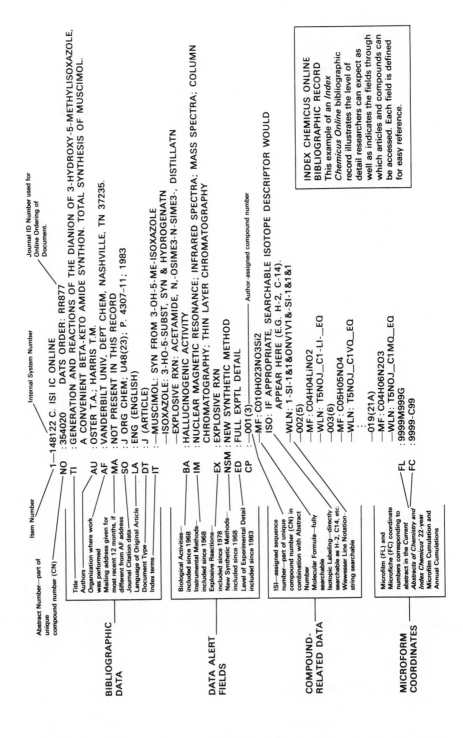

— Explosive reactions alert: all articles reporting explosive reaction are indexed.
— New synthetic method: new reactions, new synthesis or total synthesis of natural products are reported.
— Level of experimental detail: full, partial, or none tell if the article contains enough detail to repeat the synthesis.

This database, original in its conception, does not replace Chemical Abstracts (which is unique for its field of coverage), but allows the chemist to look to very specific topics or possible reaction paths or intermediates. In this point of view it illustrates some of the new tendencies of database producers, and if complete, it will constitute a very powerful tool.

7.3.5   Citation indices [22, 24]

The citation index arises from the paper in question being cited in articles appearing in the 2 500 most cited journals. It is multidisciplinary. There is also a database associated with the citation index, related to the 2 500 journals. This database is also multidisciplinary but does not contain patents or reports. It can be consulted only through title words, authors, and journals. The use of the citation index is different from that of the classical bibliographies as it allows forward recording of the literature.

Quick bibliographies   These can be constructed when two or three key articles are known. In these cases, we simply ask who has cited a key article, the citation index rapidly gives the main articles related to the subject at hand.

Mapping citation network   It is very important, when you are engaged in laboratory research, to understand the mapping citation network of the main articles in the field and also of your own papers. This gives a good overview of the various applications, fields and people who use your results. In some cases it can be surprising or helpful but, often, the result is only moderately helpful [25].

In fact, the impact of the primary journals in chemistry shows that the best journal gives, on average, only a few citations per article per year. If you also think that the lifetime of an article is very short, the total number of citations can be disappointing. However, in the opinion of some, this may be a reflection of the true state of affairs.

Citation indices are expensive and they have to be manipulated, online, by experienced people, otherwise, errors can occur. The concept of the citation index was developed by E. Garfield. For more information note reference [20].

VII.8   DOWNLOADING

Nowadays, downloading is becoming a very useful and popular tool. It allows the capabilities of a terminal to search online a database, and of a microcomputer to store and reread the files created with a word processor or other programs.

Downloading gives rise to ASCII text files. These files are stored on diskette or on hard disk. Many programs can be used, and most of them are commercial. Most of the work processors can handle ASCII text files.

Some hosts will ask special permission to perform downloading, others will not mind. The format of the references to be downloaded can be different, for instance, from the one generally used in the online print process. ESA (European Space Agency) will allow you to use the format X, which is a special format where the beginning, the end, and the various fields of a reference will be marked by special characters.

Downloading is now giving rise to statistical programs which, in local use, will analyze the file created from the downloading of the main reference file. For instance, Patstat of Derwent gives you the opportunities to analyze a patent search in 'different ways: time, year, classes, countries, punch code'. . . . The results are presented as bar charts, pie charts, etc.

With small computers (e.g. Apple II/c with the Access II communication program), it can be very easy to develop one's own software to perform statistical searches on various fields. The best, in this case, is to select the fields from which the statistical data arise, and to store them on diskette with, for instance, the PRINT SELECT command of the host SDC. This will allow more information to be put in the allocated memory volume, and also save time.

Obviously, downloading will become increasingly important in the near future, and will help people to save time and typing.

SDI and downloading   A combination has been developed by Biosis which provides for IBM and Apple microcomputers by placing part of their database on diskettes. The information is selected according to a certain number of divisions (sections, topics, etc.). The software allows one to search in-house the database, and the files can be used in combination with other programs, such as word processors, etc.

VIII.9   COST OF ONLINE ACCESS TO CHEMICAL INFORMATION

9.1   Invoicing structure

It would take much space to develop in detail the costing structure of chemical information. Thus, we will give only the main points.

Often you do not need to have subscriptions, you simply sign an agreement with the host computer, and pay only when you use the service.

Basically the price can be divided into several items.

9.1.1   Communication network

Communication network, from your office to the computer, on the basis of a price per connect hour plus a charge for the number of characters sent or received. Normally you will use to do this a telecomunication network in time sharing, and the cost will be far more important than a usual telephone call.

9.1.2 Connect time to the computer

The computer possesses its own files where general information is stored. The price per hour is generally low. But, the time spent on the database to be searched will be charged according to the producer.

9.1.3 References printed

If you print the references online, you will be charged a certain amount for each reference (do not forget also that you will use time spent on the database, and be charged accordingly). If you print the references offline they will be sent to you by mail; you will save time but will be charged a little bit more for each reference. Depending on the database and the number if references to be printed, you must know when you wish to having printing done offline or online.

9.2 Textual files

These files are the least expensive, leading most of the time to bibliographic references. The most expensive are patent files, the business files giving rise to a series of numbers. In chemistry, an exception must be made for the use of chemical dictionaries, which are textual files dealing with structural aspects of the molecules. These files, because of the number of inverted fields involved, are more expensive to use.

The textual file, in 1985, will cost between 50 to 150 dollars per connect hour, plus the cost of references (about 15 to 25 cents each). [26, 27].

The use of some files can be restricted to people who already subscribe to the printed product, or who have special subscriptions. (For more details see the information given by the various host computers or database producers.)

9.3 Structural files

They have been developed for chemistry and patents. They are more expensive than the textual files. There are prices per connect hour, number of RN selected, number of questions asked, etc. [28–29].

The answers must generally be associated with the textual bibliographic files, to have the references dealing with the selected structures. In this case, the overall cost per question is very important.

9.4 Equipment

This is not a problem, since most small computers can be used as terminals. They must be equipped with a serial card, a modem and a printer. Dumb terminals also exist, but microcomputers, because of the downloading must be preferred if possible [30].

Portable terminals are also available, and if the country is equipped with videotex, most home terminals can be used (if the host computer is also available through videotex).

9.5 The SDI (The Selective Dissemination of Information)

Most host computers have the capability to store in their memory a set profile (a set of questions and orders). This profile can be executed in two ways: manually or automatically. This is the second solution which is generally used, and the profile is run by the computer each time the databases involved are updated.

In many laboratories and research centers, it is common to have several profiles per researcher. The cost is composed of storage of the profile and offline prints. Before the creation of a profile, it is necessary to test it for several weeks and when the results match with the expected coverage, to transform the search in SDI. The SDI can be cancelled at any time by the user from its terminal.

VIII.10 CONCLUSION

Chemistry is often the crossroad between various disciplines. There is a strong temptation for the chemist to move to an increasingly specialized area where he will remain, against all odds, hopefully unique and comfortable. However, during this time, life will be passing by, and others who need chemistry will develop their own areas. To stay aware and be ready to face all situations, the chemist needs a great deal of information, not only to progress but also to survive.

Chemical information databases, if well used, are an invaluable way of gathering information, to learn, compare, classify, and attain the necessary knowledge to design new experiments, or to make good decisions [31, 32].

Like other chemical disciplines, chemical information should be considered as an indispensable tool: you would not consider giving away NMR or UV or IR! The time has gone where one person was able to retain in his memory all the main facts of his speciality; times are changing and so are the techniques. Chemical information, in conjunction with online access and home computers, will perhaps be the key to new ways forward and applications in chemistry.

VIII.11 REFERENCES

[1]   H. Dou, D. Doré and P. Hassanaly, in *Connaître et utiliser les Banques de données* (in French), Infotecture, Paris (1981).

[2]   H. Bechtel, *Nachristen fur dokumentation*, **35**, 100–107 (1984).

[3] S. H. Wilen, in *The Use of Chemical Literature*, ACS Audio Course (1978).

[4] J. T. Dickman, M. P. O'Hara and O. B. Ramsay, in *Chemical Abstracts, an Introduction to its Effective Use*, CAS (1979).

[5] P. Hassanaly, H. Dou, J. C. Bonnet, R. Attias and A. Lemagny, 4th Symposium Online Information Meeting, London (1980), *Online Review Supplement*, 307–316 (1980).

[6] C. Gueuinier, 7th Symposium Online Information Meeting, London, *Online Review Supplement*, 95–98 (1983).

[7] S. C. Stephen, *Online*, **8**, 44–49 (1984).

[8] G. K. Ostrum and D. K. Diane, National Online Meeting, New York, Proceedings, 445–454 (1982).

[9] E. Vinatzer-Chanal, in *The Application of Mini- and Micro-computers in Information, Documentation and Libraries*, 195–204 (1982).

[10] D. W. Mehlman, National Online Meeting, New York, Proceedings, 349–363 (1982).

[11] C. R. Zeidner, J. O. Amoss and R. C. Haines, National Online Meeting, New York, Proceedings, 575–586 (1982).

[12] P. G. Diitmar, N. A. Farmer, W. Fisanick, R. C. Haines and J. Mockus, *Journal of Chemical Information and Computer Sciences*, **23**, 93–102 (1983).

[13] N. A. Farmer and M. P. O'Hara, *Database*, **3**, 10–25 (1980).

[14] K. Kobayashi, Y. Kobota and M. Harada, *Sumimoto Kagayo (Osaka)*, **1**, 61–66 (1984).

[15] K. Ohoyama, *Kemikaru Enjiniyaringu*, **29**, 855–867 (1984).

[16] S. Tokizane, *Kemikaru Enjiniyaringu*, **29**, 841–854 (1984).

[17] G. S. Revesz and P. A. Cassidy, *Chemyech*, **14**, 18–25 (1984).

[18] H. Dou and P. Hassanaly, 5th International Online Information Meeting, London, 65–71 (1981).

[19] H. Dou and P. Hassanaly, 7th International Online Information Meeting, London, 175–183 (1983).

[20] E. Garfield, *Current Contents*, **41**, 5–14 (1981).
E. Garfield, *Current Contents*, **2**, 5–8 (1980).

[21] R. Erb and K. Van Der Meer, *Chem. Mag.*, 151–153 (1985).

[22] S. Tokizane, *Petrotech*, **8**, 147–151 (1985).

[23] E. Garfield, *Current Contents*, **19**, 5–13 (1980).

[24] E. Garfield, *Citation Indexing, its Theory and Application in Sciences Technology and Humanities*, John Wiley, New York, 274 pp (1979).

[25] H. Dou and P. Hassanaly, *Inf. Chim.*, **231**, 231–241 (1982).

[26] R. Sancho, *Revista Espanola de Documentacion Cientifica*, **4**, 219–229 (1981).

[27] E. Zass, *Nachrichten fur Dokumentation*, **33**, 129–137 (1982).

[28] A. Girard and M. Moureau, 4th International Online Information Meeting, London, 335–343 (1980).

[29] P. F. Urbach, *Journal of Chemical Information and Computer Sciences*, **24**, 1–3 (1984).

[30]   C. Maguire, Communicating Information Proceedings of the 43rd
       Asis Annual Meeting, Anaheim, **17**, 239–241 (1980).
[31]   J. P. Courtal, *Documentaliste*, **22**, 102–107 (1985).
[32]   P. Hassanaly and H. Dou, *Bulletin des Bibliothèques de France*, **27**,
       399–401 (1982).

RECENT LITERATURE ON THE USE OF COMPUTERS IN CHEMISTRY

K. Ebert and H. Ederer, *Computer Anwendungen in der Chemie, eine Enfüeh-rung in die Arbeiten mit Keinrechnern*, Verlag Chemie, Weinheim, F.R.G., 2nd ed., 457 pp. (1984).

Inuzuka, Kozo, *Learning Chemistry by Microcomputers*. (*Maikon de Kagaku o Manahu*). Tokyo Kagaku Dojin, Tokyo, Japan, 197 pp. (1984).

Computer Education for Chemists. Based on a symposium presented at the 184th National Meeting of the ACS, Kansas City, Missouri, Sept. 1982. (P. Lykos, Ed.); J. Wiley, New York, N.Y., 223 pp. (1984).

Proceedings of the 1984 summer computer simulation conference, July 23–25,. 1984. Boston, MA. Vol. 2. (W. D. Wade, Ed.); Soc. Comput. Simul., La Jolla, CA, 567 pp. (1984).

T. Mamuro and Y. Sudo, *BASIC Programming for Understanding Chemistry and Physics* (*Kagaku to Btsuri ga Wakaru Beshikku Proguramingu*), Nikkan Kogyo, Shinbunsha, Tokyo, Japan, 174 pp. (1984).

Methods and Applications in Cristallographic Computing. Papers presented at the international summer school on crystallographic computing, held at Tokyo, Japan, August 1983. (S. R. Hall and T. Ashida, Eds.); Oxford University Press. U.K., 506 pp. (1984).

T. Yoshimura, *Simulation of Chemical Experiments by the use of BASIC*. (*BASIC ni Yoru Kagaku Jikken Shimyureshon*). Koritsu shuppan Co., Ltd., Tokyo, Japan, 221 pp. (1985).

Computer Handling of Generic Chemical Structures. Proceedings of a conference at the University of Sheffield, (U.K.). (J. M. Barnard, Ed.); Gower Publ. Co., Aldershot, U.K., 230 pp. (1984).

P. Benedek and F. Olti, *Computer Aided Chemical Thermodynamics of Gases and Liquids: Theory, Models, Programs*. J. Wiley, New York, N.Y., 731 pp. (1985).

Proceedings of the 2nd International Conference on Vector and Parallel Processors in Computational Science. [1984, Oxford (U.K.) in: Comput. Phys. Commun. 1985: 37(1–3)] (I. S. Duff and J. K. Reid, Eds.); North Holland: Amsterdam, Neth., 386 pp. (1985).

S. Ohe, *Chemical Engineering Design Using Personal Computer*. (*Pasakon ni yoru kemikaru enjiniaringu dezain*). Kodansha Ltd., Tokyo, Japan, 186 pp. (1985).

Computerized Control and Operation of Chemical Plants [1981, Vienna, Austria in: Comput. Chem. Eng., 1985, 8(5)]. (K. Czeija, Ed.); Pergamon Press: New York, N.Y., 60 pp. (1985).

T. Yoshimura, *Introduction to Chemistry Dry Lab Using BASIC* (*BASIC ni Yoru Kagaku Drairabo Nyumon*) Kyoritsu shuppan: Tokyo, Japan, 157 pp. (1984).

Computer Applications in Fire Protection: Analysis, Modeling, Design, Pt. 1 [in: Fire Saf. J., 1985, 9(1)] (P. Di Nenno, Ed.); Elsevier Sequoia: Lausanne, Switz., 135 pp. (1985).

Managing Laboratory Information: the Computer Options. Technical Insights, Inc., Technical Insights: Fort Lee, N.J. 1985, 226 pp.

Computers in Chemistry. Practice Oriented Introduction (*Computer in der Chemie. Praxisorientierte Einfuenrung*) (E. Ziegler, Ed.); Springer: Berlin, F.R.G., 280 pp. (1984).

O. Kikuchi, *Chemistry by means of BASIC* (*BASIC ni yoru kagaku*) Kyoritsu shuppan, Tokyo, Japan, 222 pp. (1984).

Programming BASIC for Chemical Engineering (*BASIC ni Yoru Kagaku Kogaku Puroguramingu*) The Society of Chemical Engineers, Japan. Baifukan Co., Ltd., Tokyo, Japan, 272 pp. (1985).

First National Colloquium of the Centre National de l'Information Chimique on Information of Chemistry, 13th and 14th of November 1984, CNIC, PARIS, Fr., 256 pp. (1984).

Computer-Aided Molecular Design. Proc. 2 day conf. 1984, Oyez Sci., Tech. Serv. London. (1985).

L. J. Malon, *BASIC Concepts of Chemistry, Computerized Version*, 2nd Ed., (J. Wiley, New York, N.Y.), 603 pp. (1985).

J. Nentwig, M. Kreuder and K. Morgenstern, *Teaching Program, Chemistry I.* (*Lehrprogramm, Chemie I*) VCH Verlaggesellschaft: Weinheim, F.R.G., 654 pp. (1985).

J. E. Ash, P. A. Chubb, S. E. Ward, S. M. Welford and P. Willett, *Communication Storage and Retrieval of Chemical Information.* Ellis Horwood, Ltd., Chichester, U.K., 295 pp. (1985).

H. Schulz, *From CA to CAS On-line: the Data Collections of Chemica, Abstracts Service and Their Use.* (*Von CA bis CAS: die Datensammlunger. des Chemical Abstracts Service und deren Nutzung*) VCH Verlagsgesellschaft; Weinheim, F.R.G., 170 pp. (1985).

The Role of Data in Scientific Progress, Proceedings of the 9th international CODATA conference, Jerusalem, Israel, 24–28 June, 1984. (P. S. Glaeser, Ed.); North-Holland; Amsterdam, Neth., 549 pp. (1985).

ACS Symposium Series, Vol. 265: Computers in the Laboratory, Current Practice and Future Trends. (J. G. Liscouski, Ed.); ACS. Washington, D.C., 124 pp. (1984).

Applications of Artificial Intelligence in Chemistry, T. Pierce, and B. Hohme, Eds.); ACS Symposium Series, ACS, Washington, D.C. (1986).

NOTE

Due to the lack of space, all the examples have been compiled, specially for the commands and computer answers. For instance, the example 1 should be read:

Example 1 (see p. 345)

SS 5 /C? User: 90-14-2/RN

PROG:
SS 5 PSTG (43); SS 6 /C? USER: 5 AND FR./LO

To be read

SS 5 /C?
USER:
90-14-2/RN

PROG:
SS 5 PSTG (43)

SS 6 /C?
USER:
5 AND FR./LO

Concerning all the databases, it is not possible to take into account all the products, for instance since we wrote this chapter new databases have been added, for instance JAPIO, dealing with the English Abstract Journal of japanese patents (kokai). If the reader wants a more exhaustive coverage of the databases available online, she or he must consult the Directory of Online Databases, or database CUADRA, which gives the same informations online.

Subject Index

Author Index

Index

Index

Index

Index

Index

Index

Index

Index

Index

Index

Index

Index

Index

Index

Index